T0324161

Sustainable Production Technology in Food

Sustainable Production Technology in Food

Edited by

JOSÉ MANUEL LORENZO
Galician Meat Technology Center, Galicia Technology Park, Ourense, Spain
Food Technology Area, Faculty of Sciences of Ourense,
Vigo University, Ourense, Spain

PAULO EDUARDO SICHETTI MUNEKATA
Galician Meat Technology Center, Galicia Technology Park, Ourense, Spain

FRANCISCO J. BARBA
Department of Preventive Medicine and Public Health, Food Science, Toxicology and
Forensic Medicine, Faculty of Pharmacy, Universitat de València, València, Spain

ACADEMIC PRESS
An imprint of Elsevier

Academic Press is an imprint of Elsevier
125 London Wall, London EC2Y 5AS, United Kingdom
525 B Street, Suite 1650, San Diego, CA 92101, United States
50 Hampshire Street, 5th Floor, Cambridge, MA 02139, United States
The Boulevard, Langford Lane, Kidlington, Oxford OX5 1GB, United Kingdom

Copyright © 2021 Elsevier Inc. All rights reserved.

No part of this publication may be reproduced or transmitted in any form or by any means, electronic or mechanical, including photocopying, recording, or any information storage and retrieval system, without permission in writing from the publisher. Details on how to seek permission, further information about the Publisher's permissions policies and our arrangements with organizations such as the Copyright Clearance Center and the Copyright Licensing Agency, can be found at our website: www.elsevier.com/permissions.

This book and the individual contributions contained in it are protected under copyright by the Publisher (other than as may be noted herein).

Notices

Knowledge and best practice in this field are constantly changing. As new research and experience broaden our understanding, changes in research methods, professional practices, or medical treatment may become necessary.

Practitioners and researchers must always rely on their own experience and knowledge in evaluating and using any information, methods, compounds, or experiments described herein. In using such information or methods they should be mindful of their own safety and the safety of others, including parties for whom they have a professional responsibility.

To the fullest extent of the law, neither the Publisher nor the authors, contributors, or editors, assume any liability for any injury and/or damage to persons or property as a matter of products liability, negligence or otherwise, or from any use or operation of any methods, products, instructions, or ideas contained in the material herein.

Library of Congress Cataloging-in-Publication Data
A catalog record for this book is available from the Library of Congress

British Library Cataloguing-in-Publication Data
A catalogue record for this book is available from the British Library

ISBN 978-0-12-821233-2

For information on all Academic Press publications
visit our website at https://www.elsevier.com/books-and-journals

Publisher: Charlotte Cockle
Acquisitions Editor: Megan R. Ball
Editorial Project Manager: Barbara Makinster
Production Project Manager: Sreejith Viswanathan
Cover Designer: Victoria Pearson

Typeset by SPi Global, India

Working together to grow libraries in developing countries

www.elsevier.com • www.bookaid.org

Contents

Contributors

Rubén Agregán
Galician Meat Technology Center, Galicia Technology Park, Ourense, Spain

Francisco J. Barba
Department of Preventive Medicine and Public Health, Food Science, Toxicology and Forensic Medicine, Faculty of Pharmacy, Universitat de València, València, Spain

Houda Berrada
Department of Preventive Medicine and Public Health, Food Science, Toxicology and Forensic Medicine, Faculty of Pharmacy, Universitat de València, València, Spain

Benjamin Bohrer
Department of Animal Sciences, The Ohio State University, Columbus, OH, United States

Tomislav Bosiljkov
Faculty of Food Technology and Biotechnology, University of Zagreb, Zagreb, Croatia

Mladen Brnčić
Faculty of Food Technology and Biotechnology, University of Zagreb, Zagreb, Croatia

Louis Chonco
Research Unit of University Hospital Complex of Albacete, Albacete, Spain

Patricia Costa
Universidade Católica Portuguesa, CBQF, Centro de Biotecnologia e Química Fina, Laboratório Associado, Escola Superior de Biotecnologia, Porto, Portugal

Vilaine Corrêa da Silva
Laboratory of Applied Virology, Department of Microbiology, Immunology and Parasitology, Federal University of Santa Catarina (UFSC), Florianópolis, SC, Brazil

Pablo G. del Río
Department of Chemical Engineering, Faculty of Science, University of Vigo, Ourense, Spain

Rubén Domínguez
Galician Meat Technology Center, Galicia Technology Park, Ourense, Spain

Mohamed A. Farag
Department of Chemistry, School of Sciences and Engineering, The American University in Cairo; Pharmacognosy Department, College of Pharmacy, Cairo University, New Cairo, Egypt

Xi Feng
Department of Nutrition, Food Science, and Packaging, San Jose State University, San Jose, CA, United States

Emilia Ferrer
Department of Preventive Medicine and Public Health, Food Science, Toxicology and Forensic Medicine, Faculty of Pharmacy, Universitat de València, València, Spain

Gislaine Fongaro
Laboratory of Applied Virology, Department of Microbiology, Immunology and Parasitology, Federal University of Santa Catarina (UFSC), Florianópolis, SC, Brazil

Mohammed Gagaoua
Food Quality and Sensory Science Department, Teagasc Ashtown Food Research Centre, Dublin, Ireland

Andrés García
Animal Science Techniques Applied to Wildlife Management Research Group, Hunting Resources Research Institute; Hunting and Livestock Resources Section, Regional Development Institute; Department of Agroforestry Science and Technology and Genetics, Higher Technical School of Agricultural and Forestry Engineers, Castilla-La Mancha University, Albacete, Spain

Beatriz Gullón
Department of Chemical Engineering, Faculty of
Science, University of Vigo (Campus Ourense),
Ourense, Spain

Patricia Gullón
Nutrition and Bromatology Group, Department of
Analytical and Food Chemistry, Faculty of Food Science
and Technology, University of Vigo, Ourense, Spain

Anet Režek Jambrak
Faculty of Food Technology and Biotechnology,
University of Zagreb, Zagreb, Croatia

Damir Ježek
Faculty of Food Technology and Biotechnology,
University of Zagreb, Zagreb, Croatia

Sven Karlović
Faculty of Food Technology and Biotechnology,
University of Zagreb, Zagreb, Croatia

Tomás Landete-Castillejos
Animal Science Techniques Applied to Wildlife
Management Research Group, Hunting
Resources Research Institute; Hunting
and Livestock Resources Section, Regional
Development Institute; Department of
Agroforestry Science and Technology and
Genetics, Higher Technical School of
Agricultural and Forestry Engineers, Castilla-La
Mancha University, Albacete, Spain

Juliano De Dea Lindner
Food Technology and Bioprocess Research Group,
Department of Food Science and Technology,
UFSC, Florianópolis, SC, Brazil

José Manuel Lorenzo
Galician Meat Technology Center, Galicia Technology
Park; Food Technology Area, Faculty of Sciences of
Ourense, Vigo University, Ourense, Spain

Herbert L. Meiselman
Herb Meiselman Training and Consulting, Rockport,
MA, United States

Marília Miotto
Food Microbiology Laboratory, UFSC, Florianópolis,
SC, Brazil

Jane M. Misihairabgwi
Department of Biochemistry and Microbiology,
School of Medicine, Faculty of Health Sciences,
University of Namibia, Windhoek, Namibia

Paulo Eduardo Sichetti Munekata
Galician Meat Technology Center, Galicia Technology
Park, Ourense, Spain

Iwona Niedźwiedź
Department of Microbiology, Biotechnology and
Human Nutrition, University of Life Sciences in
Lublin, Lublin, Poland

Marinela Nutrizio
Faculty of Food Technology and Biotechnology,
University of Zagreb, Zagreb, Croatia

Diana Oliveira
Universidade Católica Portuguesa, CBQF, Centro
de Biotecnologia e Química Fina, Laboratório
Associado, Escola Superior de Biotecnologia,
Porto, Portugal

Noelia Pallarés
Department of Preventive Medicine and Public Health,
Food Science, Toxicology and Forensic Medicine,
Faculty of Pharmacy, Universitat de València,
València, Spain

Mirian Pateiro
Galician Meat Technology Center, Galicia Technology
Park, Ourense, Spain

Ricardo N. Pereira
CEB - Centre of Biological Engineering, University of
Minho, Braga, Portugal

Cristina Pérez-Santaescolastica
Galician Meat Technology Center, Galicia Technology
Park, Ourense, Spain

Magdalena Polak-Berecka
Department of Microbiology, Biotechnology and
Human Nutrition, University of Life Sciences in
Lublin, Lublin, Poland

Rui M. Rodrigues
CEB - Centre of Biological Engineering, University of
Minho, Braga, Portugal

David Rodríguez-Lázaro
Microbiology Division, Faculty of Sciences, University of Burgos, Burgos, Spain

Martina P. Serrano
Animal Science Techniques Applied to Wildlife Management Research Group, Hunting Resources Research Institute; Hunting and Livestock Resources Section, Regional Development Institute; Department of Agroforestry Science and Technology and Genetics, Higher Technical School of Agricultural and Forestry Engineers, Castilla-La Mancha University, Albacete, Spain

Doris Sobral Marques Souza
Laboratory of Applied Virology, Department of Microbiology, Immunology and Parasitology, Federal University of Santa Catarina (UFSC); Food Technology and Bioprocess Research Group, Department of Food Science and Technology, UFSC, Florianópolis, SC, Brazil

António J. Teixeira
CEB - Centre of Biological Engineering, University of Minho, Braga, Portugal

Igor Tomasevic
Department of Animal Source Food Technology, Faculty of Agriculture, University of Belgrade, Belgrade, Serbia

Theodoros Varzakas
Department of Food Science and Technology, Faculty of Agriculture and Food, University of the Peloponnese, Kalamata, Greece

António A. Vicente
CEB - Centre of Biological Engineering, University of Minho, Braga, Portugal

Min Wang
Department of Preventive Medicine and Public Health, Food Science, Toxicology and Forensic Medicine, Faculty of Pharmacy, Universitat de València, València, Spain

Lujuan Xing
College of Food Science and Technology, Nanjing Agricultural University, Nanjing, China

Sol Zamuz
Galician Meat Technology Center, Galicia Technology Park, Ourense, Spain

Wangang Zhang
College of Food Science and Technology, Nanjing Agricultural University, Nanjing, China

Jianjun Zhou
Department of Preventive Medicine and Public Health, Food Science, Toxicology and Forensic Medicine, Faculty of Pharmacy, Universitat de València, València, Spain

Preface

Food production has always been a vital activity for mankind, but the intensification of human activities and technological development shaped the environment, society, and the economy in order to follow the pace of global population growth.

Today, technology has a role beyond the intensification of food production by assisting in the progression of this sector towards a more sustainable system. The *Sustainable Production Technology in Food* book provides a comprehensive view about the current technological status, developments, related regulations to produce food in the context of sustainability as well as the consumers and these share of the food market.

This book is an up-to-date source of information for researchers, academics, and professionals working in sustainable food production and in the technological advances to improve this sector. The book introduces the current scenario of food production and the fundaments of technological advances and sustainability. The consumer demand and market of sustainable food product assist in the structure the background is also covered in the following chapter. Then, dedicated chapters provide up-to-date information about the use of technologies in crop and animal production, developments of current processing technologies, and the innovative processing technologies (biopreservation technologies, ultrasound, cold plasma, nanotechnology, and high-pressure processing, for instance). Finally, a chapter is centered in the regulatory scenario of sustainable food production.

- Provides up-do-date knowledge about technological advances to improve the sustainability in the food sector.
- Addresses the use of emerging and green technologies in food production.
- Comprehensively cover the technologic advances throughout the food productive chain.

CHAPTER 1

Modern Food Production: Fundaments, Sustainability, and the Role of Technological Advances

CRISTINA PÉREZ-SANTAESCOLASTICA[A] • PAULO EDUARDO SICHETTI MUNEKATA[A] • MIRIAN PATEIRO[A] • RUBÉN DOMÍNGUEZ[A] • JANE M. MISIHAIRABGWI[B] • JOSÉ MANUEL LORENZO[A,C]
[a]Galician Meat Technology Center, Galicia Technology Park, Ourense, Spain, [b]Department of Biochemistry and Microbiology, School of Medicine, Faculty of Health Sciences, University of Namibia, Windhoek, Namibia, [c]Food Technology Area, Faculty of Sciences of Ourense, Vigo University, Ourense, Spain

1.1 INTRODUCTION

With the ever-growing global population, food production, which entails the transformation of raw materials into prepared food products, is a major priority. According to the Food and Agriculture Organization (FAO), by 2050, the population is estimated to increase by 30%, so food production should increase by 70% (FAO, 2009). Increasingly diverse consumer demands have oriented food production trends, with food manufacturers striving for economic viability and sustainability in these modern consumer markets. Commensurate with current consumer demands, which include the demand for healthy foods and the use of safe, environmentally friendly technologies, the food industry has directed efforts towards the development of innovative food production technologies and novel foods (Chemat et al., 2020; Granato et al., 2020; Vargas-Ramella et al., 2020).

According to the regulations of the European Union, "novel food" pertains to any food that has not been significantly used for human consumption in the Union prior to15 May 1997, when the first novel food legislation took effect (European Regulation, 2015). This regulation encompasses 10 food categories, foods from non-European cultures, as well as innovative foods or foods made differently from the traditional ones, either by the incorporation of new ingredients or by the use of new technologies in their production.

As shown in Fig. 1.1, many factors drive the development of new food products, opening up various research avenues. The emergence of these new foods has been the result of changes in consumer preferences in recent years. Almost 29% of global greenhouse gases are linked to agriculture and food production. Nearly half of this value comes from livestock production, since currently, about 70% of agricultural land is used for livestock (Bailey, Froggatt, & Wellesley, 2014). Moreover, almost 92% of the freshwater is used for agriculture and food production, contaminating freshwater resources, leading to climate change, and affecting natural biodiversity (Gerber et al., 2013). The great concern for the environment and animal welfare has caused a rejection of meat products. The agricultural initiatives that were carried out, such as the use of fertilizers and antibiotics to promote growth triggered imbalances in nature as well as the incidence of diseases, increasing consumer rejection of animal based foods (Lymbery, 2014).

In light of the shift in consumer preferences, the food industry has been forced to search production strategies that favor the development of meat alternatives, to investigate new sources of protein and to improve techniques that allow the addition of health-benefiting ingredients or remove harmful constituents from food (Asgar, Fazilah, Huda, Bhat, & Karim, 2010; Das et al., 2020; López-Pedrouso, Lorenzo, Gullón, Campagnol, & Franco, 2021). Moreover, lifestyle changes gave impetus to consumer desire for convenient, ready to eat foods or foods requiring minimal preparation time. Faced with this, the food industry has had to develop different production techniques as well as new packages that allow convenience and comfortable use for consumers (Domínguez et al., 2018; Horita et al., 2018; Lorenzo, Batlle, & Gómez, 2014;

Copyright © 2021 Elsevier Inc. All rights reserved.

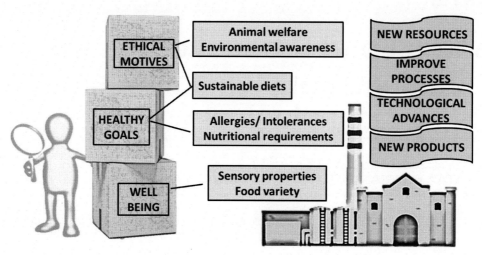

FIG. 1.1 Consumer concerns and the response observed in the objectives of the food industry.

Lorenzo, Domínguez, & Carballo, 2017; Pateiro et al., 2019; Santeramo et al., 2017; Umaraw et al., 2020). Furthermore, health concerns have boosted the demand for products with increased nutritional properties, since over the years the term nutrition has evolved, and no longer means only providing energy or promoting growth, but also is seen as a means of preventing disease as well as improving physical and mental health as seen in several studies.

Some of the health strategies adopted by the industry for the development of new products consist of limiting energy content of total fats, replacing saturated fats with unsaturated fats, eliminating trans fatty acids and increasing the content of n-3 fatty acids from fish oil or vegetable sources (Barros et al., 2020; da Silva et al., 2019; de Carvalho et al., 2019; de Oliveira Fagundes et al., 2017; Domínguez, Agregán, Gonçalves, & Lorenzo, 2016; Domínguez, Pateiro, Agregán, & Lorenzo, 2017; Domínguez, Pateiro, Munekata, Campagnol, & Lorenzo, 2017; Franco, Munekata, et al., 2020; Franco, Martins, et al., 2020; Heck et al., 2017). Additionally, strategies such as reducing the content of salt and refined carbohydrates and development of new formats that enhance the consumption of fruits and vegetables have been adopted (Cofrades, Benedí, Garcimartin, Sánchez-Muniz, & Jimenez-Colmenero, 2017; da Silva et al., 2020; Domínguez, Pateiro, Pérez-Santaescolástica, Munekata, & Lorenzo, 2017; Lorenzo et al., 2015; Nachtigall et al., 2019). Among the long list of strategies, functional foods have been the object of several studies for a long time, being understood as foods that provide health benefits, or have the potential

to prevent diseases (Griffiths, Abernethy, Schuber, & Williams, 2009).

1.2 NOVEL FOOD TRENDS

This section aims to provide an overview of the many newly developed, innovative foods or food produced using emerging technologies on which the food industry has focused its attention. Animal protein food sources due to the continuing world population growth and consequent rise in global food production demand, conventional animal protein food sources, which include beef, pork, lamb, goat and chicken meat, may become insufficient to meet consumer demands (Kearney, 2010). Livestock farming has been criticized for its huge detriments to the environment. Owing to the limited number of suitable areas available for extensive livestock production, forests have been converted into livestock ranches, making livestock production a major cause of global deforestation and consequent loss of biodiversity and climate change (Food and Agriculture Organization of the United Nations, 2012; Herrero et al., 2016). Additionally, livestock production has been associated with high water consumption and pollution, and high greenhouse gas emissions. There have also been growing concerns over the slaughter of animals, which has been perceived as being cruel and inhumane. In view of these concerns, some consumers have advocated for cutting down or eliminating the consumption of animal protein food sources, giving impetus to exploration of environmentally friendly, sustainable alternative protein food sources, which include insects, cultured meat, mycoproteins, algae and aquatic foods (Parodi et al., 2018).

1.2.1 Insects

Entomophagy, the consumption of insects, has great potential as a replacement for consumption of conventional animal protein sources, considering its comparative nutritional, economic and ecological advantages. With respect to nutritional quality and health, insects are 23.5%, 26.7%, and 41.1% richer in protein than beans, lentils and soybean, respectively (Blásquez, Manuel, Moreno, Hugo, & Camacho, 2012), with high protein digestibility in the range of 75%–98% for most insects (Teffo, Toms, & Eloff, 2007). Most insect proteins contain all the essential amino acids (Zielińska, Baraniak, Karaś, Rybczyńska, & Jakubczyk, 2015). Some researchers have highlighted that bioactive proteins and peptides derived from some edible insects possess antioxidant capacity (Yang et al., 2013), angiotensin-I converting enzyme (ACE) inhibitory activity (Wu, Jia, Yan, Du, & Gui, 2015), antiviral, antibacteria, antiinflammatory and pain-alleviating effects (Lin & Li, 2008; Xiaodong & Bo, 2005) and could enhance mineral bio absorption (Sasaki, Yamada, & Kato, 2000). Others have posited that some bioactive carbohydrates enhance immunity, wound-healing and defense effects in parasitic infections or allergy (He, Tong, Huang, & Zhou, 1999; Long, Ying, Zhao, Tao, & Xin, 2007). Regarding the lipid profile, insects present relatively high omega-3 fatty acids (Van Huis et al., 2013) low saturated fatty acid contents (Rumpold & Schlüter, 2013a, 2013b). Furthermore, insects provide a good source of minerals and vitamins (Rumpold & Schlüter, 2013a, 2013b). The benefits of insect farming are not only attributed to their nutritional quality, but also due to the reduced requirements for land and water (an essential limited resource in many regions of the world), and reduced emission of greenhouse gases. Moreover, insect farming is simple, with no in-depth training required, transportation is easy, and products that are not consumed by humans can be used as livestock feed. Since insects are cold-blooded animals, the feed conversion ratio is higher than that for livestock, and due to their short life cycles, investment can be returned faster and financial returns are higher (Mlcek, Rop, Borkovcova, & Bednarova, 2014).

Entomophagy has been traditionally practiced mostly in Asia, Africa, and South America for thousands of years (Bodenheimer, 2013), but is still uncommon in Western societies due to cultural customs (Sunwaterhouse et al., 2016). The majority of people who

reject insects as food consider them unclean (House, 2016; Megido, Haubruge, & Francis, 2018) despite most of the edible insects being herbivores which feed on plant leaves or wood (Gullan & Cranston, 2014). Based on their predominant diet, insects could actually be considered cleaner than crabs or lobsters, which eat waste that sometimes comes from contaminated water (Mitsuhashi, 2016). Over time, traditional foods such as frogs and lobsters, which were initially rejected, have become acceptable (Paoletti, 2005; Tao & Li, 2018). Consumption of edible insects may follow the same trend since the people who eat insects on a regular basis do so because of their taste (Nonaka, 2009).

1.2.2 Meat-Based Foods

Despite meat and meat products being important sources of quality nutrients in the human diet, there has been increasing concern regarding the negative health impacts of consumption of meat and its products. Meat intake has been associated with increased incidence of various diseases which include cardiovascular disease, obesity, diabetes mellitus, and hypertension. Consequently, consumer concerns with regards to health of meat and meat products have oriented research into reformulation of meat and meat products with the objective of obtaining healthier alternatives. The design and development of functional meat-based foods is generally based on reducing compounds with a detrimental effect on health, and/or increasing compounds whose presence is beneficial. These modifications, whether qualitative or quantitative, are achieved through strategies based on animal production (genetic and nutritional) and meat transformation systems (reformulation process) (de Oliveira Fagundes et al., 2017; Martins et al., 2019). Reformulation has been widely used to eliminate, reduce, increase, add and/or replace different bioactive components (Al Khawli et al., 2019; Astray, Gullón, Gullón, Munekata, & Lorenzo, 2020; Domínguez et al., 2020; Echegaray et al., 2018; Falowo et al., 2018; Gullón et al., 2020; Gullón, Astray, Gullón, Tomasevic, & Lorenzo, 2020), as is the case of meat product preparations with ingredients of plant origin (soy, nuts, oils, oats, rice, wheat, carrots, etc.), whose purpose is to improve fat content, incorporate antioxidants, prebiotics and dietary fiber, and enrich with minerals, etc. (Rocchetti et al., 2020). To extend the knowledge and to understand the high importance that the food industry has given to this area, in Table 1.1 are shown several examples of the studies that have been carried out in recent years on reformulation of meat products for various intended health benefits.

TABLE 1.1
Some Studies on Reformulation of Meat Products and Their Documented Health Benefits.

Health Benefit of Reformulation	References
To reduce salt content	López-López, Cofrades, and Jiménez-Colmenero (2009), López-López, Cofrades, Ruiz-Capillas, and Jiménez-Colmenero (2009), and López-López, Cofrades, Yakan, Solas, and Jiménez-Colmenero (2010)
To reduce fat content	Cofrades, López-López, Ruiz-Capillas, Triki, and Jiménez-Colmenero (2011), López-López, Cofrades, and Jiménez-Colmenero (2009), and López-López, Cofrades, Ruiz-Capillas, et al. (2009)
To replace saturated fats with monounsaturated or polyunsaturated fats	Delgado-Pando et al. (2011), López-López, Cofrades, and Jiménez-Colmenero (2009), and López-López, Cofrades, Ruiz-Capillas, et al. (2009)
To elevate omega-3 fatty acids content	López-López, Cofrades, and Jiménez-Colmenero (2009) and López-López, Cofrades, Ruiz-Capillas, et al. (2009)
To provide an extra content of minerals	López-López, Cofrades, and Jiménez-Colmenero (2009), López-López, Cofrades, Ruiz-Capillas, et al. (2009), and López-López et al. (2010)
To enhance protein content	Cofrades, López-López, and Jiménez-Colmenero (2012) and Cofrades et al. (2011)

1.2.3 Cultured Meat

An alternative to reduce the negative effects of meat production without sacrificing the advantages of meat is the in vitro growth of animal cell meat (cultured meat). This approach is in the development stages, and it will take years to be commercially available (Mattick & Allenby, 2013). However, although food products derived from cloned animals are considered safe for human consumption in the United States, they are not yet allowed in the European Union (Bonny, Gardner, Pethick, & Hocquette, 2015).

Non animal protein food sources as the world's population increases, the demand for protein increases, and although consumers are more aware of the environmental, animal welfare and health problems which arise from meat production, only a small part of the world's population follow a vegetarian or vegan diet. Despite increasing awareness of associated negative concerns, meat is a traditionally consumed product in many cultures, being considered healthy and nutritious (Schösler, De Boer, & Boersema, 2012; Verbeke et al., 2010). Therefore, the food industry has been forced to find new sources of protein, and has developed strategies to reduce meat consumption. Vegetable protein has been studied as a good alternative to meat protein due to the wide variety of sources, such as legumes and oilseeds. Additionally, cereals and fungi have been explored as alternative protein sources. Among these alternative protein sources of nonanimal origin, cereals, legumes and soybeans have been the most used (Day, 2013; De Boer, Schösler, & Aiking, 2014; Van Der

Weele, Feindt, Van Der Goot, Van Mierlo, & van Boekel, 2019). The use of vegetable proteins provides environmental advantages since parts of the plant that are not intended for food can be used for feed or chemical products, achieving more sustainable production (Sari, Mulder, Sanders, & Bruins, 2015).

1.2.4 Meat Analogues and Meat Extenders

Meat analogues, can be defined as products that mimic meat in its functionality, bearing similar appearance, texture, and sensory attributes to meat. Production of meat analogues has been on the increase, targeted at satisfying consumers' desire for indulgent, healthy, low environmental impact, and ethical meat substitutes. The wide range of ingredients used, variety of products made and the nutritional value of meat analogues have been extensively studied in recent years (Asgar et al., 2010; Bohrer, 2017). Meat analogues can appear in different sizes (from 6 to 20 mm) and shapes (sheets, discs, cakes, strips and others) (Riaz, 2004) to resemble hamburgers, steaks, chicken burgers, sausages, slices of luncheon meat, Canadian bacon, stuffed turkey and many other meat products (Asgar et al., 2010). These products have been well received by consumers due to their healthy image (no cholesterol, low fat, and low calorie), good taste and low cost. Among the most used analogue ingredients we can outline (Egbert & Borders, 2006):

- Water, whose function is as an emulsifier as well as providing juiciness, with concomitant cost reduction.

- Textured vegetable proteins, essentially soy, wheat, and their combinations. They are used to improve the mouthfeel and to simulate the "original meat-texture" in the analogues, owing to the mouth feel and texture generated when they are hydrated during the cooking process.
- Nontextured proteins, such as soy concentrates, wheat gluten, egg white and whey. They are used as emulsifiers, improving water binding, texture, and mouth feel.
- Fats and oils. To improve flavor and texture as well as to contribute to Maillard reactions and enzymatic browning.
- Flavors, spices and coloring agents whose purpose is to mask cereal notes, to enhance meat flavors and odors, and to modify the appearance to make the product analogous to the original meat product.
- Binder agents like gums, hydrocolloids, enzymes, and starches. Mainly, they are used to achieve adequate texture, besides acting in water binding, and can provide fiber.

Meat extenders are nonmeat substances characterized by substantial protein content. Unlike meat analogues, when extenders are consumed alone, they do not resemble meat in appearance, texture, or mouth feel. Meat extenders are in the form of layers (>2 mm), minced (>2 mm) and pieces (15–20 mm), and are characterized by absorbing a great amount of water, which may constitute between 2.5 and 5 times their weight (Riaz, 2004). By improving water-binding properties, efficiency and texture are also improved. Processed meat products, in which part of the meat has been replaced by plant-based extenders, have been developed (Boland et al., 2013). In minced meat products, for instance, nonmeat proteins are often used as alternative gelling agents (Pietrasik, Jarmoluk, & Shand, 2007). Similarly, vegetable proteins such as wheat gluten or soy concentrates and isolates are used to join cuts of meat and trimmings to make chicken rolls and pressed masses (Singh, Kumar, Sabapathy, & Bawa, 2008), as well as to improve the texture and quality of meatballs, ground beef and sausages (Asgar et al., 2010). Combination of meat extenders with meat has been employed as a means of reducing the cost of meat without reducing its nutritional value.

In thick minced meats (meat patties, sausages, meat sauces, etc.), together with soy flour, textured soy protein concentrates are also used to obtain the required final texture (Asgar et al., 2010). According to the United States Department of Agriculture (USDA), textured vegetable protein products are described as food products from edible protein sources, whose structural integrity and recognizable structure makes them able to resist cooking preparation procedures for their consumption" (Textured Vegetable Protein Products (B-1), 1971). During extrusion cooking, there is a texture modification, protein denaturation, trypsin inhibitor inactivation, in addition to controlling bitter flavors (Björck & Asp, 1983; Hayakawa, Hayashi, Urushima, Kajiwara, & Fujio, 1989). The main characteristics that should be taken into account in the choice of raw materials for texturizing are the particle sizes, the quantity and quality of protein, the quantity and type of sugar, and the levels of oil and fiber (Strahm, 2006). In this regard, soy is the most used raw material for the production of textured vegetable proteins by extrusion, employing both defatted soy flour (50%–55% protein) and soy protein concentrate (65%–70% protein) or isolates of soy protein (85%–90% protein) (Golbitz & Jordan, 2006; Riaz, 2004). Other raw materials used in the extrusion process for texturizing are wheat, sunflower, peanuts, sesame, peas, and beans (Riaz, 2004; Strahm, 2006). On the other hand, rapeseed is a crop with a high nutritional value and great economic importance. In 2008, 48.4 million metric tons of rapeseed were produced worldwide (American Soybean Association, 2009). Notwithstanding, its use has been restricted to animal feed (Bos et al., 2007). It has been observed that rapeseed protein is suitable for the production of texturized products and, enzymatic modification with microbial transglutaminase can improve its gelation properties (Pinterits & Arntfield, 2008). By introducing new transglutaminase crosslinks, functional properties can be improved, extending the potential use of rapeseed as nonmeat proteins (Pietrasik et al., 2007).

1.2.5 Single-Cell Proteins

Single-cell proteins (SCP) are proteins derived from pure or mixed cultures of microorganisms such as bacteria, microalgae, yeasts, and fungi (Upadhyaya, Tiwari, Arora, & Singh, 2016). These proteins are produced when microorganisms are cultured in various agricultural wastes or appropriate media, after which biomass is harvested for consumption (Chandrani-Wijeyaratne & Tayathilake, 2000). SCP has potential as an alternative non meat protein food source, and it also contains lipids and vitamins. Worth noting is mycoprotein, a meat substitute of fungal origin, produced by a fermentation process employing the microorganism *Fusarium venenatum*. At the end of the fermentation process, the broth is heated to decrease RNA content and biomass is harvested either by centrifugation or by filtration. To make the texture more similar to meat, calcium can be added to improve intrahyphal crosslinking. Via pressure treatment, the mixed mass forms block, which are steam-heated leading to protein denaturation. Finally, to obtain the desirable fleshy texture, the mass is cooled and frozen (Hashempour-baltork, Khosravidarani, Hosseini, Parastou Farshi, & Reihani, 2020). Mycoprotein, whose texture is perceived to be similar to chicken breast, is marketed as a healthy source of protein and fiber, with lower saturated fat relative to meat.

The most important factors influencing the SCP production process include environmental protection, cost, and safety (Suman, Nupur, Anuradha, & Pradeep, 2015). The total cost of mycoprotein production depends on the fermentation substrates (Wiebe, 2004), so the total cost can be reduced by using agro-industrial by-products such as those from pea processing industries (Anupama & Ravindra, 2000), waste materials (Gabriel, Victor, & du Preez, 2014), which include molasses, starch, fruit and vegetable wastes, as well as natural gases (Bekatorou, Psarianos, & Koutinas, 2006).

It has been shown that not only the type of microorganism and the substrates influence the nutritional composition of mycoproteins, but also the collection and processing methods. (Reihani & Khosravi-darani, 2019; Upadhyaya et al., 2016). Nevertheless, the general composition of mycoprotein is around 13 g fat (mainly polyunsaturated fatty acids, highlighting linoleic and linolenic acids), 45 g protein (with biological value similar to milk proteins), 10 g carbohydrate, and 25 g fiber per 100 g dry matter (Finnigan, Needham, & Abbott, 2017). Although the amount of sodium is low, the amounts of zinc and selenium are significant (Denny, Aisbitt, & Lunn, 2008).

Concerning the environment, it can be assumed that due to the low environmental effects associated with its production, mycoprotein production can be a suitable solution for existing environmental deterioration problems. Proof of that is the observed carbon footprint values which are at least four times lower than chicken meat and ten times less than beef (Dairy UK, 2010; MacLeod et al., 2013).

1.2.6 Milk Substitutes

Milk contains macro and micronutrients important for human health. Even though dairy products are the main source of calcium, some metabolic diseases and allergies make digestion difficult, requiring the person affected to cut out these products from the diet (Pereira et al., 2012). To reduce the effects of these diseases, lactase capsules or liquid lactase have been developed. These products are added to foods or meals that contain lactose and, therefore, partial hydrolysis of the lactose present in them is achieved, decreasing symptoms

(da Cunha, Suguimoto, de Oliveira, Sivieri, & de Costa, 2008; Mattar & Mazo, 2010). However, these products have a high cost and the people who suffer from the symptoms mostly choose to eliminate milk from the usual diet. Taking into account that the incidence of lactose intolerance is 75% of the population (Gasparin, Teles, & de Araújo, 2010), the need for products and derivatives without milk has increased, and the development of new products to substitute milk has become vital. Consequently, one of the strategies adopted is the development of the so-called "plant based milk substitutes", which are drinks similar to regular milk in appearance, consisting of homogenates of plant based-extracts from nuts, such as walnuts, almonds, Brazil nuts, cashews and hazelnuts, corn, legumes like soybeans and chickpea, oilseeds such as sesame and sunflower, cereals like rice and oats or pseudo cereals such as quinoa (Sethi, Tyagi, & Anurag, 2016).

The nature of the raw material, the extraction method, as well as the storage conditions, will determine the stability and particle size of the final product (Cruz et al., 2007). However, the stability of plant-based milk is different from regular milk, as well as the sensory and nutritional attributes (Sethi et al., 2016). Since the nutritional value is different among the plant bases, some strategies are carried out, including fortifying with proteins, or mixing different varieties of plant-based milks to obtain a product with a higher nutritional value equivalent to regular milk. Also, it has been shown that partial protein hydrolysis through the addition of enzymes can improve extraction yields (Silva, Silva, & Ribeiro, 2020; Suphamityotin, 2011).

1.2.7 Marine Products

Algae have a high potential for use in the development of functional foods. This is due to the large number of bioactive compounds with positive health effects that can be extracted from them, such as high-quality proteins, minerals, vitamins, essential fatty acids, polyphenols, carotenoids, tocopherols, etc. (Agregán et al., 2017; Agregán et al., 2017; Lorenzo et al., 2017; Lorenzo, Domínguez, & Carballo, 2017; Lorenzo, Sineiro, Amado, & Franco, 2014; Marti-Quijal et al., 2019; Parniakov et al., 2018). Furthermore, their high fiber content also imparts health benefits. Algae can not only be used for incorporation as a nutritional supplement in food, but also as a natural food coloring agent (Apt & Behrens, 1999). Therefore, there are a large variety of products in which algae can be incorporated such as pastries, snacks, candy bars or chewing gum, and drinks (Liang, Liu, Chen, & Chen, 2004). The most common currently used commercial algae strains include Arthrospira, Chlorella, *D. salina* and *Aphanizomenon flos-aquae* (Spolaore, Joannis-Cassan, Duran, & Isambert, 2006). A considerable number of studies have been carried out in recent years to evaluate the effect of the incorporation of different varieties of algae or their extracts on the nutritional, texture, and organoleptic properties of diverse types of food (Table 1.2).

Despite research studies to validate the use of algae, its incorporation into food is still limited mainly due to the low powder consistency of dry biomass, its color and the fishy smell, added to the total cost of production which is very high compared to the costs of conventional protein production (Becker, 2007).

Besides the incorporation of algae (both macro and microalgae), fish oils, and proteins are also currently

TABLE 1.2

Examples of Foods in Which Algae Have Been Incorporated for Nutritional, Texture and-or Organoleptic Purposes.

Product	Algae	Reference
Meat and meat products	*Laminaria digitate*	Moroney, O'Grady, Lordan, Stanton, and Kerry (2015) and Moroney, O'Grady, O'Doherty, and Kerry (2013)
	Ulva lactuca and *Ulva rigida*	Lorenzo, Sineiro, et al. (2014)
	Sea Spaghetti (*Himanthalia elongata*)	Cofrades, López-López, Solas, Bravo, and Jiménez-Colmenero (2008), Cox and Abu-Ghannam (2013a, 2013b), Fernández-Martín, López-López, Cofrades, and Colmenero (2009); Jiménez-Colmenero et al. (2010), López-López, Cofrades, and Jiménez-Colmenero (2009), and López-López, Cofrades, Ruiz-Capillas, et al. (2009)
	Wakame (Undaria pinnatifida)	Cofrades et al. (2008), López-López et al. (2011), López-López, Cofrades, and Jiménez-Colmenero (2009), López-López, Cofrades, Ruiz-Capillas, et al. (2009), López-López et al. (2010), and Sasaki et al. (2008)
	Nori (*Porphyra umbilicalis*)	Cofrades et al. (2008), López-López, Cofrades, and Jiménez-Colmenero (2009), and López-López, Cofrades, Ruiz-Capillas, et al. (2009)
	Sea tangle (*Lamina japonica*)	Kim et al. (2010)
Seafood products	*Fucus vesiculosus*	Dellarosa, Laghi, Martinsdóttir, Jónsdóttir, and Sveinsdóttir (2015) and Wang et al. (2010)
	Cochayuyo, sea lettuce, ulte, and red luche	Ortiz, Vivanco, and Aubourg (2014)
	Nori (*Porphyra tenera*) and Hijiki (*Hijikia fusiformis*)	Ribeiro et al. (2014)
	Eucheuma	Senthil, Mamatha, and Mahadevaswamy (2005)
Bread	*Palmaria palmata*	Fitzgerald et al. (2014)
	Kappaphycus alvarezii	Mamat et al. (2014)
	Sea Spaghetti (*Himanthalia elongata*)	Cox and Abu-Ghannam (2013a, 2013b)
	Myagropsis myagroides	Lee et al. (2011)
Pasta	*Monostroma nitidum*	Chang, Chen, and Hu (2011) and Chang and Wu (2008)
	Wakame (*Undaria pinnatifida*)	Prabhasankar et al. (2009)
Milk	*Ascophyllum nodosum* and *Fucus vesiculosus*	O'Sullivan et al. (2016)
Others	Eucheuma (*Kappaphycus alvarezzi*)	Senthil, Mamatha, Vishwanath, Bhat, and Ravishankar (2011)
	Enteromorpha	Mamatha, Namitha, Senthil, Smitha, and Ravishankar (2007)

used to increase the functionality of foods such as meat, dairy, or vegetable products (Jiménez-Colmenero, 2007). As shown in Table 1.3, a great number of functional materials including polysaccharides, polyunsaturated fatty acids, minerals, vitamins, antioxidants, collagen, gelatin, enzymes and bioactive peptides, among others, can be obtained from marine fish (Al Khawli et al., 2019; Franco, Munekata, et al., 2020;

Gómez et al., 2019; Pateiro et al., 2020). Fish bones, muscles, body, internal organs, skin, and bones are currently used as a sources of these bioactive compounds (Atef & Ojagh, 2017; Barrow & Shahidi, 2007).

Interest in the use of fish protein hydrolysates (FPH) has increased because the hydrolysis processes maintain the high content of essential amino acids in addition to providing improvements in physicochemical

TABLE 1.3
Types of Functional Material From Marine Fish.

Bioactivity	Material	Fish of Origin
Antioxidant	Frame	Cod, Flounder fish, Hoki, Pollack, Sandfish, Sole, Tilapia
	Skin	Amur sturgeon, Bluefin leatherjacket, Hoki, Pollack, Snapper, Sole, Tilapia
	Gelatin	Cod
	Scale	Yellowtail
	Bone	Cod, Flying fish, Yellowtail, Tuna
	Whole	Herring
	Body	Herring
	Head	Herring, Bluefin leatherjacket, Sardinella
	Gonads	Herring
	Viscera	Mackerel, Black Pomfret, Sardinella, Nile tilapia
	Carcass	Nile tilapia
	Pectoral fin	Salmon
	Cartilage	Skate
ACE inhibitory	Frame	Cod, Flounder fish, Leatherjacket, Lizard fish, Loach, Pollack, Shark, Tilapia
	Gelatin	Cod
	Skin	Pollack, Salmon
	Scale	Sea Bream, Yellowtail
	Bone	Yellowtail
	Head	Sardinella
	Viscera	Sardinella
Ca-binding	Frame	Hoki, Pollack
	Bone	Hoki
Fe-binding	Collagen	Milkfish
Learning and memory	Collagen	Salmon
Neuroprotective	Frame	Lantern fish
Neurobehavioral	Skin	Salmon
Antiproliferative	Bone	Flying fish
Anticoagulant	Frame	Sole
Antihypertensive	Frame	Tilapia

TABLE 1.3
Types of Functional Material From Marine Fish—cont'd

Bioactivity	Material	Fish of Origin
Antidiabetic	Livers	Shark
Cryoprotective	Skin	Amur sturgeon
Long bone development	Skin	Salmon
Facial Skin Quality	Collagen	Tilapia
Human breast cancer cell	Dark muscle	Tuna

properties that are very useful in the food industry. So, improvements in important functional properties, in terms of food formulation, such as emulsifying capacity, gelling capacity, texture, solubility, water retention capacity, and oil fixing capacity, have been found (Foh, Wenshui, Amadou, & Jiang, 2012; Gbogouri, Linder, Fanni, & Parmentier, 2004; Geirsdottir et al., 2011; Kristinsson, 2007; Thiansilakul, Benjakul, & Shahidi, 2007). Moreover, FPH derived from fish like cod and mackerel have shown antioxidant activities as well as a protective effect against fading and deterioration, extending the food shelf life (Guérard & Sumaya-Martinez, 2003; Herpandi, Rosma, & Wan Nadiah, 2011; Wu, Chen, & Shiau, 2003).

1.3   NEW TECHNOLOGIES
1.3.1   Nonthermal Processing Technologies
Non thermal processing technologies are mainly used to increase the extraction of compounds through cell rupture and to prevent microbial growth, prolonging the shelf life of food, since, due to mechanical effects, the cell membrane is damaged by the phenomenon of cavitation or electroporation. The advantages of using nonthermal technologies in food processing include enabling acceleration of heat and mass transfer, shortening processing times, controlling Maillard's reactions, deactivating enzymes, improving product quality and functionality, protecting against environmental strains, and extending shelf life (Hernández-Hernández, Moreno-Vilet, & Villanueva-Rodríguez, 2019).

Among the most important currently available technologies, and those that have potential to be commercialized in the coming years, are high hydrostatic pressure, pulsed electric field, ionizing radiation technologies and cold atmospheric plasma (Jermann, Koutchma, Margas, Leadley, & Ros-Polski, 2015). Each of these techniques is briefly explained below. Together with a description of the technique called cold atmospheric plasma, which without having the same marketing potential right now,

it has been seen that could offer many advantages to the food industry. Table 1.4 shows a general comparison of all these techniques.

High hydrostatic pressures. Technology that has been used to extract active ingredients from plant sources (Shouqin, Junjie, & Changzhen, 2004) and to extend the shelf life of products. Since the process is similar to cold pasteurization, with little effect on the sensory and nutritional characteristics of the food, it has become an innovative treatment for fresh products (Chawla, Patil, & Singh, 2011). However, it can present effects on physical properties (melting point, density, viscosity, solubility, etc.), structural changes (cell deformation, protein denaturation, etc.), and on equilibrium processes (acid-base equilibrium, ionization, dissociation of weak acids, etc.) or process rates (reaction speed variations).

Pulsed electric field (PEF). The application of electrical pulses induces local structural variations and fast decomposition of the cell membrane that can be permanent or temporary called electroporation (Toepfl, Siemer, Saldaña-Navarro, & Heinz, 2014). This makes nonthermal processing useful in food preservation because it causes less degradation of nutritional and sensory characteristics than traditional technologies (Agcam, Akyildiz, & Evrendilek, 2016). It has been widely used in many applications, such as pasteurization, extraction of bioactive compounds or food components, and modification of molecules. This processing offers increased mass transfer, reduced degradation of heat-sensitive compounds, increased extraction yield and decreased intensity of extraction parameters and processing time, facilitation of the purified extract, reduced energy costs and environmental impact (Barba et al., 2015). PEFs have been successfully used in liquids and semi-solid products. However, it has been observed that it is not suitable for solid food products (Toepfl et al., 2014). Despite commercial applications of PEF technology in recent years as a method of pasteurization of heat-sensitive foods, due to limited treatment capacity and high cost, it has limited treatment.

TABLE 1.4
Comparative Overview of Nonthermal Processing Techniques.

Technique	High Hydrostatic Pressure (HHP)	Pulsed Electric Field (PEF)	Ultrasound (US)	Ionizing Radiation (IOR)	Cold Atmospheric Plasma (CAP)
Effect	On physical properties (melting point, density, viscosity, solubility, etc.). On structural changes (cell deformation, protein denaturation, etc.). On equilibrium processes (acid—base equilibrium, ionization, dissociation of weak acids, etc.) or process rates (reaction speed variations).	Local structural changes and a rapid, permanent, or temporary, breakdown of the cell membrane.	Mechanical and chemical effects on matrix food.	Lethal-effect mechanism Bacterial DNA damage, avoiding the division of cells by inhibiting DNA synthesis. Production of active molecules leading to cell lysis.	Inactivation of spores by several synergistic mechanisms of surface erosion, UV radiation, and oxidation caused by ozone and reactive species, and charged particles.
Advantages	Can be applied with packaging. Little effect on the sensory and nutritional characteristics. Kills bacteria in the raw food. Increase shelf-life. Additive free. Enhances desired attributes (digestibility).	Increases the mass transfer. Reduces the degradation of heat-sensitive compounds, Increases the extraction yield. Decreases the intensity of extraction parameters. Decreases the processing time. Facilitates the extract purification. Reduces energy costs. Reduces the environmental impact.	Reduces processing time. Higher throughput. Lower energy consumption. Green technology friendly to the environment. Economically feasible. Meets process requirements as simplicity, energy efficient and scale up.	Less processing time. Excellent penetration into foods, processing in packaged products. Avoid recontamination. Reliable. Little loss of food quality. Suitable for large-scale production. Low energy cost. Suitable for dry foods, eco-friendly without chemicals and residues generated.	Low water usage. Low operating temperatures. Low costs.
Disadvantages	High cost. Long-term investment requirements.	Not suitable for solid food products. High cost.	Degradation of food properties (flavor, color, or nutritional value).	No permitted in all countries. High capital cost. Localized risks from radiation. Poor consumer acceptance. Changes in flavor. Quality sensorial and nutritious losses.	Produce off flavor compounds and rancidity in high lipid products.

TABLE 1.4
Comparative Overview of Nonthermal Processing Techniques—cont'd

Technique	High Hydrostatic Pressure (HHP)	Pulsed Electric Field (PEF)	Ultrasound (US)	Ionizing Radiation (IOR)	Cold Atmospheric Plasma (CAP)
Applications	To extract active ingredients. To extend the shelf life of products. Manipulates the texture of fresh food.	Pasteurization. Extracting bioactive compounds or food component. Modifying molecules.	Emulsification. Homogenization. Extraction. Crystallization. Dewatering, degassing, and defoaming.	Pasteurization. Sterilization. Inactivation of microorganisms in packaging materials. Inactivation of spores and pathogenic bacteria. Delayed senescence and maturation process. Disinfestation.	Sterilization. Disinfection. Inactivation of enzymes. Altering the hydrophilic/hydrophobic properties Etching or deposition of thin films.

Ultrasound. The application of ultrasound causes mechanical and chemical effects on food due to the generation, growth, and eventual collapse of bubbles in the matrix (called acoustic cavitation). Two types of applications can be distinguished depending on the frequency and energy used: low-intensity sonication when frequencies are above 100 kHz and energies below 1 W/cm^2 and high-intensity sonication when frequencies are below 100 kHz and energies above 10 W/cm^2). High-intensity sonication produces physical, biochemical, chemical, and mechanical changes that include mechanical damage to the cell wall caused by pressure gradients, micro-transmission within the cell and disintegration of the cell wall due to the attack of chemical compounds that have been formed during cavitation (Kadam, Tiwari, Álvarez, & O'Donnell, 2015). This technology has higher efficiency, shorter processing time and lower energy consumption than traditional methods and is considered an environmentally friendly technology that is economically feasible and simple. However, it has been observed that it can degrade the taste, color, or the nutritional value of foods that have been treated (Priyadarshini, Rajauria, O'Donnell, & Tiwari, 2019). It has been most widely used in beverages and drinks, although it has also been applied for the formation of foams, emulsions, and homogenates as well as for extraction, crystallization, degassing, and dehydration processes (Jermann et al., 2015).

Ionizing radiation. Food is exposed to gamma radiation, electron beams and x-rays to improve food safety.

The lethal effect on bacterial contaminants can be attributed to direct radiation, where the bacterial DNA is damaged and its synthesis inhibited, preventing further cell division, or from indirect radiation, whereby interacting with water molecules, cell lysis occurs due to the active molecules that are originated (Pan, Sun, & Han, 2017). It has been used on fresh fruits and vegetables, frozen food, cereals, seafood, meat, spices, cheese, nuts and other foods for sterilization, pasteurization, disinfestation, delayed ripening and senescence, and inactivation of spores and pathogenic bacteria and microorganisms in packaging materials. This technology offers less processing time, good food penetration and allows the processing of packaged products without recontamination. In addition, it does not require chemicals and does not produce waste, is environmentally friendly, and requires low energy costs. However, it can cause oxidation of lipids and thus changes in taste. High doses can also lead to loss of sensory and nutritional characteristics (Olatunde & Benjakul, 2018). Due to the various limitations cited, its use is only permitted in some countries and carries a high capital cost, linked with low acceptance of consumption due to localized radiation risks.

Cold atmospheric plasma. Through electrical discharges, the cold atmospheric plasma process ionizes a certain gas to generate reactive molecular species. Cold plasma processing has been used for enzymatic inactivation, etching or deposition of thin films, alteration of hydrophilic/hydrophobic properties, sterilization, and disinfection (Priyadarshini et al., 2019). The use of cold plasma allows the inactivation of microorganisms by direct chemical interaction between the reactive species or the charged particles and the cells, or through the effect produced by UV, which breaks the DNA chain or causes damage to the membranes or cellular components. Unlike thermal processing, cold atmospheric plasma involves lower operating temperatures, lower water consumption and lower costs. However, it is not suitable for lipid-rich food products since some of the reactive species are oxidized and unusual flavors linked to rancidity appear (Li & Farid, 2016). Although the cold atmospheric plasma technology does not currently have high marketing potential, it could offer many advantages to the food industry.

1.3.2 Microencapsulation

Microencapsulation consists of trapping components inside a capsule of a few microns in diameter called microcapsules. The process involves coating ingredients with food-grade and biodegradable materials to enhance nutrition, mask off-flavors, facilitate storage, and extend shelf life. In the core of the microcapsule (inner part), there can be found different materials, among others, suspensions, or emulsions. Depending on this material and taking into account the desired characteristics of the final product, a coating material is chosen based on its physicochemical properties from a wide variety of polymers, both natural, and synthetic. Solubility, emulsifying properties, or crystallinity are some of the properties that are generally taken into account when choosing the most suitable coating material (Gharsallaoui, Roudaut, Chambin, Voilley, & Saurel, 2007).

Common techniques that have been used are freeze drying, gelation, complex coacervation, inclusion complexation, emulsification, spray drying, and electrospinning (Encina, Vergara, Giménez, Oyarzún-Ampuero, & Robert, 2016). Among them, spray drying has gained significance recently since it is characterized by its simplicity and reproducibility, as well as being an economical and easy-to-implement technique. It requires short drying times and is therefore, particularly useful in the encapsulation of heat-sensitive materials (Desai & Jin Park, 2005).

Microencapsulation, which is a method of immobilization by which a solution or dispersion of food is converted into micro-particles of dust, is currently used to deliver unique consumer sensory and nutritional demands. In the first stage, a fine dispersion or emulsion must be prepared where the core material, usually hydrophobic, is stable in the wall solution in which it is immiscible. This dispersion is heated and homogenized, and depending on the coating material, an emulsifier can be added. Finally, the emulsion is atomized through a high-temperature drying medium, although the short exposure time and the rapid speed at which the water is evaporated means that the temperature of the particles does not reach 40°C (Gharsallaoui et al., 2007). This produces an instantaneous spherical powder that not only guarantees microbiological quality since it has little water activity but also facilitates handling (Gouin, 2004). In contrast, despite the previously mentioned advantages, there is a great loss of energy in the process since it is not possible to use all the heat that goes through the drying chamber.

Some of the most interesting food ingredients that are currently microencapsulated are flavoring compounds due to their high volatility, so their encapsulation is a great advantage for food industries (Madene, Jacquot, Scher, & Desobry, 2006). On the other hand, lipid encapsulation is very useful due to the difficulty of lipid dispersion in foods. Encapsulation of polyunsaturated fatty acids, whose capacity of auto-oxidation, by which unpleasant flavors and toxic compounds are produced, plays an important role in delaying auto-oxidation, improving stability, and maintaining desired taste of food (Gómez et al., 2018).

1.3.3 Nanotechnology

Nanostructured materials are those materials in which at least one dimension falls on a nanoscale. Such materials consist mainly of bulk materials produced from nanoscale structures, thin films, and nanowires. Depending on the dimension of the elements in their structures, they can be classified as three-dimensional (dendrimers and nanocomposites), two-dimensional (thin films), one-dimensional (nanorods or nanotubes), and zero-dimensional (fullerenes, nanoclusters and quantum dots) (Pathakoti, Manubolu, & Hwang, 2017). There are two approaches to nanomaterial synthesis and the manufacture of nanostructures. These are:

- The top-down approach is to reduce nanoscale size by breaking bulk materials by milling. With milling, materials are physically broken down by applying mechanical energy, reducing their size at the nanoscale. Thus, using dry milling, smaller-sized wheat flour with higher water retention capacity (Shibata, 2002) and green tea powder with better antioxidant activity have been obtained. However, this method is not suitable for mass production since its production rate is slow, in addition to introducing internal stresses and including imperfections and contamination (Hsieh & Ofori, 2007).

- The Bottom-up approach consists of building nanomaterials from individual atoms that have the ability to naturally self-assemble and self-regulate (Sanguansri & Augustin, 2006). This approach includes techniques such as: ultrasound emulsion, in which high intensity ultrasound waves can change the characteristics of the treated material due to cavitation caused by intensive shear forces, temperature and pressure; high pressure homogenization, used to increase the stability of fat emulsions by reducing the size of the globules (Thiebaud, Dumay, Picart, Guiraud, & Cheftel, 2003); mycofluidization, a kind of homogenization in which the size is reduced employing auxiliary chambers, improving the texture of emulsions (Degant & Schwechten, 2002); and dry grinding.

There are many advantages that nanotechnology can offer in the food sector, including enhanced absorption of nutrients, stabilization of bioactive compounds, extended product life, sensory improvements, quality, and safety monitoring (Pathakoti et al., 2017). This technology allows a wide range of applications, which can be classified into two main types: nanointerior applications when nanoparticles are incorporated into the product and nanoexterior applications when nanoparticles are included into materials in touch with foods such as packaging (Santeramo et al., 2017).

Included under nanointerior applications are the preparation of oil and water nanoemulsions (Singh, 2016), flavor oil emulsions, filled glazes, the production of creams, salad dressings, yoghurts, malted drinks, syrups, and chocolate (Kentish et al., 2008). Synthetic nanostructured systems protect bioactive components during manufacturing and storage, facilitate controlled release and improve their solubility, and bioavailability. Among the most important synthetic nanostructured systems for the food industry are liposomes, polymer nanoparticles, nanoemulsions, and microemulsions. Liposomes, which are spherical vesicles surrounded by a phospholipid double-layer, have been used for the encapsulation of functional ingredients. However, recently, it has been observed that they may help protect food products from the growth of pathogenic microorganisms due to their ability to integrate food antimicrobials (Singh, Thompson, Liu, & Corredig, 2012). Besides that, lipid-based nanoencapsulation can prevent undesirable interactions with other food components, helping to improve food solubility, bioavailability, and stability. Nanostructured biopolymer particles are protein micelles with a spherical structure of 5–100 nm diameter, which can encapsulate nonpolar molecules. Nonpolar elements such as lipids, antioxidants, and vitamins can be solubilized by the micelles to form microemulsions. In the case of antioxidants, the possibility of a synergistic effect between hydrophilic and lipophilic compounds can provide greater antioxidant effectiveness. Nanoemulsions, which are colloidal dispersions of 50–1000 nm droplets, are used in the production of food products for salad dressings, flavored oils, sweeteners, customized drinks, and other processed foods (Garti, 2008), as well as in the fortification of bottled water and milk with vitamins, minerals, and antioxidants (Huang, Li, & Zhou, 2015).

Nano-exterior application presents greater advantages than other techniques, for example, providing an improved barrier with mechanical properties and heat resistance, in addition to being biodegradable (De Azeredo, 2009). It is worth noting the improved antimicrobial effects that it can contribute, in addition to the ability to detect food spoilage thanks to the use of nanosensors (McClements & Xiao, 2012). Polymers (nanocomposites), both synthetic (polystyrene, polyamide, polyolefins, or nylon) and natural (chitosan, carrageenan, or cellulose), have been used in food packaging as an alternative to conventional materials such as metals, paper, or glass (Bastarrachea, Dhawan, & Sablani, 2011; Rhim, Park, & Ha, 2013) due to their functionality and low cost (Silvestre, Duraccio, & Cimmino, 2011). Both, nanosensors and polymers, allow one to record the temperature history, time, and expiration date of foods. Recent studies have shown that nanosensors can be used for the detection of food pathogens and toxins in packaging (Bastarrachea et al., 2011). As evidence of this, the study by Gfeller, Nugaeva, and Hegner (2005) showed that a low-cost nanobioluminescent sensor reacted with microbes to produce a visual sheen on the food indicating such contamination. Therefore, in terms of packaging, nanotechnology allows the prevention of spoilage and/or nutritional loss of food by ensuring food safety and therefore a longer shelf life.

1.3.4 3D Printing

Additive manufacturing (AM), free form solid (FFS) or, as it is better known, 3D printing is a robotically controlled process where a design program creates a 3D model that is sent to the printer and the product is built by depositing a material layer by layer (Fig. 1.2). Three classifications of materials are used for the printing (Dankar, Haddarah, Omar, & Sepulcre, 2018):

- Printable materials, that is materials that can normally be applied with a syringe and after being deposited, their stability allows the shape to be

FIG. 1.2 Printed food.(Images were obtained from: https://www.naturalmachines.com/dish-gallery (Natural Machines Co., 2020).)

maintained without further processing. Some examples are cake coating, hydrogels, soft cheese, hummus, and chocolate.

• Daily foods whose nature does not allow them to be printed and additives need to be incorporated. These are foods such as meat, rice, vegetables, and fruit.

• Other green and sustainable materials like the metabolites, flavoring compounds and enzymes from agricultural residues and other food processes that are part of biologically active foods.

This technology makes it possible for foods to be personalized according to specific composition, taste, texture, and nutritional requirements based on the age, occupation, sex, or lifestyle of the target consumer. Considering the large percentage of people over the age of 50 who have problems with chewing and swallowing food and must eat unpalatable products in the form of mash, this new technology may provide them with new opportunities. The food industry through 3D printing could provide innovative textured foods that in addition to their soft texture, are nutritious (Aguilera & Park, 2016). This innovative technology could be a good educational tool for children, who, in most cases, are reluctant to eat products such as vegetables and fruits, eating only nonnutritious snacks, thus promoting risks in their future health (Hamilton, Alici, & in het Panhuis, M., 2018). A change in the appearance of some raw materials would allow the elimination of obstacles in their use. Similarly, foods that are not socially accepted for cultural reasons but whose high nutritional value has been demonstrated, as in the case of insects, could try to be incorporated into the diet by adding other food products as ingredients to gradually create greater acceptance.

On the other hand, 3D printing allows at an industrial level to replace, reduce or automate steps in the production process, reducing in some cases the workforce which reduces errors and increases the efficiency of the process by reducing costs (Sun, Zhou, Huang, Fuh, & Hong, 2015). Likewise, since it is not necessary a massive production, only requested food manufacture, the amount of waste generated is reduced and water and energy consumption are reduced. This makes it a technique that causes a lower carbon footprint and is therefore more ecological (Kietzmann, Pitt, & Berthon, 2015).

1.4 CONCLUSION

Major factors underlying modern food production include the requirement for increased food production to meet up with the growing global population. Conventional food production has been criticized for intensive use of land and energy, high water consumption, generation of a great amount of waste, contributing to global greenhouse gas emissions, health problems arising from nutritional imbalance, such as obesity and cardiovascular disease, forcing the food industry to make changes. In recent years, the attention of researchers around the world has focused on finding alternatives to solve the problems that the food industry has been faced with, taking into consideration shifting consumer demands. Gradually, a wide variety of new foods has emerged which, on the one hand, provide nutritional benefits and on the other hand, offer a wide range of products with different sensory characteristics, making the food market close to being a personalized market for each type of consumer.

In line with the trends in food production, the technology used traditionally has had to improve and even redesign or to explore new types of equipment., Due to the environmental problems and the legal regulations that have been enforced to stop them, the design of new technologies has not only had to satisfy the requirements of the desirable manufactured product, but also to do so in a sustainable and eco-friendly way. Although many advances have been made in food production, this chapter was focused on some of the modern foods and novel technologies that have received the most attention, either because of their novelty, or because of their functionality. However, this is only the first step and much remains to be done.

ACKNOWLEDGMENTS

Acknowledgements to INIA for granting Cristina Pérez Santaescolástica with a predoctoral scholarship (grant number CPD2015-0212). Paulo E. S. Munekata acknowledges postdoctoral fellowship support from the Ministry of Economy and Competitiveness (MINECO, Spain) "Juan de la Cierva" program (FJCI-2016-29486). The authors thank GAIN (Axencia Galega de Innovación) for supporting this review (grant number IN607A2019/01). Jose M. Lorenzo is member of the HealthyMeat network, funded by CYTED Ciencia y Tecnología para el Desarrollo (ref. 119RT0568).

REFERENCES

Agcam, E., Akyildiz, A., & Evrendilek, G. A. (2016). A comparative assessment of long-term storage stability and quality attributes of orange juice in response to pulsed electric fields and heat treatments. *Food and Bioproducts Processing, 99*, 90–98.

Agregán, R., Munekata, P. E., Domínguez, R., Carballo, J., Franco, D., & Lorenzo, J. M. (2017). Proximate composition, phenolic content and in vitro antioxidant activity of aqueous extracts of the seaweeds *Ascophyllum nodosum, Bifurcaria bifurcata* and *Fucus vesiculosus*. Effect of addition of the extracts on the oxidative stability of canola oil unde. *Food Research International, 99*, 986–994.

Agregán, R., Munekata, P. E. S., Franco, D., Dominguez, R., Carballo, J., & Lorenzo, J. M. (2017). Phenolic compounds from three brown seaweed species using LC–DAD–ESI–MS/MS. *Food Research International, 99*, 979–985.

Aguilera, J. M., & Park, D. J. (2016). Texture-modified foods for the elderly: Status, technology and opportunities. *Trends in Food Science & Technology Journal, 57*, 156–164.

Al Khawli, F., Pateiro, M., Domínguez, R., Lorenzo, J. M., Gullón, P., Kousoulaki, K., et al. (2019). Innovative green technologies of intensification for valorization of seafood and their by-products. *Marine Drugs, 17*(12), 689.

American Soybean Association, A. (2009). *Soy stats guide.*

Anupama, & Ravindra, P. (2000). Value-added food: Single cell protein. *Biotechnology Advances, 18*, 459–479.

Apt, K. E., & Behrens, P. W. (1999). Commercial developments in microalgal biotechnology. *Journal of Phycology, 35*, 215–226.

Asgar, M. A., Fazilah, A., Huda, N., Bhat, R., & Karim, A. A. (2010). Nonmeat protein alternatives as meat extenders and meat analogs. *Comprehensive Reviews in Food Science and Food Safety, 9*(5), 513–529.

Astray, G., Gullón, P., Gullón, B., Munekata, P. E. S., & Lorenzo, J. M. (2020). *Humulus lupulus* L. as a natural source of

functional biomolecules. *Applied Sciences (Switzerland), 10*(15), 1–18.

Atef, M., & Ojagh, S. M. (2017). Health benefits and food applications of bioactive compounds from fish byproducts: A review. *Journal of Functional Foods, 35*, 673–681.

Bailey, R., Froggatt, A., & Wellesley, L. (2014). *Livestock—Climate change's forgotten sector global public opinion on meat and dairy consumption*. Chatham House.

Barba, F. J., Parniakov, O., Pereira, S. A., Wiktor, A., Grimi, N., Boussetta, N., et al. (2015). Current applications and new opportunities for the use of pulsed electric fields in food science and industry. *Food Research International, 77*, 773–798.

Barros, J. C., Munekata, P. E. S., de Carvalho, F. A. L., Pateiro, M., Barba, F. J., Domínguez, R., et al. (2020). Use of tiger nut (*Cyperus esculentus* L.) oil emulsion as animal fat replacement in beef burgers. *Foods, 9*(1), 44.

Barrow, C., & Shahidi, F. (2007). *Marine nutraceuticals and functional foods*. CRC Press.

Bastarrachea, L., Dhawan, S., & Sablani, S. S. (2011). Engineering properties of polymeric-based antimicrobial films for food packaging. *Food Engineering Reviews, 3*(2), 79–93.

Becker, E. W. (2007). Micro-algae as a source of protein. *Biotechnology Advances, 25*, 207–210.

Bekatorou, A., Psarianos, C., & Koutinas, A. A. (2006). Production of food grade yeasts. *Food Technology & Biotechnology, 44*(3), 407–415.

Björck, I., & Asp, N. G. (1983). The effects of extrusion cooking on nutritional value—A literature review. *Journal of Food Engineering, 2*, 281–308.

Blásquez, J. R., Manuel, J., Moreno, P., Hugo, V., & Camacho, M. (2012). Could grasshoppers be a nutritive meal? *Food and Nutrition Sciences, 3*(02), 164–175.

Bodenheimer, F. S. (2013). *Insects as human food: A chapter of the ecology of man*. Springer.

Bohrer, B. M. (2017). Nutrient density and nutritional value of meat products and non-meat foods high in protein. *Trends in Food Science & Technology, 65*, 103–112.

Boland, M. J., Rae, A. N., Vereijken, J. M., Meuwissen, M. P., Fischer, A. R., Boekel, V. M. A., et al. (2013). The future supply of animal-derived protein for human consumption. *Trends in Food Science & Technology, 29*(1), 62–73.

Bonny, S. P. F., Gardner, G. E., Pethick, D. W., & Hocquette, J. (2015). What is artificial meat and what does it mean for the future of the meat industry? *Journal of Integrative Agriculture, 14*(2), 255–263.

Bos, C., Airinei, G., Mariotti, F., Benamouzig, R., Bérot, S., Evrard, J., et al. (2007). The poor digestibility of rapeseed protein is balanced by its very high metabolic utilization in humans. *Journal of Nutrition, 137*(3), 594–600.

Chandrani-Wijeyaratne, S., & Tayathilake, A. N. (2000). Characteristics of two yeast strain (*Candida tropicalis*) isolated from Caryotaurens (Khitul) toddy for single cell protein production. *Journal of the National Science Foundation of Sri Lanka, 28*, 79–86.

Chang, H. C., Chen, H. H., & Hu, H. H. (2011). Textural changes in fresh egg noodles formulated with seaweed powder and full or partial replacement of cuttlefish paste. *Journal of Texture Studies, 42*(1), 61–71.

Chang, H. C., & Wu, L. C. (2008). Texture and quality properties of Chinese fresh egg noodles formulated with green seaweed (*Monostroma nitidum*) powder. *Journal of Food Science, 73*(8), S398–S404.

Chawla, R., Patil, G. R., & Singh, A. K. (2011). High hydrostatic pressure technology in dairy processing: A review. *Journal of Food Science and Technology, 48*(3), 260–268.

Chemat, F., Abert Vian, M., Fabiano-Tixier, A.-S., Nutrizio, M., Režek Jambrak, A., Munekata, P. E. S., et al. (2020). A review of sustainable and intensified techniques for extraction of food and natural products. *Green Chemistry, 22*(8), 2325–2353.

Cofrades, S., Benedí, J., Garcimartin, A., Sánchez-Muniz, F. J., & Jimenez-Colmenero, F. (2017). A comprehensive approach to formulation of seaweed-enriched meat products: From technological development to assessment of healthy properties. *Food Research International Journal, 99*, 1084–1094.

Cofrades, S., López-López, I., & Jiménez-Colmenero, F. (2012). Applications of seaweed in meat-based functional foods. In *Handbook of marine macroalgae* (pp. 491–499). New York: Wiley.

Cofrades, S., López-López, I., Ruiz-Capillas, C., Triki, M., & Jiménez-Colmenero, F. (2011). Quality characteristics of low-salt restructured poultry with microbial transglutaminase and seaweed. *Meat Science, 87*(4), 373–380.

Cofrades, S., López-López, I., Solas, M. T., Bravo, L., & Jiménez-Colmenero, F. (2008). Influence of different types and proportions of added edible seaweeds on characteristics of low-salt gel/emulsion meat systems. *Meat Science, 79*(4), 767–776.

Cox, S., & Abu-Ghannam, N. (2013a). Incorporation of *Himanthalia elongata* seaweed to enhance the phytochemical content of breadsticks using Response Surface Methodology (RSM). *International Food Research Journal, 20*(4), 1537–1545.

Cox, S., & Abu-Ghannam, N. (2013b). Enhancement of the phytochemical and fibre content of beef patties with *Himanthalia elongata* seaweed. *International Journal of Food Science & Technology, 48*(11), 2239–2249.

Cruz, N., Capellas, M., Hernández, M., Trujillo, A. J., Guamis, B., & Ferragut, V. (2007). Ultra high pressure homogenization of soymilk: Microbiological, physicochemical and microstructural characteristics. *Food Research International, 40*, 725–732.

da Cunha, M. E. T., Suguimoto, H. H., de Oliveira, A. N., Sivieri, K., & de Costa, M. R. (2008). Lactose intolerance and technological alternatives. *Journal of Health Sciences, 10*(2), 83–88.

da Silva, S. L., Amaral, J. T., Ribeiro, M., Sebastião, E. E., Vargas, C., de Lima Franzen, F., et al. (2019). Fat replacement by oleogel rich in oleic acid and its impact on the technological, nutritional, oxidative, and sensory properties of Bologna-type sausages. *Meat Science, 149*, 141–148.

da Silva, S. L., Lorenzo, J. M., Machado, J. M., Manfio, M., Cichoski, A. J., Fries, L. L. M., et al. (2020). Application of arginine and histidine to improve the technological and sensory properties of low-fat and low-sodium bologna-type sausages produced with high levels of KCl. *Meat Science, 159*, 107939.

Dairy UK, D. (2010). *The carbon trust, 2010. Guidelines for the carbon footprinting of dairy products in the UK.*

Dankar, I., Haddarah, A., Omar, F. E. L., & Sepulcre, F. (2018). 3D printing technology: The new era for food customization and elaboration. *Trends in Food Science & Technology, 75*(July 2017), 231–242.

Das, A. K., Das, A., Nanda, P. K., Madane, P., Biswas, S., Zhang, W., et al. (2020). A comprehensive review on antioxidant dietary fibre enriched meat-based functional foods. *Trends in Food Science & Technology, 99*, 323–336.

Day, L. (2013). Proteins from land plants-potential resources for human nutrition and food security. *Trends in Food Science & Technology, 32*(1), 25–42.

De Azeredo, H. M. C. (2009). Nanocomposites for food packaging applications. *Food Research International, 42*, 1240–1253.

De Boer, J., Schösler, H., & Aiking, H. (2014). "Meatless days" or "less but better"? Exploring strategies to adapt Western meat consumption to health and sustainability challenges. *Appetite, 76*, 120–128.

de Carvalho, F. A. L., Lorenzo, J. M., Pateiro, M., Bermúdez, R., Purriños, L., & Trindade, M. A. (2019). Effect of guarana (*Paullinia cupana*) seed and pitanga (*Eugenia uniflora* L.) leaf extracts on lamb burgers with fat replacement by chia oil emulsion during shelf life storage at 2 °C. *Food Research International, 125*, 108554.

de Oliveira Fagundes, D. T., Lorenzo, J. M., dos Santos, B., Fagundes, M., Heck, R., Cichoski, A., et al. (2017). Pork skin and canola oil as strategy to confer technological and nutritional advantages to burgers. *Czech Journal of Food Sciences, 35*(4), 352–359.

Degant, O., & Schwechten, D. (2002). *Wheat flour with increased water binding capacity and process and equipment for its manufacture.* German Patent DE10107885A1.

Delgado-Pando, G., Cofrades, S., Ruiz-Capillas, C., Solas, M. T., Triki, M., & Jiménez-Colmenero, F. (2011). Low-fat frankfurters formulated with a healthier lipid combination as functional ingredient: Microstructure, lipid oxidation, nitrite content, microbiological changes and biogenic amine formation. *Meat Science, 89*(1), 65–71.

Dellarosa, N., Laghi, L., Martinsdóttir, E., Jónsdóttir, R., & Sveinsdóttir, K. (2015). Enrichment of convenience seafood with omega-3 and seaweed extracts: Effect on lipid oxidation. *LWT-Food Science and Technology, 62*(1), 746–752.

Denny, A., Aisbitt, B., & Lunn, J. (2008). Mycoprotein and health. *Nutrition Bulletin, 33*(4), 298–310.

Desai, K. G. H., & Jin Park, H. (2005). Recent developments in microencapsulation of food ingredients. *Drying Technology, 23*(7), 1361–1394.

Domínguez, R., Agregán, R., Gonçalves, A., & Lorenzo, J. M. (2016). Effect of fat replacement by olive oil on the physico-chemical properties, fatty acids, cholesterol and tocopherol content of pâté. *Grasas y Aceites*, *67*(2), e133.

Domínguez, R., Barba, F. J., Gómez, B., Putnik, P., Bursać Kovačević, D., Pateiro, M., et al. (2018). Active packaging films with natural antioxidants to be used in meat industry: A review. *Food Research International*, *113*, 93–101.

Domínguez, R., Gullón, P., Pateiro, M., Munekata, P. E. S., Zhang, W., & Lorenzo, J. M. (2020). Tomato as potential source of natural additives for meat industry. A review. *Antioxidants*, *9*(1), 73.

Domínguez, R., Pateiro, M., Agregán, R., & Lorenzo, J. M. (2017). Effect of the partial replacement of pork backfat by microencapsulated fish oil or mixed fish and olive oil on the quality of frankfurter type sausage. *Journal of Food Science and Technology*, *54*(1), 26–37.

Domínguez, R., Pateiro, M., Munekata, P. E. S., Campagnol, P. C. B., & Lorenzo, J. M. (2017). Influence of partial pork backfat replacement by fish oil on nutritional and technological properties of liver pâté. *European Journal of Lipid Science and Technology*, *119*(5), 1600178.

Domínguez, R., Pateiro, M., Pérez-Santaescolástica, C., Munekata, P. E. S., & Lorenzo, J. M. (2017). Salt reduction strategies in meat products made from whole pieces. In *Strategies for obtaining healthier foods* (pp. 267–289). Nova Science Publishers.

Echegaray, N., Gómez, B., Barba, F. J., Franco, D., Estévez, M., Carballo, J., et al. (2018). Chestnuts and by-products as source of natural antioxidants in meat and meat products: A review. *Trends in Food Science and Technology*, *82*, 110–121.

Egbert, R., & Borders, C. (2006). Achieving success with meat analogs. *Food Technology (Chicago)*, *60*(1), 28–34.

Encina, C., Vergara, C., Giménez, B., Oyarzún-Ampuero, F., & Robert, P. (2016). Conventional spray-drying and future trends for the microencapsulation of fish oil. *Trends in Food Science and Technology*, *56*, 46–60.

European Regulation. (2015). *European Regulation (EU) 2015/2283 of the European Parliament of and of the Council of 25 November 2015 concerning concerning novel foods and novel food ingredients, Pub. L. No. 2015/2283, 2015*. Official Journal of the European Union.

Falowo, A. B., Mukumbo, F. E., Idamokoro, E. M., Lorenzo, J. M., Afolayan, A. J., & Muchenje, V. (2018). Multifunctional application of Moringa oleifera Lam. in nutrition and animal food products: A review. *Food Research International*, *106*, 317–334.

FAO. (2009). *How to feed the world in 2050. High-level expert forum*. Rome: FAO.

Fernández-Martín, F., López-López, I., Cofrades, S., & Colmenero, F. J. (2009). Influence of adding Sea Spaghetti seaweed and replacing the animal fat with olive oil or a konjac gel on pork meat batter gelation. Potential protein/alginate association. *Meat Science*, *83*(2), 209–217.

Finnigan, T., Needham, L., & Abbott, C. (2017). Mycoprotein: A healthy new protein with a low environmental impact. In *Sustainable protein sources* (pp. 305–326). Academic Press.

Fitzgerald, C., Gallagher, E., Doran, L., Auty, M., Prieto, J., & Hayes, M. (2014). Increasing the health benefits of bread: Assessment of the physical and sensory qualities of bread formulated using a renin inhibitory *Palmaria palmata* protein hydrolysate. *LWT-Food Science and Technology*, *56*(2), 398–405.

Foh, M. B. K., Wenshui, X., Amadou, I., & Jiang, Q. (2012). Influence of pH shift on functional properties of protein isolated of tilapia (*Oreochromis niloticus*) muscles and of soy protein isolate. *Food Bioprocess Technology*, *5*, 2192–2200.

Food and Agriculture Organization of the United Nations. (2012). *FAO statics*. Rome: FAO.

Franco, D., Martins, A. J., López-Pedrouso, M., Cerqueira, M. A., Purriños, L., Pastrana, L. M., et al. (2020). Evaluation of linseed oil oleogels to partially replace pork backfat in fermented sausages. *Journal of the Science of Food and Agriculture*, *100*(1), 218–224.

Franco, D., Munekata, P. E. S., Agregán, R., Bermúdez, R., López-Pedrouso, M., Pateiro, M., et al. (2020). Application of pulsed electric fields for obtaining antioxidant extracts from fish residues. *Antioxidants*, *9*, 90.

Gabriel, A., Victor, N., & du Preez, J. C. (2014). Cactus pear biomass, a potential lignocellulose raw material for single cell protein production (SCP): A review. *International Journal of Current Microbiology and Applied Sciences*, *3*(7), 171–197.

Garti, N. (2008). *Delivery and controlled release of bioactives in foods and nutraceuticals*. Elsevier.

Gasparin, F. S. R., Teles, J. M., & de Araújo, S. C. (2010). Allergy to cow milk protein versus lactose intolerance: Differences and similarities. *Revista Saúde e Pesquisa*, *3*(1), 107–114.

Gbogouri, G. A., Linder, M., Fanni, J., & Parmentier, M. (2004). Influence of hydrolysis degree on the functional properties of Salmon byproducts hydrolysates. *Food Chemistry and Toxicology Influence*, *69*(8), 615–622.

Geirsdottir, M., Sigurgisladottir, S., Hamaguchi, P. Y., Thorkelsson, G., Johannsson, R., Kristjansson, H. G., et al. (2011). Enzymatic hydrolysis of blue whiting (*Micromesistius poutassou*); functional and bioactive properties. *Journal of Food Science*, *76*(1), 14–20.

Gerber, P. J., Steinfeld, H., Henderson, B., Mottet, A., Opio, C., Dijkman, J., et al. (2013). *Tackling climate change through livestock: A global assessment of emissions and mitigation opportunities*. Food and Agriculture Organization of the United Nations (FAO).

Gfeller, K. Y., Nugaeva, N., & Hegner, M. (2005). Micromechanical oscillators as rapid biosensor for the detection of active growth of *Escherichia coli*. *Biosensors and Bioelectronics*, *21*, 528–533.

Gharsallaoui, A., Roudaut, G., Chambin, O., Voilley, A., & Saurel, R. (2007). Applications of spray-drying in microencapsulation of food ingredients: An overview. *Food Research International*, *40*, 1107–1121.

Golbitz, P., & Jordan, J. (2006). Soyfoods: Market and products. In M. Riaz (Ed.), *Soy applications in food* (pp. 1–21). New York: CRC Press.

Gómez, B., Barba, F. J., Domínguez, R., Putnik, P., Bursać Kovačević, D., Pateiro, M., et al. (2018). Microencapsulation of antioxidant compounds through innovative technologies and its specific application in meat processing. *Trends in Food Science and Technology, 82*, 135–147.

Gómez, B., Munekata, P. E. S., Gavahian, M., Barba, F. J., Martí-Quijal, F. J., Bolumar, T., et al. (2019). Application of pulsed electric fields in meat and fish processing industries: An overview. *Food Research International, 123*, 95–105.

Gouin, S. (2004). Microencapsulation: Industrial appraisal of existing technologies and trends. *Trends in Food Science & Technology, 15*(7–8), 330–347.

Granato, D., Barba, F. J., Bursać Kovačević, D., Lorenzo, J. M., Cruz, A. G., & Putnik, P. (2020). Functional foods: Product development, technological trends, efficacy testing, and safety. *Annual Review of Food Science and Technology, 11*(1), 93–118.

Griffiths, J. C., Abernethy, D. R., Schuber, S., & Williams, R. L. (2009). Opinions and perspectives functional food ingredient quality: Opportunities to improve public health by compendial standardization. *Journal of Functional Foods, 1*(1), 128–130.

Guérard, F., & Sumaya-Martinez, M.-T. (2003). Antioxidant effects of protein hydrolysates in the reaction with glucose. *Journal of the American Oil Chemists' Society, 80*(5), 467–470.

Gullan, P. J., & Cranston, P. S. (2014). *The insects: An outline of entomology.* John Wiley & Sons.

Gullón, P., Astray, G., Gullón, B., Tomasevic, I., & Lorenzo, J. M. (2020). Pomegranate peel as suitable source of high-added value bioactives: Tailored functionalized meat products. *Molecules, 25*(12), 1–18.

Gullón, B., Gagaoua, M., Barba, F. J., Gullón, P., Zhang, W., & Lorenzo, J. M. (2020). Seaweeds as promising resource of bioactive compounds: Overview of novel extraction strategies and design of tailored meat products. *Trends in Food Science & Technology, 100*, 1–18.

Hamilton, C., Alici, G., & in het Panhuis, M. (2018). 3D printing vegemite and marmite: Redefining "breadboards". *Journal of Food Engineering, 220*, 83–88.

Hashempour-baltork, F., Khosravi-darani, K., Hosseini, H., Parastou Farshi, S., & Reihani, F. S. (2020). Mycoproteins as safe meat substitutes. *Journal of Cleaner Production, 253*, 119958.

Hayakawa, I., Hayashi, N., Urushima, T., Kajiwara, Y., & Fujio, Y. (1989). Texturization of whole soybean in a twin-screw extruder. *Journal of the Faculty of Agriculture, Kyushu University, 33*(3–4), 213–220.

He, J., Tong, Q., Huang, X., & Zhou, Z. (1999). Nutritional composition analysis of moths of Dendrolimus houi Lajonquiere. *Kunchong Zhishi, 36*(2), 83–86.

Heck, R. T., Vendruscolo, R. G., de Araújo Etchepare, M., Cichoski, A. J., de Menezes, C. R., Barin, J. S., et al. (2017). Is it possible to produce a low-fat burger with a healthy n-6/n-3 PUFA ratio without affecting the technological and sensory properties? *Meat Science, 130*, 16–25.

Hernández-Hernández, H. M., Moreno-Vilet, L., & Villanueva-Rodríguez, S. J. (2019). Current status of emerging food processing technologies in Latin America: Novel non-thermal processing. *Innovative Food Science and Emerging Technologies, 58*, 102233.

Herpandi, N. H., Rosma, A., & Wan Nadiah, W. (2011). The tuna fishing industry: A new outlook on fish protein hydrolysates. *Comprehensive Reviews in Food Science and Food Safety, 10*, 195–207.

Herrero, M., Henderson, B., Havlík, P., Thornton, P. K., Conant, R. T., Smith, P., et al. (2016). Greenhouse gas mitigation potentials in the livestock sector. *Nature Climate Change, 6*(5), 452–461.

Horita, C. N., Baptista, R. C., Caturla, M. Y. R., Lorenzo, J. M., Barba, F. J., & Sant'Ana, A. S. (2018). Combining reformulation, active packaging and non-thermal post-packaging decontamination technologies to increase the microbiological quality and safety of cooked ready-to-eat meat products. *Trends in Food Science and Technology, 72*, 45–61.

House, J. (2016). Consumer acceptance of insect-based foods in the Netherlands: Academic and commercial implications. *Appetite, 107*, 47–58.

Hsieh, Y. H. P., & Ofori, J. A. (2007). Innovations in food technology for health. *Asia Pacific Journal of Clinical Nutrition, 16*, 65–73.

Huang, J. Y., Li, X., & Zhou, W. (2015). Safety assessment of nanocomposite for food packaging application. *Trends in Food Science & Technology, 2*, 187–199.

Jermann, C., Koutchma, T., Margas, E., Leadley, C., & Ros-Polski, V. (2015). Mapping trends in novel and emerging food processing technologies around the world. *Innovative Food Science and Emerging Technologies, 31*, 14–27.

Jiménez-Colmenero, F. (2007). Healthier lipid formulation approaches in meat-based functional foods. Technological options for replacement of meat fats by non-meat fats. *Trends in Food Science and Technology, 18*(11), 567–578.

Jiménez-Colmenero, F., Cofrades, S., López-López, I., Ruiz-Capillas, C., Pintado, T., & Solas, M. T. (2010). Technological and sensory characteristics of reduced/low-fat, low-salt frankfurters as affected by the addition of konjac and seaweed. *Meat Science, 84*(3), 356–363.

Kadam, S. U., Tiwari, B. K., Álvarez, C., & O'Donnell, C. P. (2015). Ultrasound applications for the extraction, identification and delivery of food proteins and bioactive peptides. *Trends in Food Science & Technology, 46*(1), 60–67.

Kearney, J. (2010). Food consumption trends and drivers. *Philosophical Transactions of the Royal Society B: Biological Sciences, 365*(1554), 2793–2807.

Kentish, S., Wooster, T. J., Ashokkumar, M., Balachandran, S., Mawson, R., & Simons, L. (2008). The use of ultrasonics for nanoemulsion preparation. *Innovative Food Science and Emerging Technologies, 9*(2), 170–175.

Kietzmann, J., Pitt, L., & Berthon, P. (2015). Disruptions, decisions, and destinations: Enter the age of 3-D printing and additive manufacturing. *Business Horizons, 58*(2), 209–215.

Kim, H. W., Choi, J. H., Choi, Y. S., Han, D. J., Kim, H. Y., Lee, M. A., … Kim, C. J. (2010). Effects of sea tangle (*Lamina japonica*) powder on quality characteristics of breakfast sausages. *Food Science of Animal Resources, 30*(1), 55–61.

Kristinsson, H. G. (2007). Aquatic food protein hydrolysates. In *Maximising the value of marine by-products* Woodhead Publishing Limited.

Lee, C. J., Song, E. J., Kim, K. B. W. R., Jung, J. Y., Kwak, J. H., Choi, M. K., … Ahn, D. H. (2011). Effect of gamma irradiation on immune activity and physicochemical properties of *Myagropsis myagroides* water extract. *Korean Journal of Fisheries and Aquatic Sciences, 44*(1), 50–57.

Li, X., & Farid, M. (2016). A review on recent development in non-conventional food sterilization technologies. *Journal of Food Engineering, 182*, 33–45.

Liang, S., Liu, X., Chen, F., & Chen, Z. (2004). Current microalgal health food R & D activities in China. *Hydrobiologia, 512*, 45–48.

Lin, Q., & Li, X. (2008). Effect of scorpion venom analgesic active peptide extracted from *Buthus martensii* Karschon evoked potential in the thalamic posterior nucleus group in rats. *Neural Regeneration Research, 3*(4), 453–455.

Long, S., Ying, F., Zhao, H., Tao, M., & Xin, Z. (2007). Studies on alkaline solution extraction of polysaccharide from silkworm pupa and its immunomodulating activities. *Forest Research, 20*(6), 782–786.

López-López, I., Cofrades, S., Cañeque, V., Díaz, M. T., López, O., & Jiménez-Colmenero, F. (2011). Effect of cooking on the chemical composition of low-salt, low-fat Wakame/olive oil added beef patties with special reference to fatty acid content. *Meat Science, 89*(1), 27–34.

López-López, I., Cofrades, S., & Jiménez-Colmenero, F. (2009). Low-fat frankfurters enriched with *n*–3 PUFA and edible seaweed: Effects of olive oil and chilled storage on physicochemical, sensory and microbial characteristics. *Meat Science, 83*(1), 148–154.

López-López, I., Cofrades, S., Ruiz-Capillas, C., & Jiménez-Colmenero, F. (2009). Design and nutritional properties of potential functional frankfurters based on lipid formulation, added seaweed and low salt content. *Meat Science, 83*(2), 255–262.

López-López, I., Cofrades, S., Yakan, A., Solas, M. T., & Jiménez-Colmenero, F. (2010). Frozen storage characteristics of low-salt and low-fat beef patties as affected by Wakame addition and replacing pork backfat with olive oil-in-water emulsion. *Food Research International, 43*(5), 1244–1254.

López-Pedrouso, M., Lorenzo, J. M., Gullón, B., Campagnol, P. C. B., & Franco, D. (2021). Novel strategy for developing healthy meat products replacing saturated fat with oleogels. *Current Opinion in Food Science, 40*, 40–45.

Lorenzo, J. M., Agregán, R., Munekata, P. E. S., Franco, D., Carballo, J., Şahin, S., et al. (2017). Proximate composition and nutritional value of three macroalgae: *Ascophyllum nodosum, Fucus vesiculosus* and *Bifurcaria bifurcata*. *Marine Drugs, 15*(11), 360.

Lorenzo, J. M., Batlle, R., & Gómez, M. (2014). Extension of the shelf-life of foal meat with two antioxidant active packaging systems. *LWT—Food Science and Technology, 59*(1), 181–188.

Lorenzo, J. M., Bermúdez, R., Domínguez, R., Guiotto, A., Franco, D., & Purriños, L. (2015). Physicochemical and microbial changes during the manufacturing process of dry-cured lacón salted with potassium, calcium and magnesium chloride as a partial replacement for sodium chloride. *Food Control, 50*, 763–769.

Lorenzo, J. M., Domínguez, R., & Carballo, J. (2017). Control of lipid oxidation in muscle food by active packaging technology. In R. Banerjee, A. K. Verma, & M. W. Siddiqui (Eds.), *Natural antioxidants: Applications in foods of animal origin* (1st ed., pp. 343–382). Boca Raton, FL: Apple Academic Press Inc.

Lorenzo, J. M., Sineiro, J., Amado, I. R., & Franco, D. (2014). Influence of natural extracts on the shelf life of modified atmosphere-packaged pork patties. *Meat Science, 96*(1), 526–534.

Lymbery, P. (2014). *Farmageddon: The true cost of cheap meat*. Bloomsbury Publishing.

MacLeod, M., Gerber, P., Mottet, A., Tempio, G., Falcucci, A., Opio, C., et al. (2013). *Greenhouse gas emissions from pig and chicken supply chains—A global life cycle assessment*. Food and Agriculture Organization of the United Nations.

Madene, A., Jacquot, M., Scher, J., & Desobry, S. (2006). Flavour encapsulation and controlled release—A review. *International Journal of Food Science and Technology, 41*, 1–21.

Mamat, H., Matanjun, P., Ibrahim, S., Amin, S. F. M., Hamid, M. A., & Rameli, A. S. (2014). The effect of seaweed composite flour on the textural properties of dough and bread. *Journal of Applied Phycology, 26*(2), 1057–1062.

Mamatha, B. S., Namitha, K. K., Senthil, A., Smitha, J., & Ravishankar, G. A. (2007). Studies on use of Enteromorpha in snack food. *Food Chemistry, 101*(4), 1707–1713.

Martins, A. J., Lorenzo, J. M., Franco, D., Vicente, A. A., Cunha, R. L., Pastrana, L. M., et al. (2019). Omega-3 and polyunsaturated fatty acids-enriched hamburgers using sterol-based oleogels. *European Journal of Lipid Science and Technology, 121*(11), 1900111.

Marti-Quijal, F. J., Zamuz, S., Tomašević, I., Gómez, B., Rocchetti, G., Lucini, L., et al. (2019). Influence of different sources of vegetable, whey and microalgae proteins on the physicochemical properties and amino acid profile of fresh pork sausages. *LWT, 110*, 316–323.

Mattar, R., & Mazo, D. F. C. (2010). Intolerância à lactose: mudança de paradigmas com a biologia molecular. *Revista da Associação Médica Brasileira, 56*(2), 230–236.

Mattick, C., & Allenby, B. (2013). The future of meat. *Issues in Science and Technology, 30*(1), 64–71.

McClements, D. J., & Xiao, H. (2012). Potential biological fate of ingested nanoemulsions: Influence of particle characteristics. *Food & Function, 3*, 202–220.

Megido, R. C., Haubruge, É., & Francis, F. (2018). Insects, the next European foodie craze? In *Edible insects in sustainable food systems*. Cham: Springer.

Mitsuhashi, J. (2016). *Edible insects of the world*. CRC Press.

Mlcek, J., Rop, O., Borkovcova, M., & Bednarova, M. (2014). A comprehensive look at the possibilities of edible insects as food in Europe—A review. *Polish Journal of Food and Nutrition Sciences*, 64(3), 147–157.

Moroney, N. C., O'Grady, M. N., Lordan, S., Stanton, C., & Kerry, J. P. (2015). Seaweed polysaccharides (laminarin and fucoidan) as functional ingredients in pork meat: An evaluation of anti-oxidative potential, thermal stability and bioaccessibility. *Marine Drugs*, 13(4), 2447–2464.

Moroney, N. C., O'Grady, M. N., O'Doherty, J. V., & Kerry, J. P. (2013). Effect of a brown seaweed (*Laminaria digitata*) extract containing laminarin and fucoidan on the quality and shelf-life of fresh and cooked minced pork patties. *Meat Science*, 94(3), 304–311.

Nachtigall, F. M., Vidal, V. A. S., Pyarasani, R. D., Domínguez, R., Lorenzo, J. M., Pollonio, M. A. R., et al. (2019). Substitution effects of NaCl by KCl and CaCl$_2$ on lipolysis of salted meat. *Foods*, 8(12), 595.

Nonaka, K. (2009). Feasting on insects. *Entomological Research*, 39(5), 304–312.

Olatunde, O. O., & Benjakul, S. (2018). Nonthermal processes for shelf-life extension of seafoods: A revisit. *Comprehensive Reviews in Food Science and Food Safety*, 17(4), 892–904.

Ortiz, J., Vivanco, J. P., & Aubourg, S. P. (2014). Lipid and sensory quality of canned Atlantic salmon (*Salmo salar*): Effect of the use of different seaweed extracts as covering liquids. *European Journal of Lipid Science and Technology*, 116(5), 596–605.

O'Sullivan, A. M., O'Grady, M. N., O'Callaghan, Y. C., Smyth, T. J., O'Brien, N. M., & Kerry, J. P. (2016). Seaweed extracts as potential functional ingredients in yogurt. *Innovative Food Science & Emerging Technologies*, 37, 293–299.

Pan, Y., Sun, D. W., & Han, Z. (2017). Applications of electromagnetic fields for nonthermal inactivation of microorganisms in foods: An overview. *Trends in Food Science and Technology*, 64, 13–22.

Paoletti, M. G. (2005). *Ecological implications of Minilivestock: Potential of insects, rodents, frogs and sails*. CRC Press.

Parniakov, O., Toepfl, S., Barba, F. J., Granato, D., Zamuz, S., Galvez, F., et al. (2018). Impact of the soy protein replacement by legumes and algae based proteins on the quality of chicken rotti. *Journal of Food Science and Technology*, 55(7), 2552–2559.

Parodi, A., Leip, A., Slegers, P. M., Ziegler, F., Herrero, M., Tuomisto, H., et al. (2018). Future foods: Towards a sustainable and healthy diet for a growing population. *Nature Sustainability*, 1(12), 782–789.

Pateiro, M., Domínguez, R., Bermúdez, R., Munekata, P. E. S., Zhang, W., Gagaoua, M., et al. (2019). Antioxidant active packaging systems to extend the shelf life of sliced cooked ham. *Current Research in Food Science*, 1, 24–30.

Pateiro, M., Munekata, P. E. S., Domínguez, R., Wang, M., Barba, F. J., Bermúdez, R., et al. (2020). Nutritional profiling and the value of processing by-products from Gilthead Sea bream (*Sparus aurata*). *Marine Drugs*, 18(2), 101.

Pathakoti, K., Manubolu, M., & Hwang, H. (2017). Nanostructures: Current uses and future applications in food science. *Journal of Food and Drug Analysis*, 25(2), 245–253.

Pereira, M. C. S., Brumano, L. P., Kamiyama, C. M., Pereira, J. P. F., Rodarte, M. P., & de Pinto, M. A. O. (2012). Low-lactose dairy: A necessity for people with lactose maldigestion and a niche market. *Revista Do Instituto de Laticínios Cândido Tostes*, 67(389), 57–65.

Pietrasik, Z., Jarmoluk, A., & Shand, P. J. (2007). Effect of non-meat proteins on hydration and textural properties of pork meat gels enhanced with microbial transglutaminase. *LWT—Food Science and Technology*, 40, 915–920.

Pinterits, A., & Arntfield, S. D. (2008). Improvement of canola protein gelation properties through enzymatic modification with transglutaminase. *LWT—Food Science and Technology*, 41, 128–138.

Prabhasankar, P., Ganesan, P., Bhaskar, N., Hirose, A., Stephen, N., Gowda, L. R., … Miyashita, K. J. F. C. (2009). Edible Japanese seaweed, wakame (*Undaria pinnatifida*) as an ingredient in pasta: Chemical, functional and structural evaluation. *Food Chemistry*, 115(2), 501–508.

Priyadarshini, A., Rajauria, G., O'Donnell, C. P., & Tiwari, B. K. (2019). Emerging food processing technologies and factors impacting their industrial adoption. *Critical Reviews in Food Science and Nutrition*, 59(19), 3082–3101.

Reihani, S. F. S., & Khosravi-darani, K. (2019). Influencing factors on single-cell protein production by submerged fermentation: A review. *Electronic Journal of Biotechnology*, 37, 34–40.

Rhim, J. W., Park, H. M., & Ha, C. S. (2013). Bio-nanocomposites for food packaging applications. *Progress in Polymer Science*, 38(10–11), 1629–1652.

Riaz, M. (2004). Texturized soy protein as an ingredient. In R. Yada (Ed.), *Proteins in food processing* (pp. 517–557). England: Woodhead Publishing Limited.

Ribeiro, I. S., Shirahigue, L. D., Ferraz de Arruda Sucasas, L., Anbe, L., da Cruz, & Oetterer, M. (2014). Shelf life and quality study of minced tilapia with Nori and Hijiki seaweeds as natural additives. *The Scientific World Journal*, 2014, 485287.

Rocchetti, G., Pateiro, M., Campagnol, P. C. B., Barba, F. J., Tomasevic, I., Montesano, D., et al. (2020). Effect of partial replacement of meat by carrot on physicochemical properties and fatty acid profile of Turkey fresh sausages: A chemometric approach. *Journal of the Science of Food and Agriculture*, 100, 4968–4977.

Rumpold, B. A., & Schlüter, O. K. (2013a). Nutritional composition and safety aspects of edible insects. *Molecular Nutrition & Food Research*, 57(5), 802–823.

Rumpold, B. A., & Schlüter, O. K. (2013b). Potential and challenges of insects as an innovative source for food and feed production. *Innovative Food Science and Emerging Technologies*, 17, 1–11.

Sanguansri, P., & Augustin, M. A. (2006). Nanoscale materials development—A food industry perspective. *Trends in Food Science and Technology*, 17, 547–556.

Santeramo, F. G., Carlucci, D., De Devitiis, B., Seccia, A., Stasi, A., Viscecchia, R., et al. (2017). Emerging trends in European food, diets and food industry. *Food Research International, 104*, 39–47.

Sari, Y. W., Mulder, W. J., Sanders, J. P. M., & Bruins, M. E. (2015). Towards plant protein refinery: Review on protein extraction using alkali and potential enzymatic assistance. *Biotechnology Journal, 10*, 1138–1157.

Sasaki, K., Ishihara, K., Oyamada, C., Sato, A., Fukushi, A., Arakane, T., … Mitsumoto, M. (2008). Effects of fucoxanthin addition to ground chicken breast meat on lipid and colour stability during chilled storage, before and after cooking. *Asian-Australasian Journal of Animal Sciences, 21*(7), 1067–1072.

Sasaki, M., Yamada, H., & Kato, N. (2000). Consumption of silk protein, sericin elevates intestinal absorption of zinc, iron, magnesium and calcium in rats. *Nutrition Research, 20*(10), 1505–1511.

Schösler, H., De Boer, J., & Boersema, J. J. (2012). Can we cut out the meat of the dish? Constructing consumer-oriented pathways towards meat substitution. *Appetite, 58*, 39–47.

Senthil, A., Mamatha, B. S., Vishwanath, P., Bhat, K. K., & Ravishankar, G. A. (2011). Studies on development and storage stability of instant spice adjunct mix from seaweed (Eucheuma). *Journal of Food Science and Technology, 48*(6), 712–717.

Senthil, M. A., Mamatha, B. S., & Mahadevaswamy, M. (2005). Effect of using seaweed (eucheuma) powder on the quality of fish cutlet. *International Journal of Food Sciences and Nutrition, 56*(5), 327–335.

Sethi, S., Tyagi, S. K., & Anurag, R. K. (2016). Plant-based milk alternatives an emerging segment of functional beverages: A review. *Journal of Food Science and Technology, 53*(9), 3408–3423.

Shibata, T. (2002). *Method for producing green tea in microfine powder*. United States Patent US6416803B1.

Shouqin, Z., Junjie, Z., & Changzhen, W. (2004). Novel high pressure extraction technology. *International Journal of Pharmaceutics, 278*(2), 471–474.

Silva, A. R. A., Silva, M. M. N., & Ribeiro, B. D. (2020). Health issues and technological aspects of plant-based alternative milk. *Food Research International, 131*, 108972.

Silvestre, C., Duraccio, D., & Cimmino, S. (2011). Food packaging based on polymer nanomaterials. *Progress in Polymer Science (Oxford), 36*(12), 1766–1782.

Singh, H. (2016). Nanotechnology applications in functional foods; opportunities and challenges. *Preventive Nutrition and Food Science, 21*(1), 1–8.

Singh, P., Kumar, R., Sabapathy, S., & Bawa, A. (2008). Functional and edible uses of soy protein products. *Comprehensive Reviews in Food Science and Food Safety, 7*(1), 14–28.

Singh, H., Thompson, A., Liu, W., & Corredig, M. (2012). Liposomes as food ingredients and nutraceutical delivery systems. In *Encapsulation technologies and delivery systems for food ingredients and nutraceuticals* Woodhead Publishing Limited.

Spolaore, P., Joannis-Cassan, C., Duran, E., & Isambert, A. (2006). Commercial applications of microalgae. *Journal of Bioscience and Bioengineering, 101*, 201–211.

Strahm, B. (2006). Meat alternatives. In M. N. Riaz (Ed.), *Soy applications in food* (pp. 135–154). New York: CRC Press.

Suman, G., Nupur, M., Anuradha, S., & Pradeep, B. (2015). Single cell protein production: A review. *International Journal of Current Microbiology and Applied Sciences, 4*(9), 251–262.

Sun, J., Zhou, W., Huang, D., Fuh, J. Y., & Hong, G. S. (2015). An overview of 3D printing technologies for food fabrication. *Food and Bioprocess Technology, 8*(8), 1605–1615.

Sun-waterhouse, D., Waterhouse, G. I. N., You, L., Zhang, J., Liu, Y., & Ma, L. (2016). Transforming insect biomass into consumer wellness foods: A review. *Food Research International, 89*, 129–151.

Suphamityotin, P. (2011). Optimizing enzymatic extraction of cereal milk using response surface methodology. *Songklanakarin Journal of Science and Technology, 33*(4), 389–395.

Tao, J., & Li, Y. O. (2018). Edible insects as a means to address global malnutrition and food insecurity issues. *Food Quality and Safety, 2*(1), 17–26.

Teffo, L. S., Toms, R. B., & Eloff, J. N. (2007). Preliminary data on the nutritional composition of the edible stink-bug, *Encosternum delegorguei* Spinola, consumed in Limpopo province, South Africa. *South African Journal of Science, 103*(11–12), 434–436.

Textured Vegetable Protein Products (B-1). (1971). *FNS notice (No. 219)*.

Thiansilakul, Y., Benjakul, S., & Shahidi, F. (2007). Food chemistry compositions, functional properties and antioxidative activity of protein hydrolysates prepared from round scad (*Decapterus maruadsi*). *Food Chemistry, 103*, 1385–1394.

Thiebaud, M., Dumay, E., Picart, L., Guiraud, J. P., & Cheftel, J. C. (2003). High-pressure homogenisation of raw bovine milk. Effects on fat globule size distribution and microbial inactivation. *International Dairy Journal, 13*, 427–439.

Toepfl, S., Siemer, C., Saldaña-Navarro, G., & Heinz, V. (2014). Overview of pulsed electric fields processing for food. In *Emerging technologies for food processing* (pp. 90–114). Academic Press.

Umaraw, P., Munekata, P. E. S., Verma, A. K., Barba, F. J., Singh, V. P., Kumar, P., et al. (2020). Edible films/coating with tailored properties for active packaging of meat, fish and derived products. *Trends in Food Science and Technology, 98*, 10–24.

Upadhyaya, S., Tiwari, S., Arora, N. K., & Singh, D. P. (2016). Microbial protein: A valuable component for future food security. In *Microbes and environmental management* (pp. 259–279). New Delhi: Studium Press.

Van Der Weele, C., Feindt, P., Van Der Goot, A. J., Van Mierlo, B., & van Boekel, M. (2019). Meat alternatives: An integrative comparison. *Trends in Food Science and Technology, 88*, 505–512.

Van Huis, A., Van Itterbeeck, J., Klunder, H., Mertens, E., Halloran, A., Muir, G., et al. (2013). *Edible insects future prospects for food and feed security*. F. and A. O. of the U. Nations.

Vargas-Ramella, M., Pateiro, M., Barba, F. J., Franco, D., Campagnol, P. C. B., Munekata, P. E. S., et al. (2020). Microencapsulation of healthier oils to enhance the physicochemical and nutritional properties of deer pâté. *LWT, 125*, 109223.

Verbeke, W., Van Wezemael, L., De Barcellos, M. D., Kügler, J. O., Hocquette, J.-F., Ueland, Ø., et al. (2010). European beef consumers' interest in a beef eating-quality guarantee insights from a qualitative study in four EU countries. *Appetite, 54*, 289–296.

Wang, T., Jónsdóttir, R., Kristinsson, H. G., Thorkelsson, G., Jacobsen, C., Hamaguchi, P. Y., & Ólafsdóttir, G. (2010). by *Fucus vesiculosus* extract and fractions. *Food Chemistry, 123*(2), 321–330.

Wiebe, M. G. (2004). Quorn TM myco-protein—Overview of a successful fungal product. *Mycologist, 18*(1), 17–20.

Wu, H., Chen, H., & Shiau, C. (2003). Free amino acids and peptides as related to antioxidant properties in protein hydrolysates of mackerel (*Scomber austriasicus*). *Food Research International, 36*, 949–957.

Wu, Q., Jia, J., Yan, H., Du, J., & Gui, Z. (2015). Peptides A novel angiotensin-I converting enzyme (ACE) inhibitory peptide from gastrointestinal protease hydrolysate of silkworm pupa (*Bombyx mori*) protein: Biochemical characterization and molecular docking study. *Peptides, 68*, 17–24.

Xiaodong, Y., & Bo, L. I. (2005). Bioactive peptide from bee venom for adjuvant-induced arthritis in rats. *Chinese Journal of Tissue Engineering Research, 9*(15), 242–243.

Yang, R., Zhao, X., Kuang, Z., Ye, M., Luo, G., Xiao, G., et al. (2013). Optimization of antioxidant peptide production in the hydrolysis of silkworm (*Bombyx mori* L.) pupa protein using response surface methodology. *Journal of Food Agriculture and Environment, 11*, 952–956.

Zielińska, E., Baraniak, B., Karaś, M., Rybczyńska, K., & Jakubczyk, A. (2015). Selected species of edible insects as a source of nutrient composition. *Food Research International, 77*, 460–466.

CHAPTER 2

Consumer and Market Demand for Sustainable Food Products

SOL ZAMUZ[A] • PAULO EDUARDO SICHETTI MUNEKATA[A] • HERBERT L. MEISELMAN[B] • WANGANG ZHANG[C] • LUJUAN XING[C] • JOSÉ MANUEL LORENZO[A,D]

[a]Galician Meat Technology Center, Galicia Technology Park, Ourense, Spain, [b]Herb Meiselman Training and Consulting, Rockport, MA, United States, [c]College of Food Science and Technology, Nanjing Agricultural University, Nanjing, China, [d]Food Technology Area, Faculty of Sciences of Ourense, Vigo University, Ourense, Spain

2.1 SUSTAINABILITY AND SUSTAINABLE FOOD PRODUCTS DEFINITION

Currently it is usual to hear the word *sustainability* word in the mass media. But what is it? What do they refer to when they talk about sustainable food products? Sustainability refers to meeting the needs of the present without compromising the ability of future generations to meet their own. The available resources must not be exhausted indiscriminately and natural resources must be protected. This involves a balance between economic growth, environmental care, and social well-being. Therefore, the *sustainable food system* (Fig. 2.1) is one that ensures food security and nutrition for everyone, so that it does not compromise the economic, social, and environmental foundations for future generations. It is one which would ensure that we can continue to feed without depleting resources (water, land, etc.) and without destroying the environment (biodiversity, pollution). In order words, this system must:

(1) Support financially as far as possible to achieve a high degree of autonomy (economic sustainability)
(2) Integrate social equity and combat precariousness criteria as far as possible (social sustainability)
(3) Preserve as far as possible the planet (environmental sustainability)
(4) Ensure that people maximize their long-term health as far as possible (nutritional sustainability)

According to Fischer and Garnett (2016) and Masson-Delmotte et al. (2019), current food production is destroying the environment: it contributes 20%–30% of greenhouse gas emissions (GHG) from people; is the leading cause of deforestation; accounts for 70% of all human water use; is a major source of water pollution; and unsustainable fishing practices deplete stocks of species. The entire food chain (production, processing, distribution, consumption, and waste) contributes to increasing these problems. As global population grows, it demands more resource intensive food. Therefore, it is necessary to change the current food system, to elaborate a global policy ensuring sustainable food at reasonable market prices, and to introduce changes in consumption patterns.

At the 2010 International Scientific Symposium "Biodiversity and Sustainable Diets: United Against Hunger" organized jointly by FAO and Biodiversity International, a definition of *sustainable food products* was agreed (FAO, 2012): "those food products with low environmental impacts which contribute to food and nutrition security and to healthy life for present and future generations. Sustainable products are protective and respectful of biodiversity and ecosystems, culturally acceptable, accessible, economically fair, and affordable; nutritionally adequate, safe, and healthy; while optimizing natural and human resources." Thus, the three characteristics that have to meet the sustainable food products are:

(1) Quality: those use quality raw material, are harvested at optimum ripening point, are safe and produced following healthier-orientated strategies to reduce and replace components associated with an increased risk of developing diseases such as synthetic additives, *trans* or saturated fats, hormones, or antibiotics.
(2) Environmentally friendly: food grown on site or as close as possible to consumers, seasonal, ecological, local food and manufactured (elaboration, packaging, and transport) under energy and resource efficient conditions.

Sustainable Production Technology in Food. https://doi.org/10.1016/B978-0-12-821233-2.00008-3
Copyright © 2021 Elsevier Inc. All rights reserved.

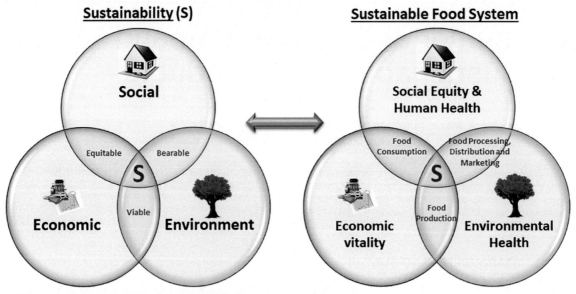

FIG. 2.1 Sustainability and components of sustainable food system.

(3) Social equity: manufactured and distributed food with conditions which ensure appropriate prices for producers and from companies that contribute to preserve the heritage and maintain biodiversity and cultural traditions. Fair trade food products are also included.

2.2 SUSTAINABLE DEVELOPMENT GOALS AND EUROPE'S FOOD SYSTEM

In 2015 all United Nations (UN) Member States adopted 17 Sustainable Development Goals (SDGs) as a universal call to action to end poverty, protect the planet and ensure that all people enjoy peace and prosperity by 2030 (2030 Agenda). The SDGs are integrated, they recognize that development must balance social, economic, and environmental sustainability and all society is needed to reach these targets (Table 2.1). The SDGs were born at the United Nations Conference on Sustainable Development in Rio de Janeiro in 2012 and they replaced the Millennium Development Goals (MDGs), which started a global effort in 2000 to tackle the indignity of poverty.

A report by the Intergovernmental Panel on Climate Change (IPCC) noted that the food system is a large emitter of greenhouse gases and air pollution

and a major consumer of energy and water and about one third of food is wasted (Masson-Delmotte et al., 2019). In developing countries, the major wastes occur in production and post-harvest by comparison with industrialized countries where the most wastes occur in processing, markets, distribution, and consumption. Therefore, Europe's food system must be transformed to achieve the SDGs to ensure the European Union's long-term sustainability goal of "living well, within the limits of the planet" by 2050. The Official Journal of the European Union published the opinion of the European Economic and Social Committee (EESC) on "Civil society's contribution to the development of a comprehensive food policy in the EU" (European Economic and Social Committee, 2018). EESC called for the development of a comprehensive food policy whose objectives are: provide healthy diets from sustainable food systems, value the nutritional and cultural importance of food, as well as its social and environmental impact, promote food waste prevention and reduction and ensure fair prices for producers so that farming remains viable, accelerate the consumer shift toward sustainability and implement a new system on sustainable food labeling due to the lack of consumer information on the environmental and social impact of food.

TABLE 2.1
List of 17 Sustainable Development Goals.

Sustainable Development Goals

	1. No poverty	End poverty in all its forms everywhere
	2. Zero hunger	End hunger, achieve food security and improve nutrition and promote sustainable agricultural
	3. Good health and well-being	Ensure healthy lives and promote well-being for all at all ages
	4. Quality education	Ensure inclusive and equitable quality education promote lifelong learning opportunities for all
	5. Gender equality	Achieve gender equality and empower all women and girls
	6. Clean water and sanitation	Ensure availability and sustainable management of water and sanitation for all
	7. Affordable and clean energy	Ensure access to affordable reliable, sustainable, and modern energy for all
	8. Decent work and economic growth	Promote sustained, inclusive, and sustainable economic growth, full and productive employment, and decent work for all
	9. Industry, innovation, and infrastructure	Build resilient infrastructure, promote inclusive and sustainable industrialization and foster innovation
	10. Reduced inequalities	Reduce inequality within and among countries
	11. Sustainable cities and communities	Make cities and human settlements inclusive, safe, resilient, and sustainable
	12. Responsible consumption and production	Ensure sustainable consumption and production patterns
	13. Climate action	Take urgent action to combat climate change and its impact
	14. Life below water	Conserve and sustainably use the oceans, seas, and marine resources for sustainable development
	15. Life on land	Protect, restore, and promote sustainable use of terrestrial ecosystems, sustainably manage forest, combat desertification, and halt and reserve land degradation and halt biodiversity

Continued

TABLE 2.1		
List of 17 Sustainable Development Goals—cont'd		
Sustainable Development Goals		
	16. Peace, justice, and strong institutions	Promote peaceful and inclusive societies for sustainable development, provide access to justice for all and build effective, accountable, and inclusive institutions at all levels
	17. Partnerships for the goals	Strengthen the means of implementation and revitalize the global partnership for sustainable development

Source: United Nations Development Programme (https://www.undp.org/content/undp/en/home/sustainable-development-goals.html).

2.3 CIRCULAR ECONOMY AND SUSTAINABLE FOOD CONSUMPTION

The linear management of the food system has caused environmental deterioration, has contributed to impoverishment, and has produced pollution from farm to fork. The food industry is aware that it can tackle environmental impacts but it is necessary to engage consumers who play an important role in the circular economy (food wastes, energy efficiency) and sustainable food consumption. The circular economy (Fig. 2.2) is a new economic model that is about de-linking the overall economic development from finite resource consumption (Kirchherr, Reike, & Hekkert, 2017). This model intends to remove the concept of waste as we know it, which involves consumption. The question is one of finding mechanisms capable of creating a new closed industrial system based on renewable energies and where all resources are consumed and no waste is generated. All values are seized and natural capital is regenerated. The food system is one way in which the human economy interacts with nature and thus the negative impacts of wastes and pollution extend into other earth and socioeconomic systems. It is overwhelmingly linear and therefore the circular economy represents drastic changes for the food system (Fassio & Tecco, 2019). The agri-food sector is one the most relevant to achieve the SDGs and a circular economy food system and has to satisfy consumer demand ensuring the security and quality of food, and the nutrition and health of people.

The "Cities and the circular economy for food" project carried out by The Ellen MacArthur Foundation (2013) defined the concept of the *food value chain* to describe the sequences of activities from farm to fork to waste. The food value chain may be divided into two parts: the agricultural system (production) and urban food system (processing, retail, preparation, consumption, and waste processing). The global food systems must face complexity changes, both in agricultural areas and consumption centers. Rural agriculture production areas need to shift towards regenerative and socially inclusive practices, while consumption centers will have to change to get healthier diets, minimize food waste and rebuild biological nutrient loops. In this sense, different actions were outlined to ensure the circular economy on the agricultural system in recent studies (Cerantola, Pinilla, & Teresa, 2018; Jurgilevich et al., 2016; Toop et al., 2017) but cities are the largest consumers and the greatest producers of wastes associated with the urban food system. Each year 1300 million tons of food are wasted (1/3 of global production), 54% in households, 29% in production, 12% in distribution, and 5% in catering (Fig. 2.3), so it is necessary that actions to achieve sustainability are not focused only on production and distribution of food but also on consumptions patterns (FAO, 2020).

2.4 CONSUMERS AND SUSTAINABLE FOOD

There is a growing trend for the consumer to make less impulsive and more thoughtful food purchases and to show greater concern for health, nutrition, animal welfare, and environmental impact. Consumers are considering the preservation of natural resources as a priority, followed by decent working conditions and accessibility for everyone to healthy and safe food when perception of sustainability was studied (Peano, Merlino, Sottile, Borra, & Massaglia, 2019). These changes are marked by demographic, attitudinal, and cultural differences. For example, younger consumers often show greater interest in sustainability (Feil, da Silva Cyrne, Sindelar, Barden, &

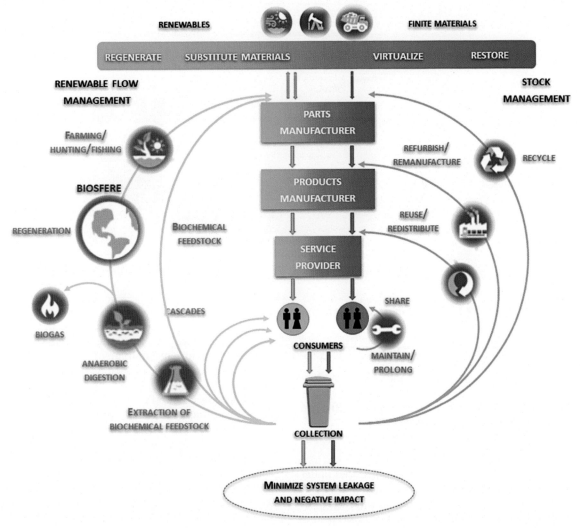

FIG. 2.2 Circular economy diagram.

Dalmoro, 2020; Hartmann & Siegrist, 2017; Nielsen & Jørgensen, 2015). The new food revolution is not marked only by flavor, nutritional properties, or food prices but also for its sustainability. A recent report of the International Trade Centre (ITC) showed that retailers in France, Germany, Italy, the Netherlands and Spain found growing consumer demand for sustainable sourced products (International Trade Centre, 2019). This growth of the recognition for greater sustainability in the food system is being built against a background of lack of awareness of the problems of sustainability and resistance to changing food patterns (de Boer & Aiking, 2018; Hartmann & Siegrist, 2017).

The EU has launched initiatives in favor of sustainable consumption to stimulate demand for better food products and to guide consumers to make better decisions (European Parliament, 2019a). EU Cities for Fair and Ethical Trade Award (www.trade-city-award.eu/) is an initiative of the European Commission to highlight innovative practices by EU cities in incentivizing sustainable trade (European Comission, 2020). The transition to a circular economy requires the responsibility of consumers to change their consumption patterns.

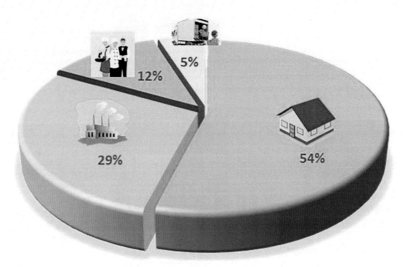

FIG. 2.3 Food wastes.

Jurgilevich et al. (2016) outlined some consumer actions to promote a circular economy for the food system:

(1) Improve menu planning: to buy only what you need.
(2) Consume seasonal food and promote local economy.
(3) Consume food with environmental certification, loose sales, and/or dehydrated products.
(4) Be well informed about nutrition and production process.

de Boer and Aiking (2018) have noted that understanding and changing food consumption must be done across the different levels of food and eating: "diets (i.e., the broad set of food items that is accepted by a population over a period of time), dishes (i.e., food items accepted on a plate in combination with each other), to ingredients (i.e., food items accepted as separate entities) and bites (i.e., single food exposures)." Thus, we cannot focus on simple agricultural products and expect changes in diets, and we cannot focus on diets and expect changes in complex prepared dishes. In order to change what we eat we must consider the whole range of levels of food (diets, dishes, ingredients, and bites), in their consumption situations or context. What works in the home might not work as well in the workplace cafeteria or school.

An additional level of food is food waste. The amount of waste generated by consumers is closely linked at our consumption patterns which vary by cultural factors and dietary habit trends. Bad eating habits

and poor management of food have led to a considerable increase in food waste. Some change in habits at home to minimize food waste are the following: prepare weekly menus, make shopping lists, choose products with the expiry date depending when they will be consumed, purchase containers in quantities appropriate to the consumption of each household, order the food by consumption priority, freeze fresh products not being consumed in a short time, calculate more accurately the ingredients, and make new recipes with reutilized food. In 2010, the "*Fundación Dieta Mediterránea*" adapted the Mediterranean diet pyramid (Fig. 2.4) to suit current food habits. The new pyramid includes social and cultural elements and besides prioritizing some food groups from others, also pays attention to the way of selecting, cooking, and eating.

Consuming more seasonal food is one of the changes proposed to achieve a more sustainable food. But what are seasonal food products? According to Macdiarmid (2014), the interpretation of seasonal food can vary depending on who is using it and the contexts in which it is being used. Two definitions are possible: (1) global seasonality when food is grown outdoor or produced during the natural growing/production period for the country or region where it is produced but is not necessarily consumed in the same place, and (2) local seasonality when food is produced and consumed in the same zone. When speaking of sustainable food, seasonal foods are purchased and consumed shortly after harvest and are produced on local farms without long distances for transport. Therefore, seasonal food

Wine in moderation and respecting social beliefs

Sweets ≤ 2s

Read meat < 2s
Processed meat ≤ 1s

Eggs 2-4s
Legumes ≥ 2s

Herbs / Spices / Garlic / Onions
(less added salt)
Variety of flavours

Olive oil
Bread / Pasta / Rice / Couscous /
Other cereals 1-2s
(preferably whole grain)

Water and herbal infusions

Biodiversity and seasonality
Traditional, local and
eco-friendly products
Culinary activities

s = serving

Potatoes ≤ 3s

White meat 2s
Fish/Seafood ≥ 2s

Dairy 2s
(preferably low fat)

Olives / Nuts / Seed 1-2s

Fruits 1-2 | Vegetables ≥ 2s
Variety of colours/textures
(cooked/raw)

Weekly

Every day

Every main
meal

Regular physical activity
Adequate rest conviviality

SERVING SIZE BASED ON FRUGALITY AND LOCAL HABITS

FIG. 2.4 Mediterranean diet pyramid.Source: https://dietamediterranea.com/

involves a local economy. More consumers are interested in knowing how their food is grown, handled, and transported and have expressed concern over quality and safety (Holt, Rumble, Telg, & Lamm, 2018).

Consumers optimize their purchase decisions in order to satisfy their own needs as fully as possible. Thus, when they accept a higher price for a product and buy it, they also reveal a higher preference for it. Consumers who purchase locally grown food are willing to pay the highest price for high-quality food grown with less environmental impact such as minimal or no pesticide/herbicide use, no-till production methods, etc. (organic food) (Zander & Hamm, 2010). Therefore, it is important that certain credence attributes exist to offer consumers full guarantees that the origin and quality of the food cover all requirements to be considered sustainable food. In this sense, geographical indications (GIs) and quality and environmental standards or other certification marks may be conducive because they are not owned but rather attributed and smaller producers have access to the marketing potential of the GI label (Giovannucci, Barham, & Pirog, 2010). Labels verified by the corresponding State enhance the trust in and credibility of the food products.

But people consume daily many food products that are typically exported from developing countries to developed countries such as coffee, sugar, etc. Fair Trade emerged to ensure that these food products meet the sustainable food goals which are defined by World Fair Trade Organization (WFTO) as a "trading partnership, based on dialogue, transparency, and respect that seek greater equity in international trade. It contributes to sustainable development by offering better trading conditions to, and securing the rights of, marginalized producers and workers" (von Meyer-Höfer, von der Wense, & Spiller, 2015). It is not only focused on exported food products but also on domestic markets and offers better trading conditions and secures the rights of marginalized producers and workers.

2.5   THE MARKETING OF SUSTAINABLE FOOD

Marketing has an influence on consumer preferences. The change of consumer food habits is leading to retailers considering sustainability key when buying from suppliers. More and more retailers are making commitments to source more sustainable food products including cutting waste, utilizing renewable commodities, and ensuring just working conditions. Therefore, policymakers should support retailers to accelerate the growing trend and suppliers on the EU

and outside of it should understand that there is a viable growing market for these food products. The EU is the second largest market after the United States for food produced in compliance with organic standards and Germany and France are the largest EU markets for these products. According to ITC data, France, Germany, Italy, the Netherlands, and Spain increased sales of sustainable food from 92% to 100% in the past 5 years (International Trade Centre, 2019). Data from the Danish Agriculture and Food Council showed Denmark had the highest market share of organic products in the world with organic food making up roughly 12% of the total retail food market in 2018. Not only is organic food widely sold in Denmark, both in mainstream discount and up-market retail stores, but organic production facilities are often as large-scale as conventional farms (Ditlevsen, Denver, Christensen, & Lassen, 2020).

Market oriented initiatives for sustainable food production processes have been expanding in recent years. There are many ways to promote sustainability in the food market. Labeling is one of the most popular instruments to communicate the sustainability of food to consumers, as producers and retailers promote their sustainable food products using different labeling strategies. The two best known and widely used sustainability labels in Germany are organic and fair trade and the impact of both labels on the global food market has grown continuously in recent decades, not only separately but also in combination (von Meyer-Höfer et al., 2015).

When talking about sustainable labeling in a broad sense, terms such as standard, label and certification are widely used, but they are not identical terms. The sustainable standards represent social and environmental requirements to reduce negative impacts of global economic activity on society and the environment and they can refer to the manufacturing process, to the properties of the final products and to the corporate management procedures. This document is normally accompanied by a verification process which represents a certification. This information is transmitted to consumers as a label. A study carried out by Kaczorowska, Rejman, Halicka, Szczebyło, and Górska-Warsewicz (2019) with urban consumers from Poland showed that sustainability labels influenced consumer buying behavior and they were very price sensitive in terms of buying products with sustainability logos. These authors concluded that the value added of sustainable products should be communicated clearly by producers in order to enable consumers to make informed choices in line with sustainable food consumption recommendations.

This labeling is voluntary and is developed at the local, national, or international level and can be managed by private or public economic entities. There are more than 400 standards across the globe and they can be consulted in different online platforms. Standard Maps is a free UN database (www.sustainabilitymap.org/standards_intro) that provides online access to a wide-range of information relative to 264 standards and it is used in 192 countries (ITC, 2020). ITC in collaborating with the Research Institute of Organic Agriculture (FiBL) and the International Institute for Sustainable Development (IISD) provides data about 14 major sustainability standards for eight agricultural products (bananas, cocoa, coffee, cotton, oil palm, soybeans, sugarcane, and tea). These data show that agriculture certification is growing in line with a consumption increase and they can be consulted in an online platform (www.sustainabilitymap.org/trends) Market Trends (Willer, Sampson, Voora, Dang, & Lernoud, 2019a, 2019b). The biggest sustainability standard for food products in terms of area and the largest variety is INFOAM—Organics International (Organic) followed by GLOBALG.A.P. In terms of numbers of producers,

INFOAM—Organics International has also the highest number followed by Fairtrade (Fig. 2.5).

INFOAM—Organics International (www.ifoam.bio/) represents the organic movement across the entire food system uniting a diverse range of stakeholders contributing to the organic vision. It sets reference standards and establishes quality assurance systems at national or regional levels and focuses on three key areas: enhancing to increase supply, raising awareness to enhance demand, and providing support for the creation of a national and international favorable policy environment. Australia has the largest agricultural organic certified area followed by Argentina and China. Apart from agricultural commodities, aquaculture, apiculture, and forestry products are certified and the leading countries for the organic market are the United States followed by Germany and France. Fairtrade International (www.fairtrade.net/) works to share the benefits of trade more equally though standards, certification, producer support, programmers and advocacy and awareness raising. It sets standards for smallholder farmers and plantations that use hired labor and trades, establishing social, economic, and

More than 750 members in over 127 countries

Leading countries organic market: the United States (43%), Germany (11%) and France (9%)

22 member organizations – three producer networks and 19 national Fairtrade organizations

Representing more than 1.7 million farmers and workers in 75 countries in Latin America and the Caribbean, Africa and the Middle East, and Asia and the Pacific

More than 400 members in over 135 countries

Certificated area: Europe (44%), Latin America (25%), Africa (12%) and North America (11%)

FIG. 2.5 Main food sustainable standards.

environmental requirements. The standards require consumers to pay a minimum price to producers for most goods. Producers receive a premium which farmers and workers decide how to invest in their businesses and communities. Fairtrade International certifies a wide range of commodities, from tropical fruit to cereals, gold, and textiles. The Global Partnership for Good Agriculture Practices (GLOBALG.P.A) (www.globalgap.org/uk_en/index.html) is an international standard that is based on an equal partnership among producers of raw agricultural food products and their buyers, such as retailers and food service organizations. It runs 40 standards for agriculture, livestock aquaculture and horticulture production. A variety of 230 different fruit and vegetables products are certified. In 2017, potatoes were the product with the largest non-certified area, followed by bananas and apples. Spain had the largest certified area followed by the United States and Italy.

EU quality policy aims at protecting the quality of food products and agricultural policy provides financial support to these types of products and their development. EU sustainable food labeling was created from the need to raise consumer awareness providing information about certain aspects of the food and/or its production method and included a European awareness system for food products linked to their geographical origin as well as traditional know-how and for products of EU's outermost regions (Table 2.2).

Food products can be granted with GIs if they have a specific link to the place where they are made which enables consumers to distinguish quality products and help producers to market their products better. There are three different GIs: Protected designation of origin (PDO) when the products have the strongest links to the place in which they are made and every part of the production, processing and preparation process must

TABLE 2.2
EU Quality Certifications for Agricultural and Other Food Products.

Label	Designation	Specifications
	Protected designation of origin (PDO)	Mandatory for food and agricultural products
		Optional for wine
	Protected geographical indication (PGI)	Mandatory for food, agricultural products
		Optional for wines
	Geographical indication of spirit drinks and aromatized wines (GI)	Spirit drinks and aromatized wines (optional)
	Traditional specialty guaranteed	Food and agricultural products (mandatory)
–	Mountain product	Agricultural and food products
	Product of EU's outermost regions	Agricultural and food products
	Organic products	Products as organic by an authorized control agency or body

Source: Certification | European Commission (https://ec.europa.eu/info/food-farming-fisheries/food-safety-and-quality/certification_en).

take place in the specific origin; Protected geographical indication (PGI) when a particular quality, reputation or other characteristic is essentially attributable to its geographical origin and at least one of the stages of production, processing or preparation takes place in the region; Geographical indication of spirit drinks and aromatized wines (GI) when the product's particular quality, reputation or other characteristic is essentially attributable to its geographical origin and at least one of the stages of distillation or preparation takes place in the region.

EU produced organic products have a specific logo (Table 2.2) which helps farmers to market their products across EU countries and it can only be used on products that have been certified by an authorized control agency. The organic food products have fulfilled strict conditions on how they must be produced, processed, transported, and stored; they contain at least 95% of organic ingredients and additionally respect further strict conditions for the remaining 5%. According to the European Commission, the organic logo must be used by all pre-packaged EU products, produced and sold as organic within the EU and can be used by imported products when they conform to EU rules as well as non-pre-packaged organic products and EU organic products commercialized in the markets of third-party countries.

From 1 January 2021, there will be an overhaul of the organics regulations (EU Regulation 834/2007, 2018) to ensure fair competition for farmers while preventing fraud and maintaining consumer trust. Organic farmers use energy and resources in a responsible way, promote animal health and contribute to maintaining biodiversity, ecological balance, and water/soil quality. The permissible farming practices are: crop rotation, ban of the use of chemical pesticides and synthetic fertilizers, strict limits on livestock antibiotics, ban genetically modified organisms, use of onsite resources for natural fertilizers and animal feed, raising livestock in a free range and use of organic fodder and tailored animal husbandry practices. Although the EU's organic farmland has increased over the years, it only represents 7% of the total agricultural area and four countries account for 54.4% of total organic area: Spain 16.9%, Italy 15.1%, France 12.9% and Germany 9.5% (European Parliament, 2019b).

Finally, EU recognizes the importance of a supply chain with good functioning that produces fair and end-to-end value for stakeholders, including consumers, retailers, manufactures, farms, and governments. The European organization involved in fair trade is the World Fair Trade Organization-Europe

(WFTO-Europe), a branch of global WFTO (wfto-europe.org/). WFTO-European is formed by 105 members across 17 European countries and they are all mission-led businesses and organizations. Netherlands followed by United Kingdom, Germany and France are the countries with the largest memberships. The main five principles that WFTO must follow are: (1) Creating opportunities for economically disadvantaged producers: reducing poverty and providing income security, market opportunities for small producers and farmers and sustainable agriculture. (2) Commitment to non-discrimination, gender equity and women's economic empowerment and freedom of association: organizations must not discriminate for any reason and will protect labor rights and promote safe and secure working environments for all workers. (3) Payment of fair price: must provide a socially acceptable remuneration for producers, be stable to enable long-term planning and trading conditions must be continuously improved. (4) Transparency and accountability: Guarantee that the system is developed to encourage the participation and policy-making of their members in each step of the supply chain, from producers to consumers. (5) Respect for environment: reducing the emission of greenhouse gases is a purpose for farmers and producers, as well as ensuring sustainable agriculture and production systems.

The food industry (including food and drinks) is the largest manufacturing sector in terms of job and added value in the EU, is a leading employer and 99% of the sector is composed of small and medium sized enterprises. It is an asset in trade with non-EU countries, and has been the main exporter sector of food and drinks in the world ahead of the United State and China (FoodDrink Europe, 2019). This sector faces challenges in both European and international markets and, as a major contribution to the EU economy and the global food system. It is committed to engage in a transition towards more sustainable food systems while remaining competitive and will have to anticipate changes in consumer and social demands. Lodorfos, Konstantopoulou, Kostopoulos, and Essien (2018) undertook a review and provided an overview of the European food and drink industry and its current state covering questions about manufacturing, consumers' purchasing behaviors, distribution, marketing, and retail, environmental, structures and economic trends. They presented some future trajectories in terms of social, consumer and regulatory trends. These authors indicated that the food and drink sector had faced a decrease in its competitiveness compared to other world food producers in terms of growth of labor productivity and added value. The current trend has led the food industry to set at boosting growth of the production and marketing of products which satisfy qualitative needs creating jobs and caring for the environment. They concluded that EU food and drinks industry may continue to maintain its competitive position if it makes efforts to accommodate the current socio-cultural, politico-regulatory, economic, and technological changes in the food and drink market being driven by changes in consumer behavior. In particular, investments in research and development, distribution channels, and also focus on government regulatory requirements for sustainable foods.

Different organizations advance the movement against high intake sugar, salt, trans-fat, and consumers demand more fresh, natural, whole ingredients and with smaller carbon footprint foods, without artificial additives and free from GMOs but without significant changes in taste and commercialized at a reasonable price. In this sense, research and technology are important supports to develop new products and to obtain new methods of organizing the food supply chain. Additionally, the technology can help food corporations to reach more consumers quicker, easier, and cheaper. EU food industry has to invest more in R&D because of technology and process innovations can give a competitive edge.

2.6   KEY FINDINGS

The food system is a strategic sector for supporting a sustainable Europe. The European food system has to move towards a more circular and sustainable economy and to adopt and promote more environmentally-friendly practices. The food industry has to work to ensure that all actors of the food supply chain (from raw material producers to consumers) are on their way to common ambitions and realistic objectives in the context of sustainability. In this sense, the food industry strives to provide consumers with products that meet their needs in terms of safety, quality, nutrition, affordability, and convenience, and to inform and guide their purchasing decisions. The European food sector shows the highest growth in sustainable products sales and it can contribute positively to the future sustainability of Europe meeting SDGs consistent with the purposes of the 2030 Agenda. This involves coherent and synergistic policies, new contracts between farmers and society, appropriate governance and new approaches to addressing consumption as well as production. (ITC, 2020).

ACKNOWLEDGMENTS

Sol Zamuz acknowledges financial support PTA program (PTA2017-14156-I). Paulo E.S. Munekata acknowledges postdoctoral fellowship support from the Ministry of Economy and Competitiveness (MINECO, Spain) "Juan de la Cierva" program (FJCI-2016-29486). Thanks to GAIN (Axencia Galega de Innovación) for supporting this study (grant number IN607A2019/01).

REFERENCES

Cerantola, N., Pinilla, O., & Teresa, M. (2018). *La economía circular en el sistema agroalimentario*. Retrieved from https://www.otroconsumoposible.es/publicacion/economia-circular.pdf.

de Boer, J., & Aiking, H. (2018). Prospects for pro-environmental protein consumption in Europe: Cultural, culinary, economic and psychological factors. *Appetite, 121*, 29–40.

Ditlevsen, K., Denver, S., Christensen, T., & Lassen, J. (2020). A taste for locally produced food—Values, opinions and sociodemographic differences among 'organic' and 'conventional' consumers. *Appetite, 147*, 104544. https://doi.org/10.1016/j.appet.2019.104544.

EU Regulation 834/2007. (2018). No 834/2007 of 28 June 2007 on organic production and labelling of organic products and repealing regulation (EEC) No 2092/91. *Official Journal of the European Union L, 189*(1), 20–27. Regulation, C. (2007).

European Comission. (2020). *EU cities for fair and ethical trade award*. Retrieved from https://www.trade-city-award.eu/.

European Economic and Social Committee. (2018). *Civil society's contribution to the development of a comprehensive food policy in the EU (own-initiative opinion)*. Retrieved from https://www.eesc.europa.eu/en/our-work/opinions-information-reports/opinions/civil-societys-contribution-development-comprehensive-food-policy-eu-own-initiative-opinion.

European Parliament. (2019a). *Sustainable consumption and production*. Retrieved from https://www.europarl.europa.eu/factsheets/en/sheet/77/sustainable-consumption-and-production.

European Parliament. (2019b). *The EU's organic food market: Facts and rules (infographic)*. Retrieved from https://www.europarl.europa.eu/news/en/headlines/society/20180404STO00909/the-eu-s-organic-food-market-facts-and-rules-infographic.

FAO. (2012). In B. Burlingame, S. Dernini, Nutrition and Consumer Protection Divisision, & FAO (Eds.), *Sustainable diets and biodiversity: Directions and solutions for policy, research and action*. Retrieved from http://www.fao.org/3/i3004e/i3004e.pdf.

FAO. (2020). *Food loss and food waste*. Retrieved from http://www.fao.org/platform-food-loss-waste/flw-data/en/.

Fassio, F., & Tecco, N. (2019). Circular economy for food: A systemic interpretation of 40 case histories in the food system in their relationships with SDGs. *Systems, 7*(3), 43. https://doi.org/10.3390/systems7030043.

Feil, A. A., da Silva Cyrne, Sindelar, F. C. W., Barden, J. E., & Dalmoro, M. (2020). Profiles of sustainable food consumption: Consumer behavior toward organic food in southern region of Brazil. *Journal of Cleaner Production, 258*, 120690.

Fischer, C. G., & Garnett, T. (2016). *Plates, pyramids, planet: Developments in national healthy and sustainable dietary guidelines: A state of play assessment.* FAO and The Food Climate Research Network. https://doi.org/978-92-5-109222-4.

FoodDrink Europe. (2019). *The path of the food and drink industry towards sustainable food systems.* Retrieved from https://www.fooddrinkeurope.eu/uploads/publications_documents/FoodDrinkEurope_-_The_path_towards_Sustainable_Food_Systems.pdf.

Giovannucci, D., Barham, E., & Pirog, R. (2010). Defining and marketing "local" foods: Geographical indications for US products. *The Journal of World Intellectual Property, 13*(2), 94–120. https://doi.org/10.1111/j.1747-1796.2009.00370.x.

Hartmann, C., & Siegrist, M. (2017). Consumer perception and behaviour regarding sustainable protein consumption: A systematic review. *Trends in Food Science & Technology, 61*, 11–25.

Holt, J., Rumble, J. N., Telg, R., & Lamm, A. (2018). Understanding consumer intent to buy local food: Adding consumer past experience and moral obligation toward buying local blueberries in Florida within the theory of planned behavior. *Journal of Applied Communications, 102*(2). https://doi.org/10.4148/1051-0834.2203.

International Trade Centre. (2019). *The European union market for sustainable products. The retail perspective on sourcing policies and consumer demand.* Geneva: ITC. Retrieved from https://www.intracen.org/uploadedFiles/intracenorg/Content/Publications/EUMarket for Sustainable Products_Report_E_WEB.pdf.

ITC. (2020). *Sustainability map.* Retrieved from https://www.sustainabilitymap.org/standards_intro.

Jurgilevich, A., Birge, T., Kentala-Lehtonen, J., Korhonen-Kurki, K., Pietikäinen, J., Saikku, L., et al. (2016). Transition towards circular economy in the food system. *Sustainability, 8*(1), 69. https://doi.org/10.3390/su8010069.

Kaczorowska, J., Rejman, K., Halicka, E., Szczebyło, A., & Górska-Warsewicz, H. (2019). Impact of food sustainability labels on the perceived product value and price expectations of urban consumers. *Sustainability (Switzerland), 11*(24), 7240. https://doi.org/10.3390/SU11247240.

Kirchherr, J., Reike, D., & Hekkert, M. (2017). Conceptualizing the circular economy: An analysis of 114 definitions. *Resources, Conservation and Recycling, 127*, 221–232. https://doi.org/10.1016/j.resconrec.2017.09.005.

Lodorfos, G., Konstantopoulou, A., Kostopoulos, I., & Essien, E. E. (2018). Food and drink industry in Europe and sustainability issues. In *The sustainable marketing concept in European SMEs* (pp. 121–140). Emerald Publishing Limited. https://doi.org/10.1108/978-1-78754-038-520180006.

Macdiarmid, J. I. (2014). Seasonality and dietary requirements: Will eating seasonal food contribute to health and environmental sustainability? In *Vol. 73. Proceedings of the nutrition society* (pp. 368–375). Cambridge University Press. https://doi.org/10.1017/S0029665113003753.

Masson-Delmotte, V., Zhai, P., Pörtner, H.-O., Roberts, D. C., Skea, J., Buendia, E. C., et al. (2019). *Climate change and land: An IPCC special report on climate change, desertification, land degradation, sustainable land management, food security, and greenhouse gas fluxes in terrestrial ecosystems.* The Intergovernmental Panel on Climate Change. Retrieved from https://www.ipcc.ch/site/assets/uploads/2019/11/SRCCL-Full-Report-Compiled-191128.pdf.

Nielsen, S. N., & Jørgensen, S. E. (2015). Sustainability analysis of a society based on exergy studies—A case study of the island of Samsø (Denmark). *Journal of Cleaner Production, 96*, 12–29.

Peano, C., Merlino, V. M., Sottile, F., Borra, D., & Massaglia, S. (2019). Sustainability for food consumers: Which perception? *Sustainability, 11*(21), 5955. https://doi.org/10.3390/su11215955.

The Ellen MacArthur Foundation. (2013). Towards the circular economy: Economic and business rationale for an accelerated transition. Retrieved from: https://www.ellenmacarthurfoundation.org/assets/downloads/publications/Ellen-MacArthur-Foundation-Towards-the-Circular-Economy-vol.1.pdf. (Accessed 9 December 2019).

Toop, T. A., Ward, S., Oldfield, T., Hull, M., Kirby, M. E., & Theodorou, M. K. (2017). AgroCycle—Developing a circular economy in agriculture. *Energy Procedia, 123*, 76–80. Elsevier Ltd. https://doi.org/10.1016/j.egypro.2017.07.269.

von Meyer-Höfer, M., von der Wense, V., & Spiller, A. (2015). Characterising convinced sustainable food consumers. *British Food Journal, 117*(3), 1082–1104. https://doi.org/10.1108/BFJ-01-2014-0003.

Willer, H., Sampson, G., Voora, V., Dang, D., & Lernoud, J. (2019a). *The state of sustainable markets 2019: Statistics and emerging trends.* Geneva. Retrieved from https://www.intracen.org/uploadedFiles/intracenorg/Content/Publications/Sustainabilemarkets2019 web.pdf.

Willer, H., Sampson, G., Voora, V., Dang, D., & Lernoud, J. (2019b). *The state of sustainable markets 2019: Statistics and emerging trends.* http://www.intracen.org/publication/Sustainable-Markets-2019/.

Zander, K., & Hamm, U. (2010). Consumer preferences for additional ethical attributes of organic food. *Food Quality and Preference, 21*(5), 495–503. https://doi.org/10.1016/j.foodqual.2010.01.006.

CHAPTER 3

Technological Advances for Sustainable Livestock Production

RUBÉN AGREGÁN[A] • PAULO EDUARDO SICHETTI MUNEKATA[A] • XI FENG[B] • BEATRIZ GULLÓN[C] • RUBEN DOMINGUEZ[A] • JOSÉ MANUEL LORENZO[A,D]
[a]Galician Meat Technology Center, Galicia Technology Park, Ourense, Spain, [b]Department of Nutrition, Food Science, and Packaging, San Jose State University, San Jose, CA, United States, [c]Department of Chemical Engineering, Faculty of Science, University of Vigo (Campus Ourense), Ourense, Spain, [d]Food Technology Area, Faculty of Sciences of Ourense, Vigo University, Ourense, Spain

3.1 INTRODUCTION

Worldwide meat consumption is increasing year by year inexorably due to the growing population (Pradère, 2014). When the quality of life improves, people also prefer to include more protein resources in their diet. This behavior pattern is accepted widely around the world (Dagevos & Voordouw, 2013). It is expected that by 2030 the demand for food of animal origin will be more than 50% compared to 2000 (van Wagenberg et al., 2017) and according to Derner et al. (2017) the animal production should increase by 70% by 2050 to satisfy consumers' demands for meat and muscle products. This increase in demands requires the changes in the current livestock production patterns, which should be able to avoid extra pressure on the ecosystem, and aim for the "sustainability." Sustainability is defined as the current and the future acceptability of a system, specifically with the availability of resources, the consequences of functioning and the morality of the action (Broom, Galindo, & Murgueitio, 2013). This proposes a challenge for the livestock industry, which should find out a solution to increase production yield, reduce the environmental impact, and also decrease costs and balance economic and social issues (van Wagenberg et al., 2017).

Livestock production systems have been adapted and changed over the decades as the constant increase in demand for muscle products as a food resource (Udo & Steenstra, 2010). In the last century, technologies were developed to increase livestock production in a more efficient manner, including artificial insemination (AI) and crossbreeding (Derner et al., 2017), which in turn can also maintain the stability of the livestock. Today, innovative technologies focus on increasing the sustainability of livestock production systems. The production

of transgenic animals is one example of novel technologies, through which the efficiency in obtaining desirable traits in livestock might be improved considerably. In fact, genetic engineering techniques have been applied not only for milk production but also to decrease its lactose or cholesterol content (Sulabh & Kumar, 2018). This transformation of animal production systems has the potential to supply meat faster, cheaper, and with less environmental impact than other existing technologies (Tait-Burkard et al., 2018).

Innovations are also taking place in animal feeding. The manufacture of livestock diets contributes to the destruction of the ecosystem. The approach of maximizing animal performance has led to deforestation and also caused global warming (Makkar & Ankers, 2014). To overcome this situation, more environmentally friendly diets are currently being studied. A very promising technology that is already introduced on farms is Precision Livestock Farming (PLF). This intelligent tool is based on the continuous measuring of determining productivity parameters on a farm, including the health and welfare of animals, production, reproduction, and the environmental impact of the activities (Berckmans, 2017). This technology has the potential of revolutionizing the livestock industries through big data, artificial intelligence, which could enhance production, reduce costs as well as increase the animal welfare (Banhazi et al., 2012).

This chapter briefly presents the environmental problems that threaten the sustainability of animal production systems and comments on the contributions of different technologies to solve this problem, including modern and those that have been developed decades ago but are still in force.

Sustainable Production Technology in Food. https://doi.org/10.1016/B978-0-12-821233-2.00005-8
Copyright © 2021 Elsevier Inc. All rights reserved.

3.2 PROBLEMS AFFECTING THE SUSTAINABILITY OF LIVESTOCK PRODUCTION SYSTEMS

3.2.1 Exploitation of Natural Resources

Livestock production systems use natural resources, such as water, land, and nutrients in soils. Among all human activities, this is the one that uses the most land, exploiting meadows, and pastures that occupy almost 26% of the total land area. On the other hand, it is also necessary to use the ground to produce the necessary fodder to feed livestock, occupying 33% of the world's farmland. Intensive land use for these tasks is causing an increase in land-use change and deforestation (Baumung, Mottet, & Teillard, 2018). The impact of the over-grazing could cause land degradation as well as a decrease in forest diversity, plant cover, plant height, and its productivity. This degradation could further cause gradual desertification and a change in the vegetation, turning grounds with perennial vegetation pastures into shrubs and bare ground (Allington & Valone, 2011). In addition to all aspects mentioned above, over-grazing decreases the yield and quality of forage (Azarnivand, Farajollahi, Bandak, & Pouzesh, 2010), which is a disadvantage to livestock. This situation has been observed in many countries. In China, livestock is identified as one of the main reasons for grassland degradation and the results of an arid environment in the northwest areas (Han et al., 2008). In Ethiopia, fertile land is lost at a rate of one billion cubic meters per year, which seriously damages agriculture and forestry stability (Mekuria et al., 2007). In Brazil, over-grazing is degrading territories in the Cerrado region, which is counted to be a quarter of the land area of the country (Pacheco, Chaves, & Nicoli, 2012).

Another natural resource widely used by livestock is freshwater. Livestock currently uses around 30% of the total amount of agricultural water in the world (Ran, Lannerstad, Herrero, Van Middelaar, & De Boer, 2016). Most of the water consumption is for the production of livestock feed (Blümmel, Haileslassie, Samireddypalle, Vadez, & Notenbaert, 2014). With the growing demand for meat origin food, the pressure of acquiring adequate water resources becomes challenged. Livestock also generates organic waste from their metabolism. These wastes, produced in significant quantities, can be used as fertilizers but also could contaminate the groundwater if it is not handled appropriately. Manure is rich in nitrogen, phosphorous, and ammoniac compounds; all of them are essential elements for the environment, but when they are added in excessive amounts, it could favor the growth of harmful algae, which causes eutrophication. Parasites, bacteria, viruses, and antibiotic residues can also reach water supplies, threatening much human health in a biological pollution manner (Bourgeois, 2012).

Industrial livestock farming is considered as one of the three most relevant activities in global warming and the main cause of water pollution and loss of biodiversity (Bourgeois, 2012).

3.2.2 Increase in Greenhouse Gases

Livestock production is one of the human activities that contribute to climate change the most (Llonch, Haskell, Dewhurst, & Turner, 2017), accounting for more than 14.5% of the total greenhouse gas (GHG) emissions released to the atmosphere (Bailey, Froggatt, & Wellesley, 2014). In fact, emissions from this activity contribute more GHG to the atmosphere than the global transport sector (Rojas-Downing, Nejadhashemi, Harrigan, & Woznicki, 2017). Livestock contributes both directly and indirectly to these emissions. The direct emission form refers to the enteric fermentation of food, breathing, and excretions (Rojas-Downing et al., 2017). Beef and dairy production account for 41% and 20% of green gas emission, respectively. However, swine and poultry only contribute 10% of total emissions (Llonch et al., 2017). Indirect emissions are referred to as those derived from forage crops, manure applications, agricultural applications, processing of livestock products, transportation, and allocation of land use for livestock production (Rojas-Downing et al., 2017). Around 44% of GHG emissions from livestock correspond to methane (CH_4) gas. The rest is nitrous oxide (NO_2) and carbon dioxide (CO_2), which account for around 27% and 29% of the total, respectively (Gerber et al., 2013). These three gases are harmful to the atmosphere, especially CH_4 and NO_2, which influences global warming (from 28 and up to 265 times) greater than CO_2 (Grossi, Goglio, Vitali, & Williams, 2019). High concentrations of these gases may be attributed to low efficiency and productivity in livestock systems (Rojas-Downing et al., 2017).

3.3 TECHNOLOGICAL ADVANCES IMPLEMENTED IN THE 20TH CENTURY

3.3.1 Artificial Insemination

In the last decades, researchers have developed new technologies to increase the efficiency of livestock production as well as reducing the environmental impact. One of these technologies that have been used for more than 60 years is Artificial insemination (AI) (Stančić & Dragin, 2011). This technology aims to improve genetic traits in order to increase productivity but also save natural resources. AI can be defined as the manual placement of semen of the male, previously collected, processed, and stored (Patel et al., 2017), which is a method different from the natural mating (Morrell, 2011). This technology belongs to the group of technologies known as "assisted reproduction technologies" (ART) (Morrell, 2011), and it is the most successful and efficient reproductive technology in the last six decades (Bertolini & Bertolini, 2009), which offers significant advantages than the natural insemination. Although the number of females fertilized with a certain amount of sperm is lower than through natural mating as a limited amount of sperms are used, the efficiency is improved with the selected sperms. In addition, sperm can be transported and stored for long periods of time, which allows it to be used in different locations and even after the male's fertile life (Patel et al., 2017). In this way, fewer males are needed for reproduction, which also reduces the number of natural resources in breeding and maintaining. Another advantage of AI is to reduce the possibility of genital disease propagation in beef cattle (e.g., brucellosis, etc.) (Patel et al., 2017). This has a direct impact on the farm's economy if a sick individual is found, which also means loss of productivity and performance. Despite all these advantages, this technology also has some drawbacks. One of them is the implementation cost excessive for many milk producers, who continue to opt for natural mating. On the other hand, there is concern that their inappropriate use may lead to breed extinction. The intensive use of genetically superior specimens produces a sharp increase in inbreeding and a loss of genetic diversity. Another major drawback in the use of AI is the problem of successfully detecting heat. This temporary imprecision is the limiting factor of reproductive performance in many dairy farms (Patel et al., 2017). However, these disadvantages won't prevent that AI is, by far, the most common method of breeding. Around 80% of dairy cattle in Europe and North America are produced through AI technology (Morrell, 2011). The reproduction technique is estimated to be responsible for approximately 50% increase in dairy products in developed countries after the 1950s (Bertolini & Bertolini, 2009). This is contributed to the work of Rendel and Robertson (1950), who proposed the progeny-testing scheme to efficiently use

of AI combined with pedigree registration and performance recording. Since then, progeny-testing schemes have been used to improve genetic traits in dairy cattle in developed countries.

Swine farming and poultry farm also benefit greatly from AI. More than 90% of the heads in Europe and North America are produced through this strategy (Morrell, 2011). Turkey is a typical example through a genetic modification to improve meat quality. AI also is used in the breeding of more than 3.3 million sheep and 0.5 million goats worldwide (Niemann & Seamark, 2018).

3.3.2 Crossbreeding to Achieve Heterosis

Crossbreeding is a mating system in which two animals from different breeds are crossed (Yadav, Singh, Sharma, & Gupta, 2018). This reproductive technique is used to combine traits of genetically unrelated breed animals, which often results in superior traits in offspring. It is also known as hybrid vigor or heterosis (Getahun, Alemneh, Akeberegn, Getabalew, & Zewdie, 2019). This superior traits in offspring phenomenon occur as a result of the increase in heterozygosity in crossbred animals (Wakchaure et al., 2015), which means a greater genetic variation with respect to an expressible trait. According to species, there are different crossbreeding systems, in which aspects, such as herd size, potential market, level of management, and facilities must be taken into account (Yadav et al., 2018).

Crossbreeding is an important practice in the beef industry, which can produce calves with enhanced reproductive, but also with an increase in survival rates, fertility ability, growth rate, lean muscle percentage, and meat quality, and resistance to disease (Yadav et al., 2018). These improvements produce greater cattle that translates into a higher profit to farmers and a benefit to the environment.

Three main types of crossing are used in beef cattle, including terminal crossing, rotational, and composite. The terminal crossing is the simplest one. In this system, all offspring are marketed and replaced by heifers purchased from outside the farm. In rotational crossing, purebred males are mated to crossbred females sired by these males and their mothers as follows: breed A sires are mated to females sired by breed B, breed B sires are mated to females sired by breed C, and breed C sires are mated to females sired by breed A (see Fig. 3.1). Composite crossing combines traits associated with factors of economical relevance of several breeds in order to obtain a composite hybrid

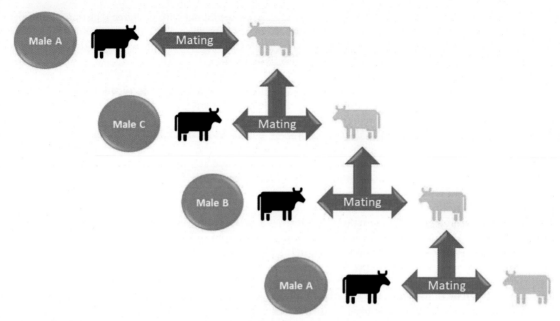

FIG. 3.1 Example of rotational crossing for beef production.

that is then maintained as a straight-bred herd (Yadav et al., 2018).

It has been shown that the use of dairy cows in cattle crossing systems mating to beef bulls improves in the meat productivity of the offspring. The crossing of dairy cows with early matured beef breeds has little positive effect on growth. However, there is an improvement in carcass conformation and a reduction of feed intake in offspring. These positive effects are observed when dairy cows are mated to late matured beef breeds, such as an increase in growth rate, kill-our proportion, and carcass muscle proportion (Keane, 2011). These genetic improvements imply a significant increase in profits through better use of resources.

Regarding swine production, terminal crossing, rotational, and rotaterminal are used. In terminal crossing, a crossbred gilt is mated to a purebred boar and their offspring are marketed. Replacement gilts are purchased or produced separately. The rotational crossing uses purebred boars rotated in a consistent order, one breed for each generation, and offspring from each generation are marketed together with replacement gilts. The third crossbreeding system used in pork production is a combination of terminal and rotational crossings, called rotaterminal crossing

(see Fig. 3.2). In this system, top females previously selected are used in a rotational crossing with maternal breed purebred boars to produce replacement gilts. Then, the replacement gilts are mated to terminal boars and their offspring are marketed (Yadav et al., 2018).

3.4 TECHNOLOGY FOR THE SUSTAINABILITY OF LIVESTOCK FARMING IN THE RECENT DECADES

3.4.1 Livestock DNA Modification: Animal Transgenesis

Modifying DNA in animals gives rise to specimens that may never have existed in nature. Their genomes have been altered to harbor genes from other species (Sulabh & Kumar, 2018). This procedure in farm animals provide a faster, cheaper, healthier, and more effective meat production than selective rearing, but also with minimal environmental impact (Tait-Burkard et al., 2018). Despite these unquestionable advantages, the production of transgenic livestock is a challenging work with a high cost. It also required to use advanced techniques in molecular biology, cell culture, reproductive biology, and biochemistry (Kues & Niemann, 2011). Important progress has been made in this field of research in the

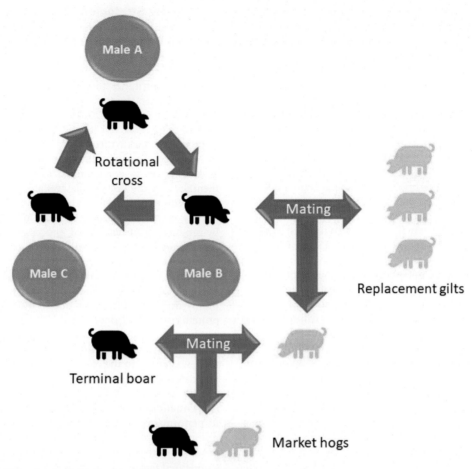

FIG. 3.2 Example of rotaterminal crossing for pork production.

latest 40 years. In the early 1980s, only a few laboratories carried out this research (Pinkert, 2014). Despite it was possible to transfer external genetic material to mice, a mouse was usually used as the animal model system, with DNA microinjection technique (Kues & Niemann, 2011) or pronuclear microinjection of genes (Murray & Maga, 2016), scientists lacked knowledge about transferred genes. Meanwhile, there were no appropriate promoters and enhancers to ensure the adequate and controlled expression of inserted transgenes, and the infectivity of the process to produce genetic engineering animals arose during this period of biotechnology. Fortunately, in recent decades, a series of progress was made, which removed the limitations, and allowed to acquire the completed genome sequence of all the major domestic species (Murray & Maga, 2016). Nowadays, genetic engineering is capable of integrating

and expressing desired genes, editing introduced genes to determinate mutation, and changing a single nucleotide, or replacing an allele through homology dependent repair (Murray & Maga, 2016).

The main contribution of the genetic modification of animals is to assist the drug products for human health or use them to serve as disease model animals. However, the most important application of this technology is in the field of animal production and welfare (Sulabh & Kumar, 2018). Regarding this issue, selective breeding has been an excellent tool to achieve larger herds and higher meat yields. From the 1960s to 2000s, this production technology resulted in a 50% size increasing and a 37% increase in lean muscle in pigs. These improvements have also been observed in poultry, where it was found that the growth rate in chickens increased with a higher meat yield. In the dairy industries, milk production increased by 67% in cattle (Tait-Burkard et al., 2018). However, the efficiency showed by this technology highly depends on the accuracy of animal selection and mating plans. In addition, it is also a time-consuming task. These problems might be solved with transgenesis, which can achieve desired traits in the animals in a shorter time and with higher precision. Cows' genes were manipulated through this technology to increase the milk-producing rate, lower lactose levels, or cholesterol contents (Sulabh & Kumar, 2018). In addition, cattle can be also manipulated to produce more meat. An important and indirect application of this technology was the genetic modification of microbes in the rumen and cecum of animals, which increased lignin degradation and improved the digestibility of dietary fibers, and also resulted in a better usage and absorption of nutrients (Segal, Knight, & Beitz, 2012).

Livestock may be also genetically modified in order to alter the fat or cholesterol content in their muscles through metabolisms. The metabolism changes in cholesterol or fatty acid absorption can reduce the amount of fat and cholesterol content in meats, eggs, cheeses, etc. In addition, there is the possibility to increase beneficial fats by genetic engineering technology, such as incorporating gene expressing omega-3 fatty acids from fishes to other animals (Sulabh & Kumar, 2018).

Livestock welfare is another crucial topic in animal production. The application of transgenic techniques can solve this issue by improving disease resistance or tolerance. Mastitis in cows can cause decreased milk production, mammary tissue damage, fever, or even death. The transgenic cows successfully show better resistance to the above illness. However, genetic engineering might not result as effective as expected, since very few genes with possible effects over disastrous animal diseases are known (Sulabh & Kumar, 2018).

Reproductive capacities and animal fertilities are also considered in increasing livestock production. Several genes with possible implications on reproductive performance and prolificacy have recently been identified (Sulabh & Kumar, 2018).

Despite all the advantages that molecular biology can offer to animal production, no genetically modified animals have moved from the laboratory to the food market. This is due to the inefficiency of producing transgenic animals with random integration following microinjection of an rDNA construct or viral transformation methods. Also, there are uncertainties in gene expression and the high cost of genetically modified animals (Van Eenennaam, 2017).

3.4.2 New Strategies in Livestock Feeding

Animal feeding is the cornerstone and its management directly or indirectly affects the entire sector. Livestock is fed to achieve the maximum yields and to bring the highest economic profits. Nevertheless, this approach generally contributes to ecosystem degradation, such as deforestation, loss of biodiversity, global warming, etc. (Makkar & Ankers, 2014). The systematic use of resources with more efficient management in livestock feeding could tackle these issues.

Numerous sustainability-focused diets are emerging as alternatives to conventional diets. These environmentally friendly diets make available resources profitable without decreasing productivity. Agro-industrial activity waste or byproduct is one possible strategy. By-products from industrial processes, such as the sugar manufacture (beet pulp), starch extraction (corn gluten feed), soybean treatment (soybean hulls), or oil extraction (soybean meal, linseed meal, corn gluten meal, cottonseed meal, and sunflower meal), are commonly used to fed livestock due to their high fiber or protein content (Correddu et al., 2020).

Other by-products from fruits and vegetables have been recently used for animal feed. Another interest in these wastes lies in the considerable amounts of bioactive compounds, especially polyphenols, tannins, and flavonoids. The addition of these compounds in low concentrations to animal diets was observed to increase its performance as well as health improvement. In addition, positive effects on the quality of the livestock products were also reported (Correddu et al., 2020). Condensed tannins, also known as proanticylanidines, have been shown to have positive effects on ruminants. Their administration at low concentrations on the diet of grazing ruminants resulted in an increased protein ingestion rate in the intestine and reduced the protein degradation rate by rumen bacteria. Conversely, excessive levels of phenolic compounds could have adverse consequences, which reduces the intake digestibility of almost all nutrients. Several studies have shown that, the inclusion of condensed tannins in sheep diets produces an increase in live weight, less carcass fat, and increased milk yield and wool production.

Phenolic compounds have also demonstrated positive effects on quality parameters of monogastric animals. The addition of seeds and grape pomace with polyphenols to pig diets improved feed to meat conversion ratio. Similarly, the supplementation of chicken diets with plant extracts rich in flavonoids and tannins increased the daily weight gain of animals and their final weight (Lipiński, Mazur, Antoszkiewicz, & Purwin, 2017). Despite all the reported benefits for both ruminants and monogastric animals, polyphenols are characterized by low bioavailability, and further research is required to optimize their use in livestock farming (Lipiński et al., 2017).

One of the biggest concerns when raising ruminants is the emission of CH_4 gas into the atmosphere and its potential effect on global warming. This environmental issue requires the development of strategies to reduce the emission of this gas and thus protect the sustainability of the breeding of livestock. Research suggests that adding certain supplements to the diet of ruminant animals may reduce CH_4 production. One of these nutritional strategies is to supplement the diets of animals with lipids. However, most of the studies carried out in this regard were relatively short in duration and there is no data available on long periods (Bayat & Shingfield, 2012). Adding tannins to the diet of ruminants has also shown to be an effective practice to limit CH_4 gas emissions (Bayat & Shingfield, 2012; Correddu et al., 2020). This anti-methanogenesis is mainly attributed to condensed tannins, which can act in two ways, directly on gas production or indirectly on hydrogen production (Bayat & Shingfield, 2012). Another way to manipulate CH_4 production in ruminants is to intervene in the microbial populations that carry out digestion. For instance, in vitro tests of non-H_2 production than H_2 producing fibrolytic bacteria reduces CH_4 gas production. Similarly, an increase in the number of capnophilic bacteria that use CO_2 and H_2 to produce organic acids, especially succinic acid, could have the same effect (Bayat & Shingfield, 2012). All these strategies represent positive approaches towards reducing CH_4 gas production in ruminant livestock. Nevertheless, until now, none of them integrates all the necessary requirements to be implemented immediately, namely, practicality, profitability, and not negatively affect meat and milk yield (Bayat & Shingfield, 2012).

Seaweeds (macroalgae) are a good source of polyphenols together with other components, such as polysaccharides, proteins, lipids, or dietary fibers. Research has shown positive effects on the health and nutrition of both humans and animals. Findings suggest that the application of seaweeds as feed additives has many benefits in livestock, including improved growth performance, increasing milk production, improving better meat quality, and stronger immune system (Rajauria, 2015). The specific composition of each one determines its function in the animals' diet. In this way, some seaweeds could contribute positively to satisfy the energy and protein requirements of livestock, while others, which contain high amounts of bioactive compounds, could be used as probiotics for improving production and health status of both monogastric and ruminant animals (Makkar et al., 2016).

The raw materials currently used in livestock feeding include higher plants, such as corn, soybeans, sorghum, oats, and barley. Most of these crops are short-lived and seasonal. In addition, they require several months to reach maturity. On the other hand, their usage as animal feeding is considered low profitability. This situation opens the door to the use of seaweeds as supplements in animal feeding, since they are a source of compounds with unique biochemical properties, grow rapidly, and provide a renewable food option. In addition, they are not seasonal, so their supply could be constantly assured and provide more security to the food market (Rajauria, 2015).

Microalgae's composition is similar to that of macroalgae. Proteins, carbohydrates, lipids, vitamins, minerals, and bioactive compounds are the main components. This composition may be affected by factors, such as species, strain, and growing conditions. Different studies reported that the inclusion of these microorganisms in the diets of both ruminant and monogastric (e.g., pigs, poultry, rabbits) animals had a positive effect on growth and meat quality, which showed to be a promising alternative to corn and soy. Unfortunately, the costs of producing microalgae are still high, which prevents it to be considered as an affordable technology in the short term (Madeira et al., 2017).

3.4.3 Towards a Smart Livestock Farming

The use of information technology is trending. Data collection and its proper management is becoming an indispensable tool for increasing productivity and generating benefits. This new concept is leading to the introduction of robots and artificial intelligence tools to improve livestock industries. PLF, precision livestock farming, is defined as a technology designed to monitor parameters in real-time manner, such as production, reproduction, etc. Measurements are taken and analyzed every second (Berckmans, 2017). Big Data technologies allow the capture of large volumes of disparate data and conduct analysis (Wolfert, Ge, Verdouw, & Bogaardt, 2017). By means of prediction models with big data analysis, the farmer is alerted of any deviation from the standard patterns, and this proceeds to their restoration as soon as possible (Tullo, Finzi, & Guarino, 2019). The continuous flow of information can be used to control processes, such as feeding times or feed intake, and monitor sounds, images, or yield parameters of animals. With this integrity system, detecting diseases at an early stage has become possible, which allows appropriate acts in place before the situation gets worse (Berckmans, 2015). Fig. 3.3 shows a scheme of farm operating with Precision Livestock Farming (PLF).

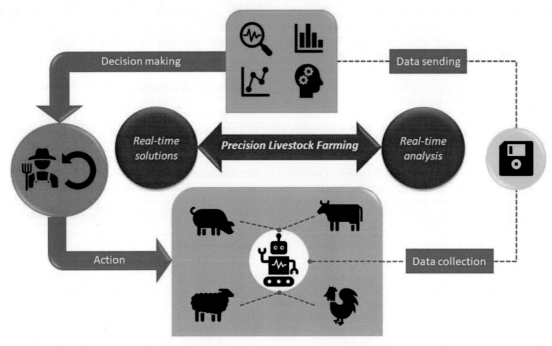

FIG. 3.3 Scheme of farm operating with Precision Livestock Farming (PLF).

The first application of PLF was an individual electronic milk meter for cows in 2004. Different technologies associated with PLF have been recently applied in farms to control parameters or processes. One of these technologies allows the feed distribution in group housing. The need to control the body weight of broiler chickens also leads to incorporate this pioneering technology in farms. An automatic animal weight monitoring system allows farmers to control productivity continuously, which ensures a high body weight of chickens while maintaining an efficient feed-conversion rate (Halachmi & Guarino, 2016).

Disease control in livestock is a constant concern for agricultural businesses. This leads to developing instruments for early detection of animal health problems. For instance, in dairy industries, the automatic milking systems record parameters that allow the detection of mastitis disease in cows (Jiang, Wang, Li, & Wang, 2017). King, LeBlanc, Pajor, Wright, and DeVries (2018) detected the disease by observing a decrease in milk yields, rumination time, milk frequency, activity, and temperature of milk in herds of commercial cows. Other health problems, such as lameness, may also be controlled with PLF instead of using subjective methods (veterinarian examination). 3D imaging technology, such as a radar micro-Doppler signatures for contactless

identification may be useful technologies for the effective early detection of anomalies (Hansen, Smith, Smith, Jabbar, & Forbes, 2018; Shrestha et al., 2018).

The air quality of livestock facilities is an important factor since it is often loaded with particles and gasses from the normal activities of animals, whose inhalation may have negative effects on both the animals and farm staff. In response to this issue, PLF can monitor ammonia emissions or dust concentration in poultry houses, which could make work easier for farmers (Tullo et al., 2019).

PLF has the potential to tackle frequent problems on livestock farms. This concept of smart farm appropriately applied could enhance animal welfare, reduce GHG emissions, facilitate product segmentation, effectively control illegal trades, and improve the economic stability in rural areas (Banhazi et al., 2012). However, only a few companies and a very limited number of farms are involved in the commercialization of PLF technologies.

3.5 CONCLUSIONS

Livestock farming is a demanding human activity with the available natural resources. It promotes the degradation of large land areas with loss of biodiversity, and consumption of water. On the other hand, the metabolic activity of animals has a great impact on the environment. Their excrements may infiltrate to the groundwater and carry possible harmful viruses and bacteria to humans. In order to meet a continuous and increasing demand for meat and meat products but also protect natural resources, innovative technologies are needed. AI has improved productivities in livestock farms for more than 60 years. The better use of male semen through this reproduction method has reduced the amount of natural resources consumption in the rearing and maintenance of animals. Crossbreeding is another technology that improves the management of available resources by farmers. This technology enhances the potential of the breed traits associated with reproduction, longevity, growth rate, meat quality, also increases the profitability of farms.

The production of genetically modified livestock is a promising technology to release pressure on the production system. This technology has the potential to obtain desirable traits more efficiently, faster, and with less environmental impact. Nevertheless, due to the high cost and the uncertainty of regulation, its application is still limited. Another cutting-edge technology is to design novel livestock diets based on available resource management, while maintaining an identical or better nutritional level than that conventional diets. PLF, as a system to monitor parameters and process data in a real-time manner, provides a possibility of immediate manipulation on farms.

ACKNOWLEDGMENTS

Thanks to GAIN (Axencia Galega de Innovación) for supporting this review (grant number IN607A2019/01). Jose M. Lorenzo is member of the HealthyMeat network, funded by CYTED (ref. 119RT0568). Paulo E. S. Munekata acknowledges postdoctoral fellowship support from Ministry of Economy and Competitiveness (MINECO, Spain) "Juan de la Cierva" program (FJCI-2016-29486). B. Gullón would like to express their gratitude to the Spanish Ministry of Economy and Competitiveness for financial support (Grant reference RYC2018-026177-I).

REFERENCES

Allington, G. R., & Valone, T. J. (2011). Long-term livestock exclusion in an arid grassland alters vegetation and soil. *Rangeland Ecology & Management, 64*(4), 424–428.

Azarnivand, H., Farajollahi, A., Bandak, E., & Pouzesh, H. (2010). Assessment of the effects of overgrazing on the soil physical characteristic and vegetation cover changes in rangelands of Hosainabad in Kurdistan province, Iran. *Journal of Rangeland Science, 1*(2), 95–102.

Bailey, R., Froggatt, A., & Wellesley, L. (2014). *Livestock—Climate change's forgotten sector: Global public opinion on meat and dairy consumption.* Chatham house report London: The Royal Institute of International Affairs.

Banhazi, T. M., Lehr, H., Black, J. L., Crabtree, H., Schofield, P., Tscharke, M., et al. (2012). Precision livestock farming: An international review of scientific and commercial aspects. *International Journal of Agricultural and Biological Engineering, 5*(3), 1–9.

Baumung, R., Mottet, A., & Teillard, F. (2018). Livestock and life on land. In A. Acosta (Ed.), *Transforming the livestock sector through the sustainable development goals* (pp. 119–127). Rome: Agriculture Organization of the United Nations (FAO).

Bayat, A., & Shingfield, K. J. (2012). Overview of nutritional strategies to lower enteric methane emissions in ruminants. *Suomen Maataloustieteellisen Seuran Tiedote, 28,* 1–7.

Berckmans, D. (2015). Smart farming for Europe: Value creation through precision livestock farming. In I. Halachmi (Ed.), *Precision livestock farming applications* (pp. 25–36). Wageningen: Wageningen Academic Publishers.

Berckmans, D. (2017). General introduction to precision livestock farming. *Animal Frontiers, 7*(1), 6–11.

Bertolini, M., & Bertolini, L. R. (2009). Advances in reproductive technologies in cattle: From artificial insemination to cloning. *Revista de la Facultad de Medicina Veterinaria y de Zootecnia, 56*(III), 184–194.

Blümmel, M., Haileslassie, A., Samireddypalle, A., Vadez, V., & Notenbaert, A. (2014). Livestock water productivity: Feed resourcing, feeding and coupled feed-water resource data bases. *Animal Production Science, 54*(10), 1584–1593.

Bourgeois, L. (2012). A discounted threat: Environmental impacts of the livestock industry. *Earth Common Journal, 2*(1).

Broom, D. M., Galindo, F. A., & Murgueitio, E. (2013). Sustainable, efficient livestock production with high biodiversity and good welfare for animals. *Proceedings of the Royal Society B: Biological Sciences, 280*(1771), 20132025.

Correddu, F., Lunesu, M. F., Buffa, G., Atzori, A. S., Nudda, A., Battacone, G., et al. (2020). Can agro-industrial by-products rich in polyphenols be advantageously used in the feeding and nutrition of dairy small ruminants? *Animals, 10*(1), 131.

Dagevos, H., & Voordouw, J. (2013). Sustainability and meat consumption: Is reduction realistic? *Sustainability: Science, Practice, and Policy, 9*(2), 60–69.

Derner, J. D., Hunt, L., Euclides Filho, K., Ritten, J., Capper, J., et al. (2017). Livestock production systems. In D. D. Briske (Ed.), *Rangeland systems* (pp. 347–372). New York: Springer.

Gerber, P. J., Steinfeld, H., Henderson, B., Mottet, A., Opio, C., Dijkman, J., et al. (2013). *Tackling climate change through livestock: A global assessment of emissions and mitigation opportunities.* Rome: Food and Agriculture Organization of the United Nations (FAO).

Getahun, D., Alemneh, T., Akeberegn, D., Getabalew, M., & Zewdie, D. (2019). Importance of hybrid vigor or heterosis for animal breeding. *Biochemistry and Biotechnology Research, 7*(1), 1–4.

Grossi, G., Goglio, P., Vitali, A., & Williams, A. G. (2019). Livestock and climate change: Impact of livestock on climate and mitigation strategies. *Animal Frontiers, 9*(1), 69–76.

Halachmi, I., & Guarino, M. (2016). Editorial: Precision livestock farming: A 'per animal' approach using advanced monitoring technologies. *Animal, 10*(9), 1482–1483.

Han, J. G., Zhang, Y. J., Wang, C. J., Bai, W. M., Wang, Y. R., Han, G. D., et al. (2008). Rangeland degradation and restoration management in China. *Rangeland Journal, 30*(2), 233–239.

Hansen, M. F., Smith, M. L., Smith, L. N., Jabbar, K. A., & Forbes, D. (2018). Automated monitoring of dairy cow body condition, mobility and weight using a single 3D video capture device. *Computers in Industry, 98*, 14–22.

Jiang, H., Wang, W., Li, C., & Wang, W. (2017). Innovation, practical benefits and prospects for the future development of automatic milking systems. *Frontiers of Agricultural Science and Engineering, 4*(1), 37–47.

Keane, M. G. (2011). *Beef cross breeding of dairy and beef cows.* Grange Beef Research Centre. Occasional Series No. 8.

King, M. T. M., LeBlanc, S. J., Pajor, E. A., Wright, T. C., & DeVries, T. J. (2018). Behavior and productivity of cows milked in automated systems before diagnosis of health disorders in early lactation. *Journal of Dairy Science, 101*(5), 4343–4356.

Kues, W. A., & Niemann, H. (2011). Advances in farm animal transgenesis. *Preventive Veterinary Medicine, 102*(2), 146–156.

Lipiński, K., Mazur, M., Antoszkiewicz, Z., & Purwin, C. (2017). Polyphenols in monogastric nutrition—A review. *Annals of Animal Science, 17*(1), 41–58.

Llonch, P., Haskell, M. J., Dewhurst, R. J., & Turner, S. P. (2017). Current available strategies to mitigate greenhouse gas emissions in livestock systems: An animal welfare perspective. *Animal, 11*(2), 274–284.

Madeira, M. S., Cardoso, C., Lopes, P. A., Coelho, D., Afonso, C., Bandarra, N. M., et al. (2017). Microalgae as feed ingredients for livestock production and meat quality: A review. *Livestock Science, 205*, 111–121.

Makkar, H. P. S., & Ankers, P. (2014). Towards a concept of sustainable animal diets. In *Animal production and health report.* Rome: Food and Agriculture Organization of the United Nations (FAO). No. 7.

Makkar, H. P. S., Tran, G., Heuzé, V., Giger-Reverdin, S., Lessire, M., Lebas, F., et al. (2016). Seaweeds for livestock diets: A review. *Animal Feed Science and Technology, 212*, 1–17.

Mekuria, W., Veldkamp, E., Haile, M., Nyssen, J., Muys, B., & Gebrehiwot, K. (2007). Effectiveness of exclosures to restore degraded soils as a result of overgrazing in Tigray, Ethiopia. *Journal of Arid Environments, 69*(2), 270–284.

Morrell, J. M. (2011). Artificial insemination: Current and future trends. In M. Manfi (Ed.), *Artificial insemination in farm animals* (pp. 1–14). London: InTech.

Murray, J. D., & Maga, E. A. (2016). Genetically engineered livestock for agriculture: A generation after the first transgenic animal research conference. *Transgenic Research, 25*(3), 321–327.

Niemann, H., & Seamark, B. (2018). The evolution of farm animal biotechnology. In H. Niemann, & C. Wrenzycki (Eds.), *Animal Biotechnology 1* (pp. 1–26). New York: Springer.

Pacheco, A. R., Chaves, R. Q., & Nicoli, C. M. L. (2012). Integration of crops, livestock, and forestry: A system of production for the Brazilian Cerrados. In K. G. Gasman (Ed.), *Eco-efficiency: From vision to reality* (pp. 51–61). Cali: Centro Internacional de Agricultura Tropical (CIAT).

Patel, G. K., Haque, N., Madhavatar, M., Chaudhari, A. K., Patel, D. K., Bhalakiya, N., et al. (2017). Artificial insemination: A tool to improve livestock productivity. *Journal of Pharmacognosy and Phytochemistry, 1*, 307–313.

Pinkert, C. A. (2014). Introduction to transgenic animal technology. In C. A. Pinkert (Ed.), *Transgenic animal technology* (pp. 3–13). Amsterdam: Elsevier.

Pradère, J. P. (2014). Links between livestock production, the environment and sustainable development. *Revue scientifique et technique/Office international des épizooties, 33*(3), 765–781.

Rajauria, G. (2015). Seaweeds: A sustainable feed source for livestock and aquaculture. In B. Tiwari, & D. Troy (Eds.), *Seaweed sustainability* (pp. 389–420). Cambridge, MA: Academic Press.

Ran, Y., Lannerstad, M., Herrero, M., Van Middelaar, C. E., & De Boer, I. J. (2016). Assessing water resource use in livestock production: A review of methods. *Livestock Science, 187*, 68–79.

Rendel, J., & Robertson, A. (1950). Estimation of genetic gain in milk yield by selection in a closed herd of dairy cattle. *Journal of Genetics, 50*, 1–8.

Rojas-Downing, M. M., Nejadhashemi, A. P., Harrigan, T., & Woznicki, S. A. (2017). Climate change and livestock: Impacts, adaptation, and mitigation. *Climate Risk Management, 16*, 145–163.

Segal, S. P., Knight, T. J., & Beitz, D. C. (2012). Use of biotechnology to increase food production and nutritional value. In T. Wilson, & D. R. Jacobs Jr., (Eds.), *Nutritional health* (pp. 505–522). Totowa: Humana Press.

Shrestha, A., Loukas, C., Le Kernec, J., Fioranelli, F., Busin, V., Jonsson, N., et al. (2018). Animal lameness detection with radar sensing. *IEEE Geoscience and Remote Sensing Letters, 15*(8), 1189–1193.

Stančić, I., & Dragin, S. (2011). Modern technology of artificial insemination in domestic animals. *Contemporary Agriculture, 60*, 204–214.

Sulabh, S., & Kumar, A. (2018). Application of transgenic animals in animal production and health. *Agricultural Reviews, 39*(2), 169–174.

Tait-Burkard, C., Doeschl-Wilson, A., McGrew, M. J., Archibald, A. L., Sang, H. M., Houston, R. D., et al. (2018). Livestock 2.0—Genome editing for fitter, healthier, and more productive farmed animals. *Genome Biology, 19*(1), 204.

Tullo, E., Finzi, A., & Guarino, M. (2019). Environmental impact of livestock farming and precision livestock farming as a mitigation strategy. *The Science of the Total Environment, 650*, 2751–2760.

Udo, H. M., & Steenstra, F. A. (2010). Intensification of smallholder livestock production, is it sustainable? In *Proceedings of the 5th international seminar on tropical animal production, community empowerment and tropical animal industry, ISTAP5, 19–22 October 2010, Yogyakarta* (pp. 19–26).

Van Eenennaam, A. L. (2017). Genetic modification of food animals. *Current Opinion in Biotechnology, 44*, 27–34.

van Wagenberg, C. P. A., De Haas, Y., Hogeveen, H., Van Krimpen, M. M., Meuwissen, M. P. M., Van Middelaar, C.

E., et al. (2017). Animal board invited review: Comparing conventional and organic livestock production systems on different aspects of sustainability. *Animal, 11*(10), 1839–1851.

Wakchaure, R., Ganguly, S., Praveen, K. P., Sharma, S., Kumar, A., Mahajan, T., et al. (2015). Importance of heterosis in animals: A review. *International Journal of Advanced Engineering Technology and Innovative Science, 1*(1), 1–5.

Wolfert, S., Ge, L., Verdouw, C., & Bogaardt, M. J. (2017). Big data in smart farming—A review. *Agricultural Systems, 153*, 69–80.

Yadav, V., Singh, N. P., Sharma, R., & Gupta, A. (2018). Crossbreeding systems of livestock. *The Pharma Innovation Journal, 7*(7), 08–13.

CHAPTER 4

Packaging Systems

RUBÉN DOMÍNGUEZ[A] • BENJAMIN BOHRER[B] • MIRIAN PATEIRO[A] •
PAULO EDUARDO SICHETTI MUNEKATA[A] • JOSÉ MANUEL LORENZO[A,C]
[a]Galician Meat Technology Center, Galicia Technology Park, Ourense, Spain, [b]Department of Animal
Sciences, The Ohio State University, Columbus, OH, United States, [c]Food Technology Area, Faculty
of Sciences of Ourense, Vigo University, Ourense, Spain

4.1 INTRODUCTION

Today's consumers demand many different types of fresh and natural food products, and at the same time consumers want those foods to be minimally-processed and to have a long shelf-life (Leimann et al., 2018). The control of several degradative and spoilage processes that are detrimental to food quality and food safety are limited without the use of food additives that are proven to be effective in both minimally-processed and further-processed foods (Lorenzo, Domínguez, & Carballo, 2017). Consequently, the influences of the consumer and their demands for additive-free (i.e., natural) food products have caused the food industry to constantly develop new strategies in the discipline of food packaging (Alizadeh-Sani, Mohammadian, Rhim, & Jafari, 2020; Anis, Pal, & Al-Zahrani, 2021; Cruz, Alves, Khmelinskii, & Vieira, 2018).

Food packaging comprises of four different, but interconnected functions summarized as the containment, the protection, the convenience, and the communication of the food product (Robertson, 2014). The primary objective of food packaging is to contain the food product (normally into desirable weight or volume proportions), which facilitates transportation, storage, and distribution. As expected, correct packaging also protects food from external contaminations and conditions that cause deteriorative chemical or microbiological processes to occur (Ghoshal, 2018). Nowadays, the incorporation of active substances in packaging materials that interact with the food product in combination with the use of specific atmospheres surrounding the food product has a significant and positive effect on shelf-life. Finally, adapting the portion size (or package size) to consumer needs and the possibility to include information on the label of the food product (e.g., product identity information, best-by-date, ingredients, nutritional composition, and nutritional claims) are two important aspects of food packaging (Berk, 2018; Cruz et al., 2018; Gupta & Dudeja, 2017).

On the other hand, almost all food products require some form of packaging, and about 99.8% of foods are sold as packaged food products (Cruz et al., 2018). Spoilage of foods during processing, distribution, and exposure in the market-setting have a significant negative impact on the food industry from an economic point of view, as well as an environmental standpoint (Domínguez et al., 2018; Lorenzo et al., 2017). The main degradative processes in foods are oxidative reactions and microbial spoilage (Domínguez et al., 2019), thus it is vital to design and select the appropriate packaging (i.e., packaging material, packaging format or technique, etc.) for a particular food product in order to minimize quality losses and food waste (Domínguez et al., 2018). In addition to packaging materials and technology, the atmosphere surrounding food plays an important role in food stability, since modified atmosphere packaging (MAP) environments could modulate the exposure to oxygen and therefore, slow oxidative processes (Domínguez et al., 2019) and delay the growth of harmful microbes (Pateiro et al., 2019).

Synthetic packaging materials (e.g., plastics and other synthetic polymers) have been used for many years in the food industry, since they have many remarkable advantages over other packaging materials, such as effective-production cost and excellent physical and barrier properties. However, challenges associated with the disposal of these materials, once they have been used, create serious environmental damage. This problem has necessitated research and development on the use of biodegradable and/or edible packaging polymers that are capable of maintaining food shelf-life without creating negative impacts on the environment (Cruz et al., 2018; Domínguez et al., 2018). Nowadays, several biopolymers are being studied worldwide for their use as food packaging materials. These bioplastics or biopolymers are biodegradable in nature and do not pollute the environment in the same way as traditional synthetic food packaging materials (Anis et al., 2021).

Sustainable Production Technology in Food. https://doi.org/10.1016/B978-0-12-821233-2.00014-9
Copyright © 2021 Elsevier Inc. All rights reserved.

Additionally, agri-food industry by-products are a renewable and interesting source of "green materials" for the production of biopolymers which in turn helps further to reduce the burden of environmental pollution (Anis et al., 2021). The biodegradable polymers could be metabolized by the microorganisms and subsequently return to nature in a short period of time (Carina, Sharma, Jaiswal, & Jaiswal, 2021). These bio-based polymers could be classified into three main categories, which are polymers that can be directly extracted from biomass (e.g., chitosan, carrageenan, etc.), renewable monomers that are chemically synthesized polymers [e.g., polylactic acid (PLA)], and polymers that are produced by microorganisms or genetically modified bacteria (e.g., polyhydroxyalkanoates) (Cruz et al., 2018).

There is a growing interest in the development of new packaging strategies and materials aimed to preserve food quality during extended periods of storage (Battisti et al., 2017; Cirillo et al., 2018). Thus, the development of sustainable and eco-friendly packaging systems, which interact with both, foods (active packaging) and consumers (intelligent packaging) are a recent and future trend in the food industry. With all this in mind, this book chapter aims to review the evolution of food packaging, including the materials and strategies employed, as well as provide a compilation of recent discoveries and future trends for the food packaging industry.

4.2   PACKAGING MATERIALS

The fundamental goals of food packaging are to cover, surround, or contain food in a cost-effective way to meet requirements for consumer desires, while also providing an environment that maintains food safety and food quality (Lee, Yam, & Piergiovanni, 2008; Marsh & Bugusu, 2007; Piergiovanni & Limbo, 2016; Robertson, 2016). In the last two centuries (detailed timeline provided in Fig. 4.1), food packaging has evolved from merely being a container for food products to an important component of branding and marketing of food products (Coles, 2003). The shape, color, and other components related to the appearance of a food package, as well as the descriptions and claims on the label of the package are all critical pieces of information to facilitate the recognition and sale of a food product. At the same time, food safety and food quality are ultimately determined by the food packaging system, and significant improvements have been made in these areas over the past few decades. The materials used for food packaging and the food atmosphere that is created are of particular importance.

Several important innovations have been made in packaging materials and packaging technologies over the last two centuries. These include the creation of new packaging materials, such as plasticized polyvinyl chloride (PVC) in 1926, polyethylene (PE) in 1936, and polypropylene (PP) in 1951, and new packaging technologies like the 1973 introduction of the Universal Product Code (UPC) using the bar code system (Berger, 2005; Coles, 2003; Ebnesajjad, 2012; Piringer & Baner, 2008; Raheem, 2013; Risch, 2009). In addition to these technologies, the emergence of modified atmosphere packaging (MAP), which was first technically recorded in 1927 (this was a modified atmosphere environment rather than modified atmosphere packaging), has been very influential for many food products, especially fresh and chilled perishable food products like raw or cooked meat, raw or cooked fish, fresh pasta, and fresh and cut produce (Goswami & Mangaraj, 2011).

In the last two decades (since the late 2000s), the requirement for sustainable and regenerative packaging materials/systems has been recognized globally. Before we begin to discuss the events that have led us to where we are today and what we anticipate as emerging trends, it is important for the evolution of food packaging materials to be explained in greater context.

4.2.1   Stoneware and Pottery Materials

The first sophisticated materials used for food packaging (beyond that of hollowed-out tree limbs and wooden baskets made from tree branches or leaves) were pottery materials. Pottery is defined as the end-product following the process of forming vessels, containers, and dishes with clay or other ceramic materials, and then firing the formed materials at high temperatures to generate a hard, durable end-product. Significant forms of pottery include earthenware, stoneware, and porcelain. Pottery is one of the earliest documented human inventions, originating before the Neolithic period (10,000 BC), with ceramic objects like the Gravettian culture Venus of Dolní Věstonice figurine discovered in the Czech Republic around 25,000 BC

Dating back to the Neolithic period (10,000 BC to 3500 BC), pottery, and particularly earthenware and stoneware, was commonly used for long-term storage of foods and beverages, as well as transportation of foods and beverages. Pottery materials offer a solid, impermeable, chemically-resistant, tamper-proof material that protects food from environmental gases (e.g., oxygen, carbon dioxide, and nitrogen) and other external factors that create food quality and microbial quality issues. While earthenware and stoneware crocks and containers were still being used for many food

Before 1850 | **1851 to 1900** | **1900s** | **1910s** | **1920s** | **1930s** | **1940s** | **1950s** | **1960s** | **1970s** | **1980s** | **1990s** | **2000s** | **2010s** | **2020s**

Before 1850s: Foods were being stored and/or thermally preserved in glass jars, tinplate canisters, and stoneware crocks.

1851 to 1900: The use of automation and the use of paper packaging materials expanded during this time period. Notable accomplishments include Albert L. Jones' 1871 patent on the use of corrugated packaging materials, Robert Gair's 1879 invention of machine-made folding cartons, William Painter's 1892 patent of the Crown cap for glass bottles, and Michael Owen's 1899 idea for fully automated bottle manufacture.

1900s: Paraffin wax coated paper milk cartons were developed in 1906.

1910s: Innovative paper containers continue to be developed. Waxed paperboard cartons were used for cream, regenerated cellulose film was developed in 1912; the "paper bottle" known as Pure-Pak was invented by John Van Wormer in 1915.

1920s: Novel retail packaging includes those by Birdseye Seafoods and their frozen foods packaged in cartons with waxed paper wrappers and the 1927 release of Du Pont's Cellophane packaging material.

1930s: The first beer can was developed in 1933 by American Can. In 1939, ethylene was first polymerized by Imperial Chemical Industries, Ltd ICI scientists Reginald Gibson and Eric Fawcett produced the solid ethylene polymer by accident.

1940s: The second world war has major impacts on food packaging. Paper and manufactured plastics are heavily used throughout the world in the retail setting since metals and glass are required for food packaging for soldiers.

1950s: New plastic processing techniques are discovered for food. This includes Giulio Natta's 1954 discovery of isotactic polypropylene, the 1956 launch of tetrahedral milk cartons constructed with low-density polyethylene extrusion coated paperboard by Tetra Pak, and the use of the first plastic sandwich bag on a roll in 1957.

1960s: Packaging innovation begins to focus efforts on product quality and preservation. Examples include the two-piece drawn and wall-ironed (DWI) can for carbonated drinks and beers, the Soudronic welded side-seam for tinplate food cans, tamper evident bottle neck shrink-sleeves, the Metal Box Company's ringpull opener, and Tetra Pak's rectangular Tetra Brik Aseptic (TBA) carton system.

1970s: Convenience becomes a priority, both for consumers and processors. Examples include the 1973 introduction of the Universal Product Code (UPC) using the bar code system and the emergence of many different ready-to-cook frozen foods that were sold in microwaveable plastic containers, bag-in-box systems, and a range of aseptic form, fill and seal (FFS) flexible packaging systems.

1980s: Packaging materials continue to advance; an example includes the co-extruded plastics that incorporate oxygen barrier plastic materials for squeezable sauce bottles. Also, The Nutrition Labeling and Education Act is passed in 1989, which requires nutrition labeling and nutrition facts on most food items.

1990s: Packaging innovation is influenced by advancements in technology. This includes digital color printing, dot-matrix laser coding, three-dimensional labeling, and oxygen scavenging film.

2000s: Investigation of more environmentally sustainable packaging becomes a priority. This includes the 2001 creation of polylactic acid (PLA) polymers, the 2006 creation of biodegradable shrink film, and societal bans on single-use plastics in several cities in the United States.

2010s: Recent packaging interventions include zero waste packaging, advanced biodegradable packaging materials, and active or intelligent packaging that provides information for the consumer.

2020s: COVID-19 is declared a global health pandemic on 11 March 2020, undoubtedly altering food packaging priorities.

FIG. 4.1 Timeline of key events related to food packaging innovation.

packaging purposes in the 19th century, today, pottery is rarely used for food packaging purposes due to its cost of production, bulk and weight, fragility, and non-transparent appearance. The creation and popularity of glassware during the Roman Empire (27 BC to 476 CE), which will be discussed more in the subsequent section, replaced pottery food packaging materials for most applications.

4.2.2 Glass Packaging Materials

When the natural and abundant raw materials silica (i.e., sand), calcium carbonate (i.e., limestone), calcium oxide (i.e., lime), soda ash, and aluminum oxide (i.e., alumina) are mixed together and melted at extremely high temperatures (around 1500°C), a new material known as molten glass is formed. Glass is structurally similar to a liquid at extremely high temperatures; however, glass behaves like a solid at ambient temperatures. The discovery of glass dates back to the year 3000 BC with archeological evidence suggesting glass was made in the coastal north region of Syria, Mesopotamia, and Egypt (Girling, 2003). The invention of the blow stick during the Roman Empire (27 BC to 476 CE) led to the manufacture of hollow glass containers, which were less expensive than pottery vessels and therefore commonly used for food packaging purposes (Girling, 2003).

There are two primary types of glass containers, which are bottles with narrow necks and jars with wide openings. An obvious challenge with both of these types of glass containers for food packaging purposes is the creation of an effective closure at the neck or opening. For the most part, glass materials are not used for the closure. Other materials such as plastic, rubber, or metal are often used to create an effective closure for glass bottles or jars.

Glass containers offer several advantageous features as packaging materials. Similar to pottery materials, glass offers a solid, impermeable, chemically-resistant, tamper-proof material that protects food from environmental gases (e.g., oxygen, carbon dioxide, and nitrogen) and other external factors that cause food quality and microbial quality issues. Glass is thermally stable, which allows for in-container sterilization and pasteurization of foods, which improves processing capabilities. Finally, glass is generally transparent and offers a smooth texture which allows foods to be displayed to the consumer, which is often preferred by consumers. From an environmental standpoint, glass is sturdy enough to be returnable or reusable, and even in cases where this is not possible, glass is fairly easy to recycle.

With the creation of multiple different types of plastic in the 20th century, glass has been replaced for many food packaging purposes. Plastics are cheaper than glass, and sometimes more convenient for consumers to use (e.g., the 1983 innovation of the squeezable plastic Heinz Ketchup bottle was easier for consumers to use than the traditional glass Heinz Ketchup bottle). Even so, glass materials are still used today for a wide range of food packaging purposes. Examples include dairy beverages, fruit beverages, alcoholic beverages (liqueurs, spirits, and beer), syrups, spices, processed baby foods, condiments (mustards, ketchups, etc.), preserves (jams, jellies, and marmalades), and processed vegetables (pickles, peppers, etc.).

4.2.3 Metal Packaging Materials

Metal materials for food packaging were first used shortly after the metal can was invented. The process of using metal materials (in the form of tin cans) for food preservation was allegedly conceived by the French engineer Philippe de Girard, but the idea eventually was discovered by British merchant Peter Durand who patented Girard's idea in 1810 (Apotheker & Markstahler, 1986; Busch, 1981). The concept of using a tin can for food preservation was based on experimental food preservation work using glass containers completed the year before by the French cook Nicholas Appert, who used glass champagne bottles to preserve broths and other food products using various levels of exposure to heat (Standage, 2009). It was actually not until the 1860s that Louis Pasteur would discover the process now known as pasteurization, which is a process where packaged or non-packaged foods are treated with mild heat to eliminate pathogens and extend shelf-life. Durand did not pursue food canning as a business, but rather sold his patent in 1812 to two Englishmen; Bryan Donkin and John Hall, who refined the process and product, and opened the world's first commercial canning factory in London, England (Apotheker & Markstahler, 1986; Busch, 1981). By 1813, this canning factory was producing tin canned goods for the United Kingdom's Royal Navy and by 1820, tin canisters or cans were being used for many military items including gunpowder, seeds, and turpentine (Apotheker & Markstahler, 1986; Busch, 1981). These tin canisters or cans were produced using tinplate, which is a low-carbon steel that is coated on both sides with a thin layer of tin.

There were several advantages to metal packaging materials when compared with stoneware and glass materials. Metal packaging containers offer a similar level of protection and preservation compared with stoneware and glass packaging containers, yet metal is usually lighter, more durable, and more versatile. While stoneware and glass are brittle and can break during transport, the risk of metal cans breaking is considerably less. Also, the openings/closures of metal canisters or cans were made from the same materials as the canisters or cans, and often times were more secure than the openings/closures of stoneware and glass packaging containers.

One notable challenge with early metal packaging is the type of metal materials that were used. Early tin cans were sealed with tin–lead alloy soldering, which could lead to lead poisoning. Today, coated steel (coated in tin or chromium compounds) and aluminum are used for most metal container and closure construction for food packaging purposes. Over the years, metal packaging materials have been used for many different purposes, many of which are still used to this day. Currently, the majority of metal food containers are used for beverage cans (approximately 78%) and processed food cans (approximately 18%), while the remaining portion is made up by aerosol and other miscellaneous products (Page, Edwards, & May, 2003).

4.2.4 Paper and Paperboard Packaging Materials

The use of paper and paper-derived materials like paperboard for food packaging dates back to the 17th century when sugar-bakers make reference to the use of "blew paper" (Kirwan, 2003; Marsh & Bugusu, 2007). This is actually many years after the manufacture of paper was first documented, as it is believed that Cai Lun (name later changed to Ts'ai Lun) invented paper in the year 105 CE (Hunter, 1978). This was described as a thin, felted material formed on flat, porous molds from macerated vegetable fiber. Many advancements have been made since then, yet paper is still roughly defined as formed materials manufactured from an interlaced network of cellulose fibers. These cellulose fibers are derived from plant sources, mainly wood from trees. The cellulose fibers are isolated using sulfate or sulfite to generate the purified pulp, and then the pulp is combined with water and placed into a press where it is flattened, dried, and cut into sheets or rolls. The sheets or rolls are then finished with the desired gloss or coating.

Today, paper is most commonly used for printing, writing, and cleaning. However, it is estimated that approximately 10% of all manufactured paper is used for packaging, and 50% of this packaging paper is used by the food industry (Kirwan, 2003). Paper and paperboard are used in many different forms and for many different applications in food packaging (Berger, 2005; Kirwan, 2003; Raheem, 2013). Table 4.1 provides

details for several different types of paper and paperboard materials used in the industry today.

The use of paper for food packaging became more common in the second half of the 19th century. This can be partially explained by the progression of the manufacturing industry, which would have enabled high speed production of paper. The other primary reason for the expansion of paper use in food packaging can be explained by the greater number of food items bought outside of the home, or off the farm. For instance, in 1820 approximately 72% of the American work force had farm occupations, this number was down to 64% of the American work force in 1850 and was further reduced to 30% of the American work force by 1920 (New York Times, 1988). The decreased percentage of the population working on farms is likely responsible for the transition from at-home storage and preservation of foods with glass jars, stoneware crocks, and tinplate canisters to paper, as packaging made with paper or paperboard was (and still is to this day) primarily found at the point of retail sale, during commercial storage of foods, or used exclusively for retail or food-service distribution of foods.

Paper and paper-derived products like paperboard have unique physical properties which enable them to be flexible yet rigid as food packages. The many different types of packaging derived from paper can vary in their appearance, strength, and other functional properties depending on the type and amount of fiber (or fibers) used and how the fibers are processed during the aforementioned manufacturing steps for paper. Paper packaging can be used with a similar level of effectiveness over a wide temperature range (i.e., from frozen food storage to high temperatures during cooking). Some paper packaging can be approved for direct contact with food products, yet packaging manufactured solely from paper is usually permeable to most food substances (e.g., water, water vapor, aqueous solutions, organic solvents, and fatty substances) and most environmental gases (e.g., oxygen, carbon dioxide, and nitrogen). This limits the ability of paper packaging to prevent quality deterioration and microbial spoilage when used for perishable foods. Additionally, paper is not heat sealable by itself which limits its ability to be used as a sole source of packaging for perishable foods. However, paper packaging can acquire barrier properties and extended functional performance, such as heat sealability for leak-proof liquid packaging, through coating and lamination with plastics, which will be discussed in the next section, and through interactions with other packaging materials such as aluminum foil or wax. Today, we commonly see paper or paperboard

TABLE 4.1
Types of Paper and Paperboards Used in Food Packaging.

	Description	Food Applications
Paper		
Kraft paper	Kraft paper is a strong, smooth paper produced with a sulfate treatment process	Commonly used for bags and wrapping (e.g., packaged flour, packaged sugar)
Sulfite paper	Sulfite paper is lighter and weaker than kraft paper, this paper is produced from sulfite pulp after lignin has been extracted in its entirety	Commonly used for bags and wrapping (e.g., bakery and confectionary items)
Greaseproof paper	Greaseproof paper is produced through a process known as beathing, in which cellulose fibers undergo longer than normal hydration periods that cause the fibers to break up and become gelatinous. The result is a paper that is resistant to oils, but not wet agents	Commonly used for bags and wrapping of oily foods (e.g., cookies, biscuits, and candy bars), and cooking/baking where it provides a non-stick surface
Glassine	Glassine is a smooth and glossy paper that is air, water, and grease resistant. It is produced with extreme hydration	Commonly used as a liner for oily foods (e.g., biscuits, cooking fats, fast foods, and baked goods)
Parchment paper	Parchment paper is essentially paper that has been coated in a layer of silicone, which is what gives it its superb nonstick quality. The silicone coating also makes it heat-resistant as well as water-resistant	Commonly used for bags and wrapping of oily foods (e.g., packaged fats like butter or lard), and cooking/baking where it provides a non-stick, heat-resistant surface
Paperboard		
Whiteboard	Whiteboard is made from several thin layers of bleached chemical pulp and can be coated with wax or laminated with polyethylene for greater heat stability	The only form of paperboard recommended for direct food contact. Commonly used as the inner layer of cartons
Solid board	Solid board has multiple layers of bleached sulfate board and possesses high levels of strength and durability	Commonly used for packaging of fruit juices and soft drinks
Chipboard	Chipboard is inexpensive and made from recycled paper and often contains blemishes and impurities from the original paper, which makes it unsuitable for direct contact with food, printing, and folding	Commonly used to make the outer layers of cartons for foods such as tea and cereals
Fiberboard	Fiberboard can be solid or corrugated. The solid type has an inner white board layer and outer kraft layer and provides good protection against impact and compression. The corrugated type, also known as corrugated board, is made with 2 layers of kraft paper with a central corrugating (or fluting) material	Commonly used for shipping bulk food and case packaging of retail food products

Information derived from Kirwan, M. J. (2003). Paper and paperboard packaging. In Coles, R., McDowell, D., & Kirwan, M. J. (Eds.). Food packaging technology (Vol. 5). CRC Press; Marsh, K., & Bugusu, B. (2007). Food packaging—Roles, materials, and environmental issues. *Journal of Food Science, 72* (3), R39–R55. https://doi.org/10.1111/j.1750-3841.2007.00301.x, and Raheem (2012).

packaging materials used as protective exterior barriers for food packages (e.g., a plastic bag of cereal inside a paperboard box or aluminum cans of soda inside a paperboard box), interior protection for food packages (e.g., parchment paper for confectionary items that are placed inside a plastic container), corrugated paperboard boxes used for storage or distribution of foods (e.g., cardboard boxes and some cartons like egg cartons), paper sacks used for storage or distribution of foods (e.g., paper sacks and the multi-layered sacks that contain flour and sugar), and paper used in combination with other packaging materials (e.g., paperboard boxes lined with plastic used for milk cartons or other multi-fold containers).

One tremendous advantage of paper packaging, especially when compared with other packaging materials, is that paper is a naturally renewable material that is also biodegradable, which will be discussed further in an upcoming section of this book chapter.

4.2.5 Plastic Packaging Materials

Plastics are a large category of synthetic, or human-made, materials that use polymers as a main ingredient (Ebnesajjad, 2012; Kirwan & Strawbridge, 2003; Marsh & Bugusu, 2007). Polymers are defined as substances that have a molecular structure consisting of a large number of similar units bonded together. Plastics are produced with two primary polymerization methods, condensation polymerization or addition polymerization. Condensation polymerization, or sometimes referred to as step-growth polymerization, is a type of polymerization in which bi-functional or multifunctional monomer units grow by condensation reactions between molecules. These condensation reactions generate by-products such as water or methanol. Addition polymerization, or sometimes referred to as polyaddition, is a type of polymerization in which polymers are formed with addition reactions (i.e., two or more molecules combine to form a larger molecule with the liberation of by-products).

The creation and development of plastics began in the 19th century (a detailed history is provided by Plastics Industry Association, 2021). Alexander Parkes, an English engineer, is credited with the invention of the first man-made plastic called Parkesine in 1855. Parkesine was manufactured following the reaction between cellulose and the solvent nitric acid. This material was essentially cellulose nitrate and could be dissolved in alcohol and hardened into an elastic material that could be molded when heated. In the next 50 years (1855–1905), the chemical modification of natural materials (e.g., cellulose, collagen, egg, blood proteins, and milk proteins) continued to be researched and developed. It was not until 1907, that the first fully synthetic plastic material was created. This material was invented by Leo Baekeland, a Belgium chemist living in the United States, who would coin the terminology plastic and be referred to many as "the father of the plastics industry." Bakelite or polyoxybenzylmethylenglycolanhydride—$(C_6H_6O \cdot CH_2O)_n$ was created from a condensation reaction of phenol with formaldehyde. Bakelite was a revolutionary material due to its many desirable qualities including its low-cost and versatility (e.g., nonflammable, heat-resistant, sturdy, smooth surface, etc.), which led to many important applications including radio and telephone casings, children's toys, jewelry, firearms, and kitchenware. In the decades following Baekeland's discovery, several other new plasticized materials were either created or further developed. Polyvinyl chloride (PVC), which was initially produced by several different individuals in the 19th century, was mastered for commercial applications and patented in 1913 by a German inventor named Friedrich Heinrich August Klatte. PVC materials were then improved in 1926 by Waldo Lonsbury Semon who was working for the B.F. Goodrich Company. The second world war helped generate further development of a number of additional polymerized plastic materials. For instance, polystyrene (PS) was first industrially manufactured in 1931, polyethylene (PE) was created in England in 1933, nylon was released for sale by DuPont in 1939, and expanded polystyrene (EPS) was created in 1941. During the war, many of these plastic materials were used exclusively for military purposes. However, following the Second World War, manufacturers of plastic materials turned their focus towards consumer goods. This led to the continued development of existing plasticized materials as well as the invention of new plastic materials like polyesters and polypropylene, both of which were introduced as consumer products in the 1950s. Today, there are many different plastic materials used for food packaging, a comprehensive list is provided in Table 4.2.

Approximately 50% of food is packaged in plastic or plastic-based materials (Kirwan & Strawbridge, 2003). There are several reasons why plastics are used for food packaging. The first reason is that plastic materials are extremely effective at containing, protecting, and preserving food and beverages. Secondly, plastic materials offer tremendous versatility. They can be easily molded into various shapes, sheets, and structures to offer unmatched design flexibility for food applications. Additionally, plastic materials are often heat sealable, which offers a high level of convenience in food applications where integration into a production facility includes forming, filling, and sealing the package. Finally, plastic materials are generally inexpensive and easy to manufacture.

The major functional disadvantages of plastic materials are that they do have some permeability to light, gases, and other external factors, and they are variable in terms of their durability during long-term storage and transportation. Safety and health concerns over the use of some materials used during the manufacture of plastic food packaging materials have also been questioned. But perhaps, the greatest disadvantage of plastic food packaging materials is from an environmental standpoint. Most plastic materials can be recycled, yet synthetic plastic materials are not biodegradable when landfilled creating a significant waste management issue.

TABLE 4.2
Types of Plastics Used in Food Packaging.

Type of Plastic	Recycling Number	Description	Food Applications
Polyethylene terephthalate (PET or PETE)	1	A type of polyester created when terephthalic acid reacts with ethylene glycol. This polyester provides a good barrier to from environmental gases (e.g., oxygen, carbon dioxide, and nitrogen) and other external factors that cause food quality and microbial quality issues	Beverage bottles, condiment jars, salad dressing bottles, medicine jars.
High-density polyethylene (HDPE)	2	The most widely used type of plastic. A type of polyolefin created from the monomer ethylene. HDPE is stiff, strong, tough, resistant to chemicals and moisture, permeable to gas, easy to process, and easy to form	Milk containers, juice containers, beverage bottles, cereal box liners, grocery bags, trash bags, soap containers
Polyvinyl chloride (PVC)	3	PVC is produced by polymerization of the vinyl chloride monomer. PVC is heavy, stiff, and ductile, with varying levels of strength. PVC has excellent resistance to chemicals, grease, and oils	Blister packages, vacuum packages, grocery bags, films, and at-home storage containers
Low-density polyethylene (LDPE)	4	A type of polyolefin created from the monomer ethylene. LDPE is flexible, strong, tough, easy to seal, and resistant to moisture	Bread bags, frozen food bags, flexible lids, squeezable food bottles coatings for paper milk cartons and paper cups, shrink/stretch wrap
Polypropylene (PP)	5	A type of polyolefin created from the monomer propylene. It is one of the most-used plastic materials (behind HDPE and LDPE). It softer, less dense, and more transparent than HDPE and LDPE, but it has a higher melting point (160°C)	Yogurt containers, margarine containers, condiment bottles, microwavable packaging, hot-filled containers
Polystyrene or Styrofoam	6	Polystyrene is created from the monomer styrene. General purpose polystyrene is clear, hard, and brittle. The temperature behavior of polystyrene can be exploited to create a foam known as Styrofoam, which has varying levels of density and strength	Foodservice items (such as cups, plates, and bowls), meat trays, rigid food containers (e.g., yogurt)
Miscellaneous/ other plastics	7	This includes many different types of plastic materials, many of which are not or seldomly used for food packaging. Examples are polyamide (i.e., nylon) polycarbonate, polyctide, acrylic, acrylonitrile, butadiene, and styrene	

Information derived from Kirwan, M.J., Strawbridge, J. W. (2003). Plastics in food packaging. In Coles, R., McDowell, D., & Kirwan, M. J. (Eds.). Food packaging technology (Vol. 5). CRC Press, Marsh, K., Bugusu, B. (2007). Food packaging—Roles, materials, and environmental issues. *Journal of Food Science, 72*(3), R39–R55. https://doi.org/10.1111/j.1750-3841.2007.00301.x, and Raheem (2012).

4.2.6 Biodegradable and Bio-Based Packaging Materials

Increasing the use of biodegradable materials for food packaging is an important goal for the global food industry. These materials are defined as materials with the ability to breakdown and return to their natural state in a short period of time following their disposal— typically in less than a year. When plastic polymers were first being developed in the 19th century, many were produced with chemical reactions between two or more naturally-occurring compounds. The resulting plastic polymer materials would often have been biodegradable. However, the use of synthetic polymers makes plastics non-biodegradable. There has been a global interest in the re-development of biodegradable polymers made with naturally occurring materials.

Recent research has focused significant attention to the creation of biodegradable polymers for food packaging (Fahmy et al., 2020; Halonen et al., 2020; Muller, González-Martínez, & Chiralt, 2017; Zhong, Godwin, Jin, & Xiao, 2020). In particular, polylactic acid (PLA), various sources of starch, and other biodegradable polymers have been investigated for their potential to replace synthetic plastics in food packaging applications.

4.2.7 Edible Food Packaging Materials

Edible food packaging is defined as food packaging that is produced with materials that are edible according to legislation, both in the initial (packaging ingredients) and in the final (packaging) forms (Cerqueira, Pereira, da Silva Ramos, Teixeira, & Vicente, 2017). Two types of edible food packaging exist—films and coatings. A film is a dried film-forming solution that is applied as a self-standing material on the food product. A coating is when the film-forming solution is applied directly on the food product and left to dry on the food surface to form a thin film. Examples of edible food packaging materials include those created with varying levels of gelatin, starch, potato, or other food materials and examples of applications include pharmaceutical capsules.

4.3 PACKAGING EFFECTS ON SHELF-LIFE

4.3.1 Modified Atmosphere Packaging (MAP)

Modified atmosphere packaging (MAP) is a packaging technique that involves changing the gaseous atmosphere surrounding a food product inside a packaging barrier and then ensuring that the packaging materials used are capable of providing a sufficient gas barrier to maintain the altered atmosphere for the duration of the product's shelf-life (Church & Parsons, 1995; Mullan & McDowell, 2003; Sivertsvik, Rosnes, & Bergslien, 2002). The history of modified atmosphere storage dates back to the year 1821 when a French pharmacy professor named Jacques Etienne Berard reported that a low-oxygen atmosphere delayed the ripening of fruit (Kirtil & Oztop, 2016). The first recorded commercial use of modified atmosphere storage dates back to 1927 when apples were stored in an elevated carbon dioxide and reduced oxygen atmosphere to extend shelf-life (Goswami & Mangaraj, 2011). By the 1930s, modified atmosphere storage (elevated carbon dioxide) was being used during long-distance

transport of several different types of foods including fruit and beef carcasses (Goswami & Mangaraj, 2011). However, MAP, by definition, was not actually utilized by the food industry until the invention of vacuum packaging in the 1950s, and gas flushed MAP was not utilized until the 1970s.

Vacuum packaging, which will be further discussed in the next section, consists of removing all of the air inside a food package, thus modifying the atmosphere inside a package. However, MAP usually refers to the addition of new gaseous compounds into the package. This process is called gas flushed MAP, which consists of injecting an inert gas such as nitrogen, carbon dioxide, carbon monoxide, or other exotic gases (e.g., noble gases) into the packaging atmosphere and then frequently removing the inert gases to eliminate the concentration of oxygen remaining inside the food package. The inert gases that are used are typically denser than oxygen, thus forcing oxygen out of the package. Several considerations that must be determined when utilizing MAP include the type of gases used, the concentration of the gaseous mixture that will be injected into the package, and the packaging materials. Carbon dioxide has been proven to elicit an antimicrobial effect when used in MAP, while nitrogen is a relatively unreactive gas and is used to displace oxygen in MAP (Church & Parsons, 1995; Farber, 1991; Mullan & McDowell, 2003). Thus, many MAP applications utilize a combination of carbon dioxide and nitrogen to elicit antimicrobial effects, while also displacing oxygen. Carbon monoxide is used exclusively in meat packaging, as this compound is capable of eliciting a change in the chemical state of the protein myoglobin, which is responsible for the red color of meat products (Djenane & Roncalés, 2018; Hunt et al., 2004; Suman & Joseph, 2013). Carbon monoxide alters the structure of myoglobin to form carboxymyoglobin, which is similar in appearance to the red color of oxygenated myoglobin. Other exotic gases, such as the noble gases helium, argon, xenon, and neon, have actually been used for some food applications (e.g., potato-based snacks, fruits, and vegetables); however, the wide-spread use and functional role of these gases in MAP are still being explored (Ghidelli & Pérez-Gago, 2018; Wilson, Stanley, Eyles, & Ross, 2019). Finally, the packaging material that is used in MAP is an important consideration. Packaging materials used in MAP are very dependent on the food product. For example, a product like potato chips/crisps may not require a transparent film, so a non-transparent polymer-based film will be sufficient as the only packaging material, while a product like fresh meat may require both a rigid polymer tray and a transparent appearance so that the consumer can view the product. Creating a transparent portion for a MAP product usually requires multiple different types of packaging materials. A common example of this would be the use of a non-transparent rigid polymer-based material to create a pre-formed tray and then a transparent polymer-based film to create a viewable exterior for consumers to view the product. A significant amount of research and development has been conducted on these films. Flexible and semi-rigid plastic and plastic laminates are most commonly used for MAP packaging films due to their gas and vapor barrier properties, optical and antifogging properties, mechanical strength yet flexibility, and heat-sealing properties (Mullan & McDowell, 2003).

To conclude, MAP offers several interesting advantages to food manufacturers and many MAP applications are still being investigated by the food industry. Of particular interest is the interactive effects of MAP with novel functional food ingredients, as well as the interactive effects of MAP with novel packaging materials.

4.3.2   Vacuum Packaging

Vacuum packaging is a method of packaging where air from the interior of a package is removed in its entirety immediately prior to sealing the package. This involves placing food products in a plastic film package (or potentially a film package made with other polymer materials), removing air from inside the package with a vacuum pump, and then sealing the package (Church & Parsons, 1995). The materials used for vacuum packaging must be strong enough to withstand the vacuum, flexible enough to form around the food product, and have a very specific level of heat sensitivity. Regarding the heat sensitivity of the material, the material must be sealed during the packaging process, but also suitable for moderate temperatures during cooking (e.g., 75 °C). This is especially the situation with the emerging popularity of *sous vide* cooking of foods.

Vacuum packaging was invented in the 1950s by a German inventor named Karl Busch, who used the discovery for vacuum packaging meat products. Today, vacuum packaging is common for long duration storage of dry foods like cereals, nuts, cured meats, cheese, and smoked fish, as well as short duration storage of fresh foods like vegetables, meats, and liquids. Vacuum packaging of foods offers several distinct advantages for the food industry, mainly associated with the reduction, or even the potential elimination, of atmospheric oxygen. This slows the rate of oxidative reactions (i.e., lipid oxidation and protein oxidation) and limits the growth of aerobic bacteria and fungi.

There are several challenges with the implementation and use of vacuum packaging in the food industry. The first challenge is associated with consumer perception of some food products. For example, consumers of meat products are accustomed to specific color cues of meat products (e.g., fresh beef is characteristically red when exposed to oxygen or stored in an aerobic environment); however, an anaerobic environment causes a change in the chemical state of the protein myoglobin, which causes temporary undesirable color cues for meat products (e.g., fresh beef appears dark red or purple in an anerobic environment) (Suman & Joseph, 2013). Another significant challenge with vacuum packaging is the requirement of synthetic packaging materials (i.e., plastic), as very few biodegradable and bio-based packaging materials are capable of meeting the aforementioned technical requirements to undergo vacuum packaging.

4.4   RECENT DISCOVERIES AND FUTURE TRENDS IN FOOD PACKAGING

There are many studies published on the topic of food packaging, and these topics range in terms of their focus and purpose. Multiple research articles, review articles, and books have documented comprehensive descriptions of different food packaging techniques, methods used in the formation of packaging materials, the development of active or intelligent packaging, and the use of bio-based materials to improve food quality and extend the shelf-life of foods. Therefore, this section of the book chapter focuses on the most recent advancements and future trends in the food packaging industry. A summary of recently published studies that have investigated the use of natural compounds (e.g., biopolymers, bioactive substances, and biosensors/natural dyes) for active and intelligent packaging applications is presented in Table 4.3.

4.4.1   Active Packaging and Edible Films/Coatings

The use of edible films and coatings in food packaging can be considered active packaging since most edible films and coatings are formulated with active ingredients (Table 4.3). Therefore, both types of packaging (active and edible) will be discussed together in this section.

Active packaging can be defined as packaging that interacts with the food product or the packaging environment, and thus modifies the condition of the packaged food product (Berk, 2018). Improved food quality, improved food safety, and the increase in the

TABLE 4.3
Summary of Active and Edible Films/Coatings, Bioactive Compounds and Natural Dyes Used in Food Packaging.

Biopolymer	Active Compound	Properties	Reference
Active and/or edible coatings/films			
Fish gelatin	Mango peels extract	Antioxidant activity	Nor Adilah, Noranizan, Jamilah, and Nur Hanani (2020)
Chitosan	ε-Polylysine	Antioxidant and antimicrobial activity	Alirezalu, Pirouzi, et al. (2021)
–	Green tea with caffeine	Antioxidant activity	Borzi et al. (2019)
Chitosan/montmorillonite	Rosemary essential oil	Antioxidant and antimicrobial activity	Souza et al. (2019)
Whey protein	Rosemary, basil, and cinnamon essential oils	Antioxidant and antimicrobial activity	Ribeiro-Santos et al. (2017)
Chitosan/gelatin	β-Carotene loaded starch nanocrystals	Antioxidant activity	Hari et al. (2018)
Cassava starch	Free and nano-encapsulated β-carotene	Antioxidant activity	Assis et al. (2018)
Alginate	*Citrus sinensis* essential oil	Antimicrobial activity	Das et al. (2020)
Chitosan	Liquid smoke	Antimicrobial activity	Desvita et al. (2020)
Methylcellulose	Jambolão extract	Antioxidant activity	da Silva Filipini et al. (2020)
ε-polylysine	Stinging nettle extract	Antioxidant and antimicrobial activity	Alirezalu, Movlan, et al. (2021)
Zein/gelatin	Oregano essential oil	Antimicrobial activity	Cai et al. (2020)
Carboxymethyl cellulose	Rosemary essential oil	Antioxidant and antimicrobial activity	Choulitoudi et al. (2017)
Watermelon rind pectin	kiwifruit peel extract	Antioxidant activity	Han and Song (2021)
Carboxymethylcellulose, bacterial cellulose fibrils and pectin	Blackberry pomace powder	Antioxidant and antimicrobial activity	Isopencu et al. (2021)
Pectin/gelatin	Virgin olive oil and grape seed oil	Antioxidant and antimicrobial activity	Khah et al. (2021)
Gelatin	Pitanga leaf extract and/or nisin	Antioxidant and antimicrobial activity	Luciano et al. (2021)
Alginate	Thyme or oregano essential oils	Antioxidant activity	Pelaes Vital et al. (2021)
Chitosan	Thyme, mugwort, centaury and oregano essential oils	Antioxidant and antimicrobial activity	Quintana et al. (2021)
Chitosan	*Allium tuberosum* root extract	Antioxidant and antimicrobial activity	Riaz et al. (2020)
Pectin/maltodextrin	Sodium chloride	Antimicrobial activity	Mohd Suhaimi et al. (2021)

Continued

TABLE 4.3
Summary of Active and Edible Films/Coatings, Bioactive Compounds and Natural Dyes Used in Food Packaging—cont'd

Biopolymer	Active Compound	Properties	Reference
Gelatin/palm wax	Lemongrass essential oil	Antioxidant and antimicrobial activity	Nurul Syahida et al. (2021)
Gelatin/chitosan	Pitanga leaf extract	Antioxidant and antimicrobial activity	Tessaro et al. (2021)
Gelatin	Propolis extract	Antioxidant and antimicrobial activity	Ucak et al. (2020)
Chitosan	*Artemisia fragrans* essential oil	Antioxidant and antimicrobial activity	Yaghoubi et al. (2021)
Gelatin/chitosan	Gallic acid and clove essential oil	Antioxidant and antimicrobial activity	Xiong et al. (2021)
Intelligent or smart packaging			
Chitin nanofiber and methylcellulose	Red barberry anthocyanin	Ammonia and pH halochromic indicator	Sani et al. (2021)
Cellulose	Synthetic acidochromic dye	Spoilage monitoring (ammonia and pH) sensor	Ding et al. (2020)
Cellulose	*Echium amoenum* anthocyanins	Freshness/spoilage monitoring (ammonia and pH) sensor	Mohammadalinejhad et al. (2020)
Cellulose nanofibers	Black carrot anthocyanins	Freshness (pH) indicator	Moradi et al. (2019)
Cassava starch	Grape skins anthocyanins	Smart pH-change indicator	Vedove et al. (2021)
chitosan/oxidized chitin nanocrystals	Black rice bran anthocyanins	Spoilage monitoring indicator	Wu et al. (2019)

shelf-life of foods are often associated with active packaging (Cruz et al., 2018).

In addition to the general definition provided for active packaging, the European Regulation [Regulation (CE), 2009] defines active materials with the application in food packaging as "materials that are intended to extend the shelf-life of foods and to maintain or improve the condition of packaged food. They are designed to deliberately incorporate components that may release substances into the packaged food or the surrounding environment or absorb some substances from food or the environment." In the same regulation, it is also specified that "active and intelligent materials and articles must be suitable and effective for the intended purpose of use, must not release into food any components that may endanger human health or cause an unacceptable change in the composition or organoleptic characteristics of food, must not mislead the consumer by labelling, presentation or advertising material" (Cruz et al., 2018). Therefore, the investigation of natural extracts and biopolymers that can be used by the food packaging industry is a priority. In fact, many of the extracts and biopolymer materials used for active packaging are

considered generally recognized as safe (GRAS) materials, which is in line with the recommendations by the international authorities. Moreover, the use of nontoxic and biodegradable protective materials to maintain levels of food quality and food safety while increasing food shelf-life helps satisfy consumer demands for minimally processed perishable food products that are manufactured without synthetic food additives or preservatives (Cruz et al., 2018; Munekata et al., 2020).

The operating principle of active packaging is that the package contains active compounds that are released from the food product or the environment that surrounds the food product (Alizadeh-Sani et al., 2020). There are countless types of active packaging that can be applied to the food industry, many of which include agents that modify the atmosphere (e.g., oxygen, ethylene or odors produced by degradation absorbers, water vapor regulators, carbon dioxide emitters, etc.). Although these technologies have been implemented with a high degree of success in the past, current trends in active packaging focus on greater development of novel antimicrobial and antioxidant agents. This is certainly logical as microbial and oxidative processes are those that cause the greatest deterioration of foods (Domínguez et al., 2019). Novel antimicrobial and antioxidant agents used in active packaging can range from synthetic compounds to natural plant extracts or essential oils (Cruz et al., 2018; Munekata et al., 2020; Pateiro et al., 2019). It should be considered that all these agents can be in the form of independent sachets incorporated into the packaging, or they can be incorporated into the polymer matrix and become part of the packaging material (Domínguez et al., 2018). With that stated, antimicrobial and antioxidant active compounds are normally incorporated in the form of an active film or coating (Berk, 2018). In any of the formats of active packaging, the release of the active compound must always be controlled and progressive in order to improve food properties (Cruz et al., 2018). In this sense, the possibility of creating multilayer films allows for improved control of the entire process (Cirillo et al., 2018; Domínguez et al., 2018).

Despite the trend towards the further development of biopolymers, plastics are currently the most widely used materials in food packaging, primarily for the aforementioned reasons listed in the previous section of this book chapter entitled *plastic packaging materials* (Section 4.2.5). In the case of active synthetic polymers, there are three main methods for the manufacture of active films (Domínguez et al., 2018; Suhag, Kumar, Petkoska, & Upadhyay, 2020). The first method is the casting procedure, in which the polymer and active compounds are dissolved in a suitable solvent. The casting procedure involves pouring the polymer/ antioxidant solution onto a surface and allowing the solvent to evaporate, thus obtaining a plastic film with the desired antioxidant characteristics. The second method is the extrusion procedure, which is the most commonly used method for polymer processing. In this case, the active compound is incorporated into the film matrix, in which the polymer is melted (normally due to high shear stress and high temperatures). Finally, the third method is the coating procedure, in which the active compound is immobilized on the surface of the film. The coating procedure has important advantages in comparison with the other techniques. Additionally, the use of biopolymers as carriers of active compounds improves the functionality of these compounds, since high temperatures are required during the production of synthetic polymers, which degrade bioactive compounds (Nor Adilah et al., 2020). The application of the biopolymer layer with active compounds is a promising strategy with great application in the food industry.

On the other hand, natural bioactive compounds can elicit a functional role in the package with or without direct contact between the food and the polymer material (i.e., film or barrier) (Borzi et al., 2019). Among the different functional compounds used for active packaging, essential oils possess promising antimicrobial activity (Pateiro et al., 2021; Sharma, Barkauskaite, Jaiswal, & Jaiswal, 2021) and antioxidant activity (Pateiro et al., 2018, 2019). Similarly, natural plant extracts possess important biological and functional properties due to their high quantity of phenolic compounds (Munekata et al., 2020, 2021), and these extracts can be obtained from several renewable sources such as agri-food by-products (Domínguez, Gullón, et al., 2020; Domínguez, Munekata, et al., 2020), berries (Domínguez, Zhang, et al., 2020; Lorenzo et al., 2018), or seeds (Munekata, Gullón, et al., 2020), among others.

Nowadays there is a growing interest in the production of packaging using renewable and environmentally-friendly materials (Carpena, Nuñez-Estevez, Soria-Lopez, Garcia-Oliveira, & Prieto, 2021; Trajkovska Petkoska, Daniloski, D'Cunha, Naumovski, & Broach, 2021). In this sense, edible films and coatings for food packaging applications are defined as materials applied over or on foods, which extend their shelf-life (Cruz et al., 2018). Edible coatings and films can be formed from bio-based polymers, such as hydrocolloids like proteins (e.g., gelatin, whey protein or corn-zein), polysaccharides (e.g., chitosan, starch, cellulose-derived polysaccharides, carrageenans, or alginates), or lipids (e.g., waxes, neutral lipids), and their combinations (Al-Tayyar, Youssef, & Al-Hindi, 2020; Domínguez et al., 2018). However, biopolymers should fulfill the physical requirements needed from packaging materials, as well as be suitable in terms of the other unique technical properties of active packaging, such as providing a barrier for (or appropriate release of) the active compound to be used in the packaging industry (Domínguez et al., 2018). Generally, polysaccharide and protein films present suitable oxygen barrier properties and fairly good mechanical characteristics, however they do have very poor water-vapor barrier properties (Leimann et al., 2018). This attribute could be improved since the use of lipid-derived polymers alone is not common (except for fruit and vegetable protection). The combination of lipid-derived polymers with protein and/or carbohydrates could improve the water-vapor barrier properties of the film (Domínguez et al., 2018). Additionally, in order to overcome the physical limitations of natural and edible polymers, plasticizers (low molecular weight compounds) are normally incorporated into the film/coating formulation, which improves their mechanical and barrier properties (Domínguez et al., 2018). As is the case with active packaging, the incorporation of antimicrobial and/or antioxidant compounds in the formulation of the edible coating is common. This would extend the shelf-life of foods by preventing oxidation and microbial degradation, but also by regulating moisture and gas exchange, aroma release, and other deteriorations (Ghoshal, 2018). The main biopolymers and bioactive compounds used for active-edible food packaging are summarized in Fig. 4.2.

Among biopolymers, linear polysaccharides derived from cellulose, such as alkyl cellulose, have excellent technological function as films, are widely available, low cost, and generally easy to work with from a manufacturing standpoint (Cirillo et al., 2018). Carboxymethyl cellulose is another interesting biopolymer to develop food coatings with, since it is biodegradable, water-soluble, and transparent when manufactured into a film or coating (Al-Tayyar et al., 2020; Panahirad et al., 2021). However, carboxymethyl cellulose has very poor water-vapor barrier activity. Thus, the combination with other biopolymers, such as collagen, starch, or chitosan is usually necessary in order to improve film properties and application of carboxymethyl cellulose.

Starches are polysaccharide polymers generated from most plant sources. Thus, starch is considered as a renewable raw material, and it actually has multiple packaging applications. Starch is an economically competitive alternative to petroleum-derived polymers, and starch-films are colorless, odorless, non-toxic, and biodegradable (Leimann et al., 2018). The combination of starch with hydrophilic compounds is usually convenient since it increases their functional properties. The main properties of starch-based materials depend on several factors such as the source of the starch (i.e., type of plant), chemical modification, pH, and other technological properties. In fact, the proportions and structure of the two main glucose polymers (amylose and amylopectin) that compose starch are vital for industrial purposes, since they directly influence the technological characteristics of starch-films (Leimann et al., 2018).

Chitosan is a biodegradable and natural material, obtained by the deacetylation of chitin (a molecule present in several living organisms). Chitosan is non-toxic, possesses natural antioxidant and antimicrobial properties, and is able to be integrated with active

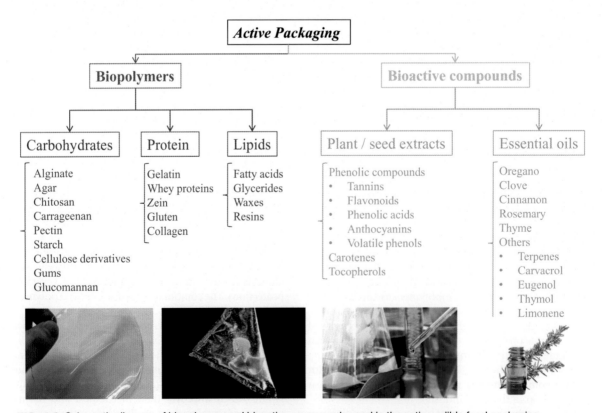

FIG. 4.2 Schematic diagram of biopolymers and bioactive compounds used in the active-edible food packaging.

compounds (Cirillo et al., 2018; Cruz et al., 2018). Due to these characteristics, chitosan is a promising bio-based material for the food industry, and its use could increase the shelf-life of foods, film manufacturing abilities, and general film characteristics (Cruz et al., 2018). However, chitosan is a non-water-soluble compound, therefore acid solutions are required to solubilize chitosan. This is mainly accomplished using organic acids at low concentrations. The films formed with chitosan are inexpensive, transparent, and water-vapor permeable, but have low permeability to oxygen and carbon dioxide (Cruz et al., 2018). Additionally, the hydrophobicity of chitosan films could be increased with the addition of lipid compounds, but this could decrease the physical and chemical characteristics of the films in their final form.

Carrageenans are linear water-soluble polymers that are generally described by their gelation abilities. Carrageenans are classified into three groups: λ-carrageenan (viscosity enhancer properties), ι-carrageenan (thermo-reversible soft gels) and k-carrageenan (strong and hard gels) (Cruz et al., 2018). The incorporation of active compounds (e.g., antioxidants or antimicrobials) and barrier effects (i.e., preventing the exchange of moisture, gases, and flavors) can help maintain or improve food quality and extend shelf-life.

Although the carbohydrate-derived biopolymers that have been described herein are currently the most used in the food industry, other coatings/films produced from agar, pectin, gums, and alginates (both alone and in combination with other biopolymers) are currently being investigated by multiple different research groups worldwide (Al-Tayyar et al., 2020; Chen et al., 2021; Senturk Parreidt, Müller, & Schmid, 2018).

Regarding protein-based biofilms—wheat proteins, which are mainly composed of gluten proteins (> 80% of total wheat protein), corn-zein (prolamin type protein), which is derived from cornstarch and the bioethanol industry, and the proteins from sunflower seed, which are derived as a by-product from the oil industry, could be potential sources of raw materials for the development of novel biopolymers (Leimann et al., 2018). Another interesting protein is gelatin, which is an animal protein obtained following hydrolysis of insoluble collagen. This is an interesting ingredient since it could be easily obtained from the food industry as a by-product. Additionally, gelatin is non-toxic, biodegradable, inexpensive, and an excellent film-forming biopolymer. Milk and whey proteins are also good alternatives to produce bio-based films or edible coatings. It is important to highlight that the protein-based biopolymers can be used as carriers of antioxidant or antimicrobial compounds for active packaging purposes. In fact, protein films are commonly used for the development of bilayer active materials with a commercial polymer (Leimann et al., 2018). But not only protein-based coatings or films are used, since the inclusion of both essential oils and natural plant extracts are also used within the field of carbohydrate edible films and coatings, and these matrices have unique functional properties (Anis et al., 2021). Thus, active, and edible coatings and films are produced with the incorporation of several different types of plant bioactive compounds (Pérez-Santaescolástica et al., 2020).

4.4.2 Intelligent and Smart Packaging

Another innovation used in food packaging is intelligent packaging (synonymous with the terminology smart packaging). This is an emerging technology that uses different sensors or indicators in order to monitor the traceability, freshness, safety, and quality of food products in real-time, and then uses this information to inform the consumers of food conditions by emitting a signal in response to changes in the food and/or packaging environment (Alizadeh-Sani, Mohammadian, Rhim, & Jafari, 2020; Firouz, Mohi-Alden, & Omid, 2021). According to the European Regulation [Regulation (CE), 2009], intelligent packaging systems are those "that provide the user with information on the conditions of the food and should not release their constituents into the food. Intelligent systems may be positioned on the outer surface of the package and may be separated from the food by a functional barrier, which is a barrier within food contact materials or articles preventing the migration of substances from behind that barrier into the food. Behind a functional barrier, nonauthorized substances may be used, provided they comply with certain criteria and their migration remains below a given detection limit (0.01 mg/kg in food)."

Therefore, unlike active packaging, intelligent packaging cannot modify or affect the conditions of the food or the environment, but simply monitors the changes that occur (Cruz et al., 2018). Several conditions of packaged food could be controlled with intelligent packaging, including time/temperature indicators (TTIs), which offer information about the food storage conditions in both real-time and between the time of manufacture and real-time, and package integrity or traceability using radio frequency identifiers (RFID) in labels or chips (Cruz et al., 2018). The TTIs are broadly used to indicate microbial spoilage and for accurate modeling of microbial growth (Firouz et al., 2021; Ghoshal, 2018). Similarly, there are also sensors to control gas composition changes, pH variations, and microbial spoilage, which display important information about food freshness and food safety (Berk, 2018; Sani et al., 2021). Another option is intelligent indicators that increase the flow of information between the consumer and the food product, such as the storage or cooking conditions, recommended preparation/consumption procedures, the expiration date, and others (Ghoshal, 2018).

Normally, intelligent packaging uses colorimetric sensors and indicators that can be easily interpreted by consumers. Nowadays, pH-sensing indicators are widely developed and could be used for several different food applications (Alizadeh-Sani et al., 2020). However, the colorimetric sensors should have a strong reaction between colorant and target analyte, the color change should be easily appreciable by the consumer, interferences with other compounds must not occur, and finally the sensors must be reproducible and reliable (Alizadeh-Sani et al., 2020). Intelligent packaging sensors could be placed on the exterior of the package (external indicators), which could be used to indicate the time, temperature or other external incidences, or in the headspace of the package (internal indicators), which could be used to indicate the changes in the atmosphere, and the release of specific substances related to deteriorative processes or microbial spoilage.

The most recent investigations about intelligent packaging are focused on the use of biodegradable polymers/films and natural dyes as biosensors since they are non-toxic, biodegradable, cost-effective, highly available, and renewable (Alizadeh-Sani et al., 2020). This responds to the trend on the part of consumers to use bio-based colorants, due to the distrust of food safety and security that synthetic colorants present. However, at this point in time, it should be noted that synthetic dyes have multiple advantages since they are more stable, have more intense color cues, and are more easily replicable (Alizadeh-Sani et al., 2020). For these reasons synthetic dyes are usually preferred over natural dyes by the food industry. There are many research studies that have investigated the use of natural colorants for intelligent packaging purposes. In this sense, anthocyanins are the most investigated natural dye (Alizadeh-Sani et al., 2020), due to their wide range of color cues. Nevertheless, not only anthocyanins have been studied, as other natural colorants such as betalains, carotenoids, and chlorophylls have also revealed excellent potential for the development of pH, gas, and temperature-responsive dyes. Moreover, it is important to highlight that several conditions (e.g., type and quantity of natural dyes, storage conditions and

oxidative stress, humidity, cultivar, etc.) could highly influence the stability and functionality of natural dyes, thus their application on an industrial scale is restricted (Alizadeh-Sani et al., 2020).

On the other hand, the use of reactive nanoparticles in intelligent food packaging could improve these devices, since nanosensors are able to react to small environmental changes, sensitive degradation products, and low levels of microbial contamination (Cruz et al., 2018). In perishable food products, the control of storage and distribution conditions is vital to ensure acceptable levels of food quality and food safety. Thus, nanosensors capable of detecting chemical compounds with high levels of sensitivity, and the release of toxins provide a real-time indication of food quality and food safety (Cruz et al., 2018).

In summary, although the incorporation of biosensor systems in packaging films is in the early stages of development, it is a promising area for future research that should also have great applicability to the food industry (Firouz et al., 2021; Ghoshal, 2018).

4.5 CONCLUSIONS

Packaging is a fundamental part of the food industry, as it plays an essential role in food quality and food safety. In fact, since ancient times, advancements in packaging technology have been an important concept for both food manufacturers and consumers. Packaging technology is very complex, encompassing the disciplines of engineering, material science, and food science and technology. Furthermore, packaging has changed from merely being a food container to becoming an ally for food, since packaging interacts with food and the environment that surrounds it, increases the shelf-life and prevents/limits its spoilage, and at the same time the packaging allows consumers to obtain information in real-time of all the conditions of the food. In addition, taking into account the different drawbacks that synthetic polymers present (e.g., significant contributor to pollution, mostly derived from non-renewable petrochemicals, use of toxic organic compounds, migration of compounds to food such as bisphenol A, etc.), there is a global necessity to further research and development efforts for natural and biodegradable polymer packaging materials.

As a general conclusion, the use of biopolymers as biodegradable and non-toxic coating/films, bioactive compounds such as essential oils and natural extracts in active packaging as well as the use of natural dyes as biosensors for intelligent packaging appear to be important future considerations for the food packaging industry. Even so, it is important to take into account the enormous variety of foods, with totally different chemical compositions, physicochemical properties, and elaboration processes, which limit a common strategy. Therefore, the correct development of a package must be designed for each food, taking into account all its peculiarities. In the same way, there are innumerable bioactive compounds that can be part of packaging (such as active packaging), each with an antioxidant or antimicrobial potential, but also with other physical properties (for example intense color) that can limit their use in the development of active packaging. In the same way, despite all the advantages mentioned for natural dyes, they are less stable and more sensitive to certain common processes of the food industry, so their use in biosensors could also be limited. Hence, a significant challenge for researchers is to identify a suitable combination of polymers, additives (i.e., plasticizers), active compounds (i.e., antioxidants and antimicrobials), and/or natural biosensors that can fulfill the requirements for the packaging of specific food products. It is also important to bear in mind that the methodologies used during manufacture of food packages must be easily scalable, in order for them to be effectively applied to the global food industry. With all this in mind, it seems clear that food packaging is evolving immensely, although there is still much to be learned, so research in this field promises great advances in the coming decades.

REFERENCES

Alirezalu, K., Movlan, H. S., Yaghoubi, M., Pateiro, M., & Lorenzo, J. M. (2021). ε-polylysine coating with stinging nettle extract for fresh beef preservation. *Meat Science, 176,* 108474. https://doi.org/10.1016/j.meatsci.2021.108474.

Alirezalu, K., Pirouzi, S., Yaghoubi, M., Karimi-Dehkordi, M., Jafarzadeh, S., & Mousavi Khaneghah, A. (2021). Packaging of beef fillet with active chitosan film incorporated with ε-polylysine: An assessment of quality indices and shelf life. *Meat Science, 176,* 108475. https://doi.org/10.1016/j.meatsci.2021.108475.

Alizadeh-Sani, M., Mohammadian, E., Rhim, J. W., & Jafari, S. M. (2020). pH-sensitive (halochromic) smart packaging films based on natural food colorants for the monitoring of food quality and safety. *Trends in Food Science and Technology, 105,* 93–144. https://doi.org/10.1016/j.tifs.2020.08.014.

Al-Tayyar, N. A., Youssef, A. M., & Al-Hindi, R. R. (2020). Edible coatings and antimicrobial nanoemulsions for enhancing shelf life and reducing foodborne pathogens of fruits and vegetables: A review. *Sustainable Materials and Technologies, 26,* e00215. https://doi.org/10.1016/j.susmat.2020.e00215.

Anis, A., Pal, K., & Al-Zahrani, S. M. (2021). Essential oil-containing polysaccharide-based edible films and coatings for food security applications. *Polymers, 13*(4), 575. https://doi.org/10.3390/polym13040575.

Apotheker, S., & Markstahler, E. (1986). *Tin cans: History and outlook. Part 2 (No. PB-95-235370/XAB).* Champaign, IL (United States): Community Recycling Center.

Assis, R. Q., Pagno, C. H., Costa, T. M. H., Flôres, S. H., & Rios, A.d. O. (2018). Synthesis of biodegradable films based on cassava starch containing free and nanoencapsulated β-carotene. *Packaging Technology and Science, 31*(3), 157–166. https://doi.org/10.1002/pts.2364.

Battisti, R., Fronza, N., Vargas Júnior, Á., da Silveira, S. M., Damas, M. S. P., & Quadri, M. G. N. (2017). Gelatin-coated paper with antimicrobial and antioxidant effect for beef packaging. *Food Packaging and Shelf Life, 11,* 115–124. https://doi.org/10.1016/j.fpsl.2017.01.009.

Berger, K. R. (2005). *A brief history of packaging. Vol. 2005.* Gainesville: Institute of Food and Agricultural Sciences, University of Florida.

Berk, Z. (2018). Food packaging. In Z. Berk (Ed.), *Food process engineering and technology* (3rd ed., pp. 625–641). https://doi.org/10.1016/B978-0-12-812018-7.00027-0.

Borzi, F., Torrieri, E., Wrona, M., & Nerín, C. (2019). Polyamide modified with green tea extract for fresh minced meat active packaging applications. *Food Chemistry, 300,* 125242. https://doi.org/10.1016/J.FOODCHEM.2019.125242.

Busch, J. (1981). An introduction to the tin can. *Historical Archaeology, 15*(1), 95–104.

Cai, J., Xiao, J., Chen, X., & Liu, H. (2020). Essential oil loaded edible films prepared by continuous casting method: Effects of casting cycle and loading position on the release properties. *Food Packaging and Shelf Life, 26,* 100555. https://doi.org/10.1016/j.fpsl.2020.100555.

Carina, D., Sharma, S., Jaiswal, A. K., & Jaiswal, S. (2021). Seaweeds polysaccharides in active food packaging: A review of recent progress. *Trends in Food Science and Technology, 110,* 559–572. https://doi.org/10.1016/j.tifs.2021.02.022.

Carpena, M., Nuñez-Estevez, B., Soria-Lopez, A., Garcia-Oliveira, P., & Prieto, M. A. (2021). Essential oils and their application on active packaging systems: A review. *Resources, 10*(1), 7. https://doi.org/10.3390/resources10010007.

Cerqueira, M. A. P. R., Pereira, R. N. C., da Silva Ramos, O. L., Teixeira, J. A. C., & Vicente, A. A. (Eds.). (2017). *Edible food packaging: Materials and processing technologies* CRC Press.

Chen, W., Ma, S., Wang, Q., McClements, D. J., Liu, X., Ngai, T., et al. (2021). Fortification of edible films with bioactive agents: A review of their formation, properties, and application in food preservation. *Critical Reviews in Food Science and Nutrition,* 1–27. https://doi.org/10.1080/10408398.2021.1881435.

Choulitoudi, E., Ganiari, S., Tsironi, T., Ntzimani, A., Tsimogiannis, D., Taoukis, P., et al. (2017). Edible coating enriched with rosemary extracts to enhance oxidative and microbial stability of smoked eel fillets. *Food Packaging and Shelf Life, 12,* 107–113. https://doi.org/10.1016/j.fpsl.2017.04.009.

Church, I. J., & Parsons, A. L. (1995). Modified atmosphere packaging technology: A review. *Journal of the Science of Food and Agriculture, 67*(2), 143–152. https://doi.org/10.1002/jsfa.2740670202.

Cirillo, G., Curcio, M., Spataro, T., Picci, N., Restuccia, D., Iemma, F., et al. (2018). Antioxidant polymers for food packaging. In A. M. Grumezescu, & A. M. Holban (Eds.), *Food packaging and preservation* (pp. 213–238). https://doi.org/10.1016/b978-0-12-811516-9.00006-3.

Coles, R. (2003). Introduction. In R. Coles, D. McDowell, & M. J. Kirwan (Eds.), *Vol. 5. Food packaging technology* CRC Press.

Cruz, R. M. S., Alves, V., Khmelinskii, I., & Vieira, M. C. (2018). New food packaging systems. In A. M. Grumezescu, & A. M. Holban (Eds.), *Food packaging and preservation* (pp. 63–85). https://doi.org/10.1016/b978-0-12-811516-9.00002-6.

da Silva Filipini, G., Romani, V. P., & Guimarães Martins, V. (2020). Biodegradable and active-intelligent films based on methylcellulose and jambolão (*Syzygium cumini*) skins extract for food packaging. *Food Hydrocolloids, 109*, 106139. https://doi.org/10.1016/j.foodhyd.2020.106139.

Das, S., Vishakha, K., Banerjee, S., Mondal, S., & Ganguli, A. (2020). Sodium alginate-based edible coating containing nanoemulsion of *Citrus sinensis* essential oil eradicates planktonic and sessile cells of food-borne pathogens and increased quality attributes of tomatoes. *International Journal of Biological Macromolecules, 162*, 1770–1779. https://doi.org/10.1016/j.ijbiomac.2020.08.086.

Desvita, H., Faisal, M., Mahidin, & Suhendrayatna. (2020). Preservation of meatballs with edible coating of chitosan dissolved in rice hull-based liquid smoke. *Heliyon, 6*(10), e05228. https://doi.org/10.1016/j.heliyon.2020.e05228.

Ding, L., Li, X., Hu, L., Zhang, Y., Jiang, Y., Mao, Z., et al. (2020). A naked-eye detection polyvinyl alcohol/cellulose-based pH sensor for intelligent packaging. *Carbohydrate Polymers, 233*, 115859. https://doi.org/10.1016/j.carbpol.2020.115859.

Djenane, D., & Roncalés, P. (2018). Carbon monoxide in meat and fish packaging: Advantages and limits. *Food, 7*(2), 12. https://doi.org/10.3390/foods7020012.

Domínguez, R., Barba, F. J., Gómez, B., Putnik, P., Bursać Kovačević, D., Pateiro, M., et al. (2018). Active packaging films with natural antioxidants to be used in meat industry: A review. *Food Research International, 113*, 93–101. https://doi.org/10.1016/j.foodres.2018.06.073.

Domínguez, R., Gullón, P., Pateiro, M., Munekata, P. E. S., Zhang, W., & Lorenzo, J. M. (2020). Tomato as potential source of natural additives for meat industry. A review. *Antioxidants, 9*(1), 73. https://doi.org/10.3390/antiox9010073.

Domínguez, R., Munekata, P. E. S., Pateiro, M., Maggiolino, A., Bohrer, B., & Lorenzo, J. M. (2020). Red beetroot. A potential source of natural additives for the meat industry. *Applied Sciences, 10*(23), 8340. https://doi.org/10.3390/app10238340.

Domínguez, R., Pateiro, M., Gagaoua, M., Barba, F. J., Zhang, W., & Lorenzo, J. M. (2019). A comprehensive review on lipid oxidation in meat and meat products. *Antioxidants, 8*(10), 429. https://doi.org/10.3390/ANTIOX8100429.

Domínguez, R., Zhang, L., Rocchetti, G., Lucini, L., Pateiro, M., Munekata, P. E. S., et al. (2020). Elderberry (*Sambucus nigra* L.) as potential source of antioxidants. Characterization, optimization of extraction parameters and bioactive properties. *Food Chemistry, 330*, 127266. https://doi.org/10.1016/j.foodchem.2020.127266.

Ebnesajjad, S. (2012). *Plastic films in food packaging: Materials, technology and applications.* Elsevier.

Fahmy, H. M., Eldin, R. E. S., Serea, E. S. A., Gomaa, N. M., AboElmagd, G. M., Salem, S. A., et al. (2020). Advances in nanotechnology and antibacterial properties of biodegradable food packaging materials. *RSC Advances, 10*(35), 20467–20484. https://doi.org/10.1039/D0RA02922J.

Farber, J. M. (1991). Microbiological aspects of modified-atmosphere packaging technology-a review. *Journal of Food Protection, 54*(1), 58–70. https://doi.org/10.4315/0362-028X-54.1.58.

Firouz, M. S., Mohi-Alden, K., & Omid, M. (2021). A critical review on intelligent and active packaging in the food industry. *Research and development. Food Research International, 141*, 110113. https://doi.org/10.1016/j.foodres.2021.110113.

Ghidelli, C., & Pérez-Gago, M. B. (2018). Recent advances in modified atmosphere packaging and edible coatings to maintain quality of fresh-cut fruits and vegetables. *Critical Reviews in Food Science and Nutrition, 58*(4), 662–679. https://doi.org/10.1080/10408398.2016.1211087.

Ghoshal, G. (2018). Recent trends in active, smart, and intelligent packaging for food products. In A. M. Grumezescu, & A. M. Holban (Eds.), *Food packaging and preservation. Handbook of food bioengineering* (pp. 343–373). https://doi.org/10.1016/b978-0-12-811516-9.00010-5.

Girling, P. J. (2003). Packaging of food in glass containers. In R. Coles, D. McDowell, & M. J. Kirwan (Eds.), *Vol. 5. Food packaging technology* CRC Press.

Goswami, T. K., & Mangaraj, S. (2011). Advances in polymeric materials for modified atmosphere packaging (MAP). In *Multifunctional and nanoreinforced polymers for food packaging* (pp. 163–242). Woodhead Publishing.

Gupta, R. K., & Dudeja, P. (2017). Food packaging. In R. K. Gupta, P. Dudeja, & S. Minhas (Eds.), *Food safety in the 21st century: public health perspective* (pp. 547–553). https://doi.org/10.1016/B978-0-12-801773-9.00046-7.

Halonen, N. J., Pálvölgyi, P. S., Bassani, A., Fiorentini, C., Nair, R., Spigno, G., et al. (2020). Bio-based smart materials for food packaging and sensors—A review. *Frontiers in Materials, 7*, 82. https://doi.org/10.3389/fmats.2020.00082.

Han, H. S., & Song, K. B. (2021). Antioxidant properties of watermelon (*Citrullus lanatus*) rind pectin films containing kiwifruit (*Actinidia chinensis*) peel extract and their application as chicken thigh packaging. *Food Packaging and Shelf Life, 28*. https://doi.org/10.1016/j.fpsl.2021.100636.

Hari, N., Francis, S., Rajendran Nair, A. G., & Nair, A. J. (2018). Synthesis, characterization and biological evaluation of chitosan film incorporated with β-carotene loaded starch nanocrystals. *Food Packaging and Shelf Life, 16*, 69–76. https://doi.org/10.1016/j.fpsl.2018.02.003.

Hunt, M. C., Mancini, R. A., Hachmeister, K. A., Kropf, D. H., Merriman, M., De Lduca, G., et al. (2004). Carbon monoxide in modified atmosphere packaging affects color, shelf life, and microorganisms of beef steaks and ground beef. *Journal of Food Science, 69*(1), FCT45–FCT52. https://doi.org/10.1111/j.1365-2621.2004.tb17854.x.

Hunter, D. (1978). *Papermaking: The history and technique of an ancient craft.* Courier Corporation.

Isopencu, G. O., Stoica-Guzun, A., Busuioc, C., Stroescu, M., & Deleanu, I. M. (2021). Development of antioxidant and antimicrobial edible coatings incorporating bacterial cellulose, pectin, and blackberry pomace. *Carbohydrate Polymer Technologies and Applications, 2*, 100057. https://doi.org/10.1016/j.carpta.2021.100057.

Khah, M. D., Ghanbarzadeh, B., Roufegarinejad Nezhad, L., & Ostadrahimi, A. (2021). Effects of virgin olive oil and grape seed oil on physicochemical and antimicrobial properties of pectin-gelatin blend emulsified films. *International Journal of Biological Macromolecules, 171*, 262–274. https://doi.org/10.1016/j.ijbiomac.2021.01.020.

Kirtil, E., & Oztop, M. H. (2016). Controlled and modified atmosphere packaging. In *Reference module in food science* (pp. 1–2). Amsterdam: Elsevier. https://doi.org/10.1016/B978-0-08-100596-5.03376-X.

Kirwan, M. J. (2003). Paper and paperboard packaging. In R. Coles, D. McDowell, & M. J. Kirwan (Eds.), *Vol. 5. Food packaging technology* CRC Press.

Kirwan, M. J., & Strawbridge, J. W. (2003). Plastics in food packaging. In R. Coles, D. McDowell, & M. J. Kirwan (Eds.), *Vol. 5. Food packaging technology* CRC Press.

Lee, D. S., Yam, K. L., & Piergiovanni, L. (2008). *Food packaging science and technology.* CRC Press.

Leimann, F. V., Gonçalves, O. H., Sakanaka, L. S., Azevedo, A. S. B., Lima, M. V., Barreiro, F., et al. (2018). Active food packaging from botanical, animal, bacterial, and synthetic sources. In A. M. Grumezescu, & A. M. Holban (Eds.), *Food packaging and preservation* (pp. 87–135). https://doi.org/10.1016/b978-0-12-811516-9.00003-8.

Lorenzo, J. M., Domínguez, R., & Carballo, J. (2017). Control of lipid oxidation in muscle food by active packaging technology. In R. Banerjee, A. K. Verma, & M. W. Siddiqui (Eds.), *Natural antioxidants: Applications in foods of animal origin* (1st ed., pp. 343–382). https://doi.org/10.1201/9781315365916.

Lorenzo, J. M., Pateiro, M., Domínguez, R., Barba, F. J., Putnik, P., Kovačević, D. B., et al. (2018). Berries extracts as natural antioxidants in meat products: A review. *Food Research International, 106*, 1095–1104. https://doi.org/10.1016/j.foodres.2017.12.005.

Luciano, C. G., Rodrigues, M. M., Lourenço, R. V., Bittante, A. M. Q. B., Fernandes, A. M., & do Amaral Sobral, P. J. (2021). Bi-layer Gelatin film: Activating film by incorporation of "Pitanga" leaf hydroethanolic extract and/or Nisin in the second layer. *Food and Bioprocess Technology, 14*(1), 106–119. https://doi.org/10.1007/s11947-020-02568-w.

Marsh, K., & Bugusu, B. (2007). Food packaging—Roles, materials, and environmental issues. *Journal of Food Science, 72*(3), R39–R55. https://doi.org/10.1111/j.1750-3841.2007.00301.x.

Mohammadalinejhad, S., Almasi, H., & Moradi, M. (2020). Immobilization of Echium amoenum anthocyanins into bacterial cellulose film: A novel colorimetric pH indicator for freshness/spoilage monitoring of shrimp. *Food Control, 113*, 107169. https://doi.org/10.1016/j.foodcont.2020.107169.

Mohd Suhaimi, N. I., Mat Ropi, A. A., & Shaharuddin, S. (2021). Safety and quality preservation of starfruit (*Averrhoa carambola*) at ambient shelf life using synergistic pectin-maltodextrin-sodium chloride edible coating. *Heliyon, 7*(2). https://doi.org/10.1016/j.heliyon.2021.e06279, e06279.

Moradi, M., Tajik, H., Almasi, H., Forough, M., & Ezati, P. (2019). A novel pH-sensing indicator based on bacterial cellulose nanofibers and black carrot anthocyanins for monitoring fish freshness. *Carbohydrate Polymers, 222*, 115030. https://doi.org/10.1016/j.carbpol.2019.115030.

Mullan, M., & McDowell, D. (2003). Modified atmosphere packaging. In R. Coles, D. McDowell, & M. J. Kirwan (Eds.), *Vol. 5. Food packaging technology* CRC Press.

Muller, J., González-Martínez, C., & Chiralt, A. (2017). Combination of poly (lactic) acid and starch for biodegradable food packaging. *Materials, 10*(8), 952. https://doi.org/10.3390/ma10080952.

Munekata, P. E. S., Gullón, B., Pateiro, M., Tomasevic, I., Domínguez, R., & Lorenzo, J. M. (2020). Natural antioxidants from seeds and their application in meat products. *Antioxidants, 9*(9), 815. https://doi.org/10.3390/antiox9090815.

Munekata, P. E. S., Pateiro, M., Bellucci, E. R. B., Domínguez, R., da Silva Barretto, A. C., & Lorenzo, J. M. (2021). Strategies to increase the shelf life of meat and meat products with phenolic compounds. *Advances in Food and Nutrition Research*. https://doi.org/10.1016/bs.afnr.2021.02.008.

Munekata, P. E. S., Rocchetti, G., Pateiro, M., Lucini, L., Domínguez, R., & Lorenzo, J. M. (2020). Addition of plant extracts to meat and meat products to extend shelf-life and health-promoting attributes: An overview. *Current Opinion in Food Science, 31*, 81–87. https://doi.org/10.1016/j.cofs.2020.03.003.

New York Times. (1988). *Farm population lowest since 1850s.* Accessed at https://www.nytimes.com/1988/07/20/us/farm-population-lowest-since-1850-s.html#:~:text=By%201850%2C%20farm%20people%20made,the%20nation's%207.7%20million%20workers.

Nor Adilah, A., Noranizan, M. A., Jamilah, B., & Nur Hanani, Z. A. (2020). Development of polyethylene films coated with gelatin and mango peel extract and the effect on the quality of margarine. *Food Packaging and Shelf Life, 26*, 100577. https://doi.org/10.1016/j.fpsl.2020.100577.

Nurul Syahida, S., Ismail-Fitry, M. R., Ainun, Z. M. A., & Nur Hanani, Z. A. (2021). Effects of gelatin/palm wax/lemongrass essential oil (GPL)-coated Kraft paper on the quality and shelf life of ground beef stored at 4 °C. *Food Packaging and Shelf Life, 28*, 100640. https://doi.org/10.1016/j.fpsl.2021.100640.

Page, B., Edwards, M., & May, N. (2003). Metal cans. In R. Coles, D. McDowell, & M. J. Kirwan (Eds.), *Vol. 5. Food packaging technology* CRC Press.

Panahirad, S., Dadpour, M., Peighambardoust, S. H., Soltanzadeh, M., Gullón, B., Alirezalu, K., et al. (2021). Applications of carboxymethyl cellulose- and pectin-based active edible coatings in preservation of fruits and vegetables: A review. *Trends in Food Science and Technology, 110,* 663–673. https://doi.org/10.1016/j.tifs.2021.02.025.

Pateiro, M., Barba, F. J., Domínguez, R., Sant'Ana, A. S., Mousavi Khaneghah, A., Gavahian, M., et al. (2018). Essential oils as natural additives to prevent oxidation reactions in meat and meat products: A review. *Food Research International, 113,* 156–166. https://doi.org/10.1016/j.foodres.2018.07.014.

Pateiro, M., Domínguez, R., Bermúdez, R., Munekata, P. E. S., Zhang, W., Gagaoua, M., et al. (2019). Antioxidant active packaging systems to extend the shelf life of sliced cooked ham. *Current Research in Food Science, 1,* 24–30. https://doi.org/10.1016/j.crfs.2019.10.002.

Pateiro, M., Munekata, P. E. S., Sant'Ana, A. S., Domínguez, R., Rodríguez-Lázaro, D., & Lorenzo, J. M. (2021). Application of essential oils as antimicrobial agents against spoilage and pathogenic microorganisms in meat products. *International Journal of Food Microbiology, 337,* 108966. https://doi.org/10.1016/j.ijfoodmicro.2020.108966.

Pelaes Vital, A. C., Guerrero, A., Guarnido, P., Cordeiro Severino, I., Olleta, J. L., Blasco, M., et al. (2021). Effect of active-edible coating and essential oils on lamb patties oxidation during display. *Food, 10*(2), 263. https://doi.org/10.3390/foods10020263.

Pérez-Santaescolástica, C., Munekata, P. E. S., Feng, X., Liu, Y., Bastianello Campagnol, P. C., & Lorenzo, J. M. (2020). Active edible coatings and films with Mediterranean herbs to improve food shelf-life. *Critical Reviews in Food Science and Nutrition.* https://doi.org/10.1080/10408398.2020.1853036.

Piergiovanni, L., & Limbo, S. (2016). *Food packaging materials* (pp. 33–49). Basel, Switzerland: Springer.

Piringer, O. G., & Baner, A. L. (2008). *Plastic packaging: Interactions with food and pharmaceuticals.* John Wiley & Sons.

Plastics Industry Association. (2021). *History of plastics.* Accessed at: https://www.plasticsindustry.org/history-plastics.

Quintana, S. E., Llalla, O., García-Risco, M. R., & Fornari, T. (2021). Comparison between essential oils and supercritical extracts into chitosan-based edible coatings on strawberry quality during cold storage. *Journal of Supercritical Fluids, 171,* 105198. https://doi.org/10.1016/j.supflu.2021.105198.

Raheem, D. (2013). Application of plastics and paper as food packaging materials—An overview. *Emirates Journal of Food and Agriculture, 25*(3), 177–188. https://doi.org/10.9755/ejfa.v25i3.11509.

Regulation (CE). (2009). *Guidance to the commission regulation (EC) No 450/2009 of 29 May 2009 on active and intelligent materials and articles intended to come into contact with food.* Retrieved from Official Journal of the European Union website https://eur-lex.europa.eu/legal-content/EN/TXT/PDF/?uri=CELEX:32009R0450&from=EN.

Riaz, A., Lagnika, C., Luo, H., Dai, Z., Nie, M., Hashim, M. M., et al. (2020). Chitosan-based biodegradable active food packaging film containing Chinese chive (*Allium tuberosum*) root extract for food application. *International Journal of Biological Macromolecules, 150,* 595–604. https://doi.org/10.1016/j.ijbiomac.2020.02.078.

Ribeiro-Santos, R., Sanches-Silva, A., Motta, J. F. G., Andrade, M., de Araújo Neves, I., Teófilo, R. F., et al. (2017). Combined use of essential oils applied to protein base active food packaging: Study in vitro and in a food simulant. *European Polymer Journal, 93,* 75–86. https://doi.org/10.1016/j.eurpolymj.2017.03.055.

Risch, S. J. (2009). Food packaging history and innovations. *Journal of Agricultural and Food Chemistry, 57*(18), 8089–8092. https://doi.org/10.1021/jf900040r.

Robertson, G. L. (2014). Food packaging. In *Encyclopedia of agriculture and food systems* (pp. 232–249). https://doi.org/10.1016/B978-0-444-52512-3.00063-2.

Robertson, G. L. (2016). *Food packaging: Principles and practice.* CRC Press.

Sani, M. A., Tavassoli, M., Hamishehkar, H., & McClements, D. J. (2021). Carbohydrate-based films containing pH-sensitive red barberry anthocyanins: Application as biodegradable smart food packaging materials. *Carbohydrate Polymers, 255,* 117488. https://doi.org/10.1016/j.carbpol.2020.117488.

Senturk Parreidt, T., Müller, K., & Schmid, M. (2018). Alginate-based edible films and coatings for food packaging applications. *Food, 7*(10), 170. https://doi.org/10.3390/foods7100170.

Sharma, S., Barkauskaite, S., Jaiswal, A. K., & Jaiswal, S. (2021). Essential oils as additives in active food packaging. *Food Chemistry, 343,* 128403. https://doi.org/10.1016/j.foodchem.2020.128403.

Sivertsvik, M., Rosnes, J. T., & Bergslien, H. (2002). Modified atmosphere packaging. In T. Ohlsson, & N. Bengtsson (Eds.), *Minimal processing technologies in the food industry* Woodhead Publishing in Food Science and Technology.

Souza, V. G. L., Pires, J. R. A., Vieira, É. T., Coelhoso, I. M., Duarte, M. P., & Fernando, A. L. (2019). Activity of chitosan-montmorillonite bionanocomposites incorporated with rosemary essential oil: From in vitro assays to application in fresh poultry meat. *Food Hydrocolloids, 89,* 241–252. https://doi.org/10.1016/j.foodhyd.2018.10.049.

Standage, T. (2009). *An edible history of humanity.* Bloomsbury Publishing PLC.

Suhag, R., Kumar, N., Petkoska, A. T., & Upadhyay, A. (2020). Film formation and deposition methods of edible coating on food products: A review. *Food Research International, 136,* 109582. https://doi.org/10.1016/j.foodres.2020.109582.

Suman, S. P., & Joseph, P. (2013). Myoglobin chemistry and meat color. *Annual Review of Food Science and Technology, 4,* 79–99. https://doi.org/10.1146/annurev-food-030212-182623.

Tessaro, L., Luciano, C. G., Quinta Barbosa Bittante, A. M., Lourenço, R. V., Martelli-Tosi, M., & José do Amaral Sobral, P. (2021). Gelatin and/or chitosan-based films activated with "Pitanga" (*Eugenia uniflora* L.) leaf hydroethanolic extract encapsulated in double emulsion. *Food Hydrocolloids, 113*, 106523. https://doi.org/10.1016/j.foodhyd.2020.106523.

Trajkovska Petkoska, A., Daniloski, D., D'Cunha, N. M., Naumovski, N., & Broach, A. T. (2021). Edible packaging: Sustainable solutions and novel trends in food packaging. *Food Research International, 140*, 109981. https://doi.org/10.1016/j.foodres.2020.109981.

Ucak, I., Khalily, R., Carrillo, C., Tomasevic, I., & Barba, F. J. (2020). Potential of propolis extract as a natural antioxidant and antimicrobial in gelatin films applied to rainbow trout (*Oncorhynchus mykiss*) fillets. *Food, 9*(11), 1584. https://doi.org/10.3390/foods9111584.

Vedove, T. M. A. R. D., Maniglia, B. C., & Tadini, C. C. (2021). Production of sustainable smart packaging based on cassava starch and anthocyanin by an extrusion process. *Journal of Food Engineering, 289*, 110274. https://doi.org/10.1016/j.jfoodeng.2020.110274.

Wilson, M. D., Stanley, R. A., Eyles, A., & Ross, T. (2019). Innovative processes and technologies for modified atmosphere packaging of fresh and fresh-cut fruits and vegetables. *Critical Reviews in Food Science and Nutrition, 59*(3), 411–422. https://doi.org/10.1080/10408398.2017.1375892.

Wu, C., Sun, J., Zheng, P., Kang, X., Chen, M., Li, Y., et al. (2019). Preparation of an intelligent film based on chitosan/oxidized chitin nanocrystals incorporating black rice bran anthocyanins for seafood spoilage monitoring. *Carbohydrate Polymers, 222*, 115006. https://doi.org/10.1016/j.carbpol.2019.115006.

Xiong, Y., Kamboj, M., Ajlouni, S., & Fang, Z. (2021). Incorporation of salmon bone gelatine with chitosan, gallic acid and clove oil as edible coating for the cold storage of fresh salmon fillet. *Food Control, 125*, 107994. https://doi.org/10.1016/j.foodcont.2021.107994.

Yaghoubi, M., Ayaseh, A., Alirezalu, K., Nemati, Z., Pateiro, M., & Lorenzo, J. M. (2021). Effect of chitosan coating incorporated with Artemisia fragrans essential oil on fresh chicken meat during refrigerated storage. *Polymers, 13*(5), 716. https://doi.org/10.3390/polym13050716.

Zhong, Y., Godwin, P., Jin, Y., & Xiao, H. (2020). Biodegradable polymers and green-based antimicrobial packaging materials: A mini-review. *Advanced Industrial and Engineering Polymer Research, 3*(1), 27–35. https://doi.org/10.1016/j.aiepr.2019.11.002.

CHAPTER 5

Pectooligosaccharides as Emerging Functional Ingredients: Sources, Extraction Technologies, and Biological Activities

PATRICIA GULLÓN[A] • PABLO G. DEL RÍO[B] • BEATRIZ GULLÓN[B] •
DIANA OLIVEIRA[C] • PATRICIA COSTA[C] • JOSÉ MANUEL LORENZO[D,E]
[a]Nutrition and Bromatology Group, Department of Analytical and Food Chemistry, Faculty of Food Science and Technology, University of Vigo, Ourense, Spain, [b]Department of Chemical Engineering, Faculty of Science, University of Vigo, Ourense, Spain, [c]Universidade Católica Portuguesa, CBQF, Centro de Biotecnologia e Química Fina, Laboratório Associado, Escola Superior de Biotecnologia, Porto, Portugal, [d]Galician Meat Technology Center, Galicia Technology Park, Ourense, Spain, [e]Food Technology Area, Faculty of Sciences of Ourense, Vigo University, Ourense, Spain

5.1 INTRODUCTION

There is an increased interest in the identification, evaluation, and commercialization of new functional products, which provide additional health benefits to both humans and animals. Products with bioactive properties such as antioxidant, antidiabetic, antihypertensive, hypocholesterolemic, etc., have been investigated over the past years. Resistant starch, xylooligosaccharides, soya oligosaccharides, mannooligosaccharides, and pectin-derived oligosaccharides are representative of these new products with improved functional properties (Gullón et al., 2013), mainly due to their capacity to modulate the gut microbiota and exert prebiotic properties within the gastrointestinal system. Pectin-derived oligosaccharides, known as POS, have been proposed as a new class of prebiotics capable of exerting a number of health-promoting effects (Hotchkiss et al., 2004). The importance of prebiotic foods lies in their active stimulation of growth of beneficial bacteria, thereby adding to potential health and nutritional benefits (Panesar, Bali, Kumari, Babbar, & Oberoi, 2014). In addition, the colonic fermentation of prebiotic oligosaccharides results in the generation of short chain fatty acids (SCFA). These SCFA present a number of beneficial effects, including, reduction of the blood glucose level, improvement of mineral absorption, regulation of lipid metabolism, decreased incidence of colonic cancer, and modulation of the immune system (Gullón et al., 2013).

POS are oligosaccharides that can be obtained by partial hydrolysis of pectins, which are structurally complex heteropolysaccharides (Míguez, Gómez, Gullón, Gullón, & Alonso, 2016). Pectins are industrially produced from peel and pulp of fruits and vegetables, mainly, citrus fruits, sugar beet, and apples and has been used for long as food additives as emulsifiers, gelling or stabilizing agents in yoghurts, jams, and other food products (Larsen et al., 2019). Moreover, POS can be produced from agriculture by-products such as fruit peels, vegetable leftovers, and sugar production side streams. Therefore, pectin-rich agricultural by-products are potential cost-effective sources of functional-POS with prebiotic activity and ability to exert a number of health benefits.

Pectin is a complex and heterogeneous polysaccharide present within the primary cell wall and intercellular regions of higher plants (Chen et al., 2013). Pectin-derived oligosaccharides, often denoted as POS (pectic oligosaccharides), can vary from almost pure homogalacturonan (HG) oligosaccharides, to arabino-oligosaccharides (AOS), galactan-rich rhamnogalacturonan I (RGI), arabinan-rich RGI, arabinogalactan (AG) and include more or less well-characterized, mixed hydrolysates depending on the sources of fruits or vegetables or agro-industrial streams. Production of POS from plant cell-wall material or industrial plant biomass residues involves several steps, including pretreatment of cell-wall material,

Sustainable Production Technology in Food. https://doi.org/10.1016/B978-0-12-821233-2.00004-6
Copyright © 2021 Elsevier Inc. All rights reserved.

extraction of pectin polysaccharides, oligosaccharide (OS) release, and separation of oligosaccharides from nondegraded pectin and other plant cell-wall material, and finally purification by chromatography (Holck, Hotchkiss, Meyer, Mikkelsen, & Rastall, 2014). Several technologies have been used to extract POS from different agri-food by-products such as enzymatic hydrolysis, hydrothermal treatment, chemical process, dynamic high-pressure microfluidization, and microwave-assisted extraction, among other (Ameer, Shahbaz, & Kwon, 2017; Chen et al., 2013; Gómez, Gullón, Yáñez, Schols, & Alonso, 2016; Gómez, Yáñez, Parajó, & Alonso, 2016). POS are nondigestible oligosaccharides and as such, they affect the host beneficially by selectively stimulating the growth and/or activity of one or a limited number of bacteria in the colon (e.g., *Bifidobacteria* and *Lactobacilli*) (Manderson et al., 2005). There is an increased interest in developing new prebiotics and functional oligosaccharides from renewable bio-resources in order to find new and more efficient, functional compounds (Holck et al., 2014). Some agricultural by-products such as apple pomace, sugar beet pulp, and berry pomace contain significant amounts of pectin (Muñoz, Rodríguez-Gutiérrez, Rubio-Senent, & Fernández-Bolaños, 2011), thus they are a potential source of bioactive POS. This book chapter offers an overview of pectin and POS sources, the main methods employed for their extraction and their main biological properties.

5.2   SOURCE OF PECTIN AND PECTIN OLIGOSACCHARIDES

Pectin is a complex heteropolymer which comprehend different structural components (see Fig. 5.1) (Gullón et al., 2013; Míguez et al., 2016):

(a) **Homogalacturonan (HG)**. This polymer is made out of 70- to 100-unit length of D-galacturonic acid (GalA), including free or esterified carboxyl groups which can be partly replaced by sugars. Acetylation and methylation degrees (DA and DM) can fluctuate depending on the origin and the growth stage of the biomass. HG comprises close to the 60% of pectin.

(b) **Rhamnogalacturonan type I (RG-I)**. It is composed by alternated units of rhamnose and GalA that, in some cases, can exhibit substituent side chains composed by arabinan, galactan, or arabinogalactan I and II. Nevertheless, xylose and glucose may be found as well. It comprises up to 7%–14% of the total pectin.

(c) **Rhamnogalacturonan type II (RG-II)**. It consists of a polymer of 7–9 GalA units with complex ramifications of different monomers.

Moreover, POS can be obtained by employing distinct kind of treatments (traditionally, enzymatic, or acid hydrolysis) over the pectin. These POS are generally made up of GalA and some units of various sugars such as rhamnose, arabinose, or galactose; however, the structure depends principally on the source and the processing treatment. Due to its composition, recent

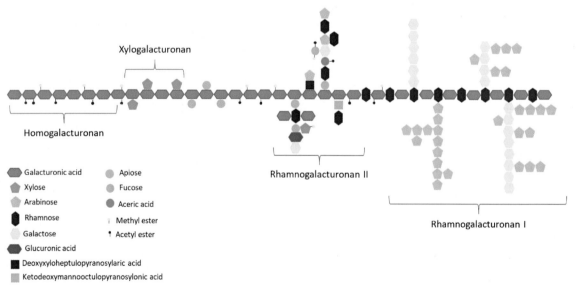

FIG. 5.1 Simplified structure of pectin.

researches classify POS as a potential and promising source of novel prebiotics with several health-promoting effects (Gómez, Míguez, Yáñez, & Alonso, 2017; Tan, Chen, Liu, Yang, & Li, 2018; Wilkowska et al., 2019; Zofou, Shu, Foba-Tendo, Tabouguia, & Assob, 2019).

In order to obtain pectin and POS, there are a lot of different fruits and vegetables that contain high levels of them, for instance apple, orange, or lemon. In spite of this, not only the pulp has high content of extractable bioactive compounds (such as pectin), but also, and often to a larger extent, their peels and other by-products (Rico, Gullón, Alonso, & Yáñez, 2020).

On the other hand, the agri-food industry generates a huge quantity of residues that creates a large loss of capital and a considerable issue in the environment (Gullón et al., 2018; Rico et al., 2020). Nonetheless, these wastes can be turned into high added-value by-products to improve the financial resources and stop the environmental drawbacks (Babbar, Dejonghe, Gatti, Sforza, & Elst, 2015; Domínguez-Rodríguez, García, Plaza, & Marina, 2019; Ismail et al., 2019). Hence, these wastes can be exploited to find new sources of bioactive compounds that are renewable and does not compete with food (Gullón et al., 2020).

Table 5.1 displays the pectin yields of a variety of by-products from the agri-food industry and the extraction methods used. The main, and more employed, sources of pectin are usually citrus peel, sugar beet pulp, potato pulp, or apple pomace (Babbar et al., 2015; Prandi et al., 2018; Tan et al., 2018). Furthermore, fruits are processed for diverse purposes, such as juice or purée, so the peels are easy to utilize for pectin production. In this case, pectin content of peels can be widely ranged between 5% to almost 33%. The lowest values correspond to salacca (5%) and passion fruit (15.2%) (Showpanish et al., 2020), whereas the highest are orange (29.4%) (Tovar et al., 2019) and tomato (32.6%) (Grassino et al., 2016). Among the fruits, citric (orange, pomelo, or lemon) peels usually show the highest pectin content, greater than 25% (Showpanish et al., 2020; Tovar et al., 2019; Wang et al., 2017). Apple pomace, which is the solid produced in cider and juice industries, can also be tapped, representing about a 20% of the processed biomass, and accounting a 15% of pectin (Wilkowska et al., 2019). Other part of the fruit that can be employed for this purpose is the endocarp of passion fruit, although it allows to achieve only about 4.5% (Talma et al., 2019). Alternatively, other discarded wastes from the agriculture industry and pruning are: broad bean pods, sweet potato residue, sunflower head, or sisal fiber (Abang Zaidel et al., 2020; Fazio et al., 2020; Maran & Priya, 2015; Tan et al., 2020).

On this basis, recent studies determined the effectiveness of various by-products for POS extraction, employing fruit wastes such as apple pomace (Wilkowska et al., 2019), lemon peel (Gómez, Gullón, et al., 2016; Gómez, Yáñez, et al., 2016), Korean citrus peel (Park et al., 2019), cranberry (Sun et al., 2019) or cas mango (Zofou et al., 2019), and other agriculture residues such as sugar beet pulp (Martínez, Gullón, Schols, Alonso, & Parajó, 2009; Martínez, Gullón, Yáñez, Alonso, & Parajó, 2009), or olive extracts (Bermúdez-Oria et al., 2019).

To summarize, diverse industrial by-products can represent an important source of pectin and POS, which have great health benefits, and can be utilized to formulate functional food by the nutraceutical industry.

5.3 POS PRODUCTION

Over the last decade, a variety of alternative methods have been proposed to extract POS from different sources (Gómez, Gullón, et al., 2016; Gómez, Yáñez, et al., 2016; Holck et al., 2014; Martínez, Yáñez, Alonso, & Parajó, 2010, 2012; Zhu et al., 2019). Methods include enzymatic hydrolysis (Gómez, Yáñez, et al., 2016; Martínez et al., 2012; Martínez, Gullón, Schols, et al., 2009; Martínez, Gullón, Yáñez, et al., 2009), hydrothermal treatment (Gómez, Gullón, et al., 2016; Martínez et al., 2010), chemical processes (Liang, Liao, Ma, Li, & Wang, 2017; Zhang, Hu, Wang, Liu, & Pan, 2018), dynamic high-pressure microfluidization (Chen et al., 2013; Chen et al., 2012; Chen, Ma, Liu, Liao, & Zhao, 2012), and microwave-assisted extraction (Ameer et al., 2017). Table 5.2 condenses information on the extraction methods previously mentioned.

Several authors have recognized that the enzymatic treatment has one of the most efficient approaches to the pectin depolymerization from different sources. Sugar beet pulp (Concha Olmos & Zúñiga Hansen, 2012; Martínez, Gullón, Schols, et al., 2009; Martínez, Gullón, Yáñez, et al., 2009), orange peel wastes (Martínez et al., 2012), lemon peel wastes (Gómez, Gullón, et al., 2016; Gómez, Yáñez, et al., 2016), apple pectin (Sabater, Ferreira-Lazarte, Montilla, & Corzo, 2019), and onion skins (Baldassarre et al., 2018) are examples of raw materials from which pectin-derived oligosaccharides have been extracted using enzymatic hydrolysis. The application of enzyme mixtures or monoactive enzymes (alone or combined) determines if the degradation of pectin is nonspecific or controlled, respectively (Holck et al., 2014). For example, Holck et al. (2011) states that a sequential use of monocomponent enzymes (pectin lyase and rhamnogalacturonan I

TABLE 5.1
Pectin Yields Extracted From Different By-products From the Food and Agricultural Industry.

Source	Parts Employed	Extraction Method	Pectin Yield (%)	Reference
Apple	Pomace	Acid (H_2SO_4, HNO_3, or HCl)	14.2–16.1	Luo, Ma, and Xu (2020)
Banana	Peel	Acid (HCl)	22.4	Ni'mah, Makhyarini, and Normalina (2020)
Broad bean	Pods	Acid (citric)	8.7–17.2	Fazio, La Torre, Dalena, and Plastina (2020)
Citron	Peel	Acid (citric)	28.3	Pasandide, Khodaiyan, Mousavi, and Hosseini (2018)
Durian	Peel	Acid (HCl)	20.24	Showpanish et al. (2020)
Eggplant	Peel	Acid (citric)	26.1	Kazemi, Khodaiyan, Hosseini, and Najari (2019)
Grapefruit	Peel	Ultrasound	23.5	Wang et al. (2017)
Jackfruit	Peel	Ultrasound and microwave	21.5	Xu et al. (2018)
Orange	Peel	Acid (H_3PO_4, HCl)	29.4	Tovar, Godínez, Espejel, Ramírez-Zamora, and Robles (2019)
Passion fruit	Peel	Acid (HNO_3)	14.4	Talma, Regis, Ferreira, Mellinger-Silva, and de Resende (2019)
Passion fruit	Peel	Acid (HCl)	15.2	Showpanish et al. (2020)
Passion fruit	Endocarp	Acid (HNO_3)	4.5	Talma et al. (2019)
Passion fruit	Pulp	Acid (HNO_3)	26.6	Talma et al. (2019)
Pomegranate	Peel	Hydrothermal	18.8–20.9	Talekar, Patti, Vijayraghavan, and Arora (2018)
Pomelo	Peel	Acid (HCl)	26.4	Showpanish et al. (2020)
Salacca	Peel	Acid (HCl)	5.0	Showpanish et al. (2020)
Sisal	Fiber	Ultrasound	29.3	Maran and Priya (2015)
Sugar beet	Pulp	–	15.0–25.0	Prandi et al. (2018)
Sunflower	Head	Acid (sodium citrate)	14.5	Tan et al. (2020)
Sweet lemon	Peel	Microwave	25.3	Rahmani, Khodaiyan, Kazemi, and Sharifan (2020)
Sweet potato	Residue	Acid (HCl)	23.5	Abang Zaidel, Ismail, Mohd Jusoh, Hashim, and Wan Azelee (2020)
Tomato	Peel	Ammonium oxalate, oxalic acid	31.9–32.6	Grassino et al. (2016)

lyase) allowed the production of homogalacturonides and rhamnogalacturonides of defined molecular size from sugar beet pectin. The literature also shows that the manufacture of POS is also dependent on temperature, time, enzyme concentration, and absence and presence of a particular enzyme (Babbar et al., 2015).

Gómez, Yáñez, et al. (2016)) confirmed that different enzymatic hydrolysis time (7.5 and 24h) yielded different pectin-derived oligosaccharides mixtures—mainly composed of oligogalacturonides—with different molar mass distribution and degree of methylation and degree of acetylation (5.5% and 8.6%, for 7.5 and

TABLE 5.2
Extraction Methods for Producing Pectin-derived Oligosaccharides From Different Sources, Experimental Conditions and Main Results Reported in the Literature.

Source	Type of Extraction	Extraction Conditions	Main Results	Reference
Orange peel waste	Enzymatic hydrolysis (Celluclast 1.5L and Viscozyme L)	37°C for 20h	GOS yield: 0.76% GalOS yield: 0.45% AraOS yield: 0.64% OgalA yield: 1.3%	Martínez et al. (2012)
Lemon peel waste	Enzymatic hydrolysis (Celluclast 1.5L, Viscozyme L and Pectinex Ultra SP-L, and Pectinase 62L)	37°C, at 150rpm, for 7.5 or 24h	*After 7.5h hydrolysis* 13.1g OGalA/100g dry matter Degree of methylation: 43.4% Degree of acetylation: 5.5% *After 24h hydrolysis* 10.8g OGalA/100g dry matter Degree of methylation: 37.4% Degree of acetylation: 8.6%	Gómez, Yáñez, et al. (2016)
Sugar beet pulp	Enzymatic hydrolysis (Celluclast 1.5L and Viscozyme L)	37°C, at 150rpm, for 4–16h	GOS yield: 5.9% GalOS yield: 2.4% AraOS yield: 9.7% OGalA yield: 8.7%	Martínez, Gullón, Yáñez, et al. (2009)
Flaxseed gum	Chemical process (Hydrogen peroxide)	120°C, 0.2M H_2O_2, for 2.0h	Acidic pyran oligosaccharide with a molecular weight of 1047Da, containing 12.31% alduronic acid	Liang et al. (2017)
Citrus peel	Chemical process (Trifluoroacetic acid and hydrogen peroxide)	Trifluoroacetic acid: 1.2M or 2M of TFA solution at 85°C, for 2.5h Hydrogen peroxide: 66.18mM H_2O_2 at 90°C for 3h 88.24mM H_2O_2, at 90°C for 4h	Galacturonic acid yield: 68.58%–89.93% using trifluoroacetic acid and 63.74%–83.26% using hydrogen peroxide Molecular weight: 3000–4000Da, 2000–3000Da and lower than 2000Da for trifluoroacetic acid and 3000–4000 and 2000–3000Da for hydrogen peroxide	Zhang et al. (2018)
Orange peel wastes	Hydrothermal	160°C	OGaU: 17.6g/L AOS: 6.6g/L GaOS: 5.52g/L	Martínez et al. (2010)
Lemon peel wastes and sugar beet pulp	Hydrothermal	160°C	*Lemon-derived POS:* OGalA: 62.4% AraOS: 13.9% GalOS: 9.4% *Sugar beet-derived POS:* AraOS: 36.2% OGalA: 33.7%	Gómez, Gullón, et al. (2016)
Potato waste	Microwave-assisted extraction	2kW single mode, ±2.5°C, ≈1min, feed flow rate of 250mL/min	Galacturonic acid yield: 40%–45%	Arrutia, Adam, Calvo-Carrascal, Mao, and Binner (2020)

Continued

TABLE 5.2
Extraction Methods for Producing Pectin-derived Oligosaccharides From Different Sources, Experimental Conditions and Main Results Reported in the Literature—cont'd

Source	Type of Extraction	Extraction Conditions	Main Results	Reference
Apple pectin	Dynamic high-pressure microfluidization	63°C, 155MPa	Pectin-derived oligosaccharides yield of 32.91%	Chen et al. (2013)
Potato pulp	Microwave-assisted alkaline extraction	1.5M KOH, microwave power of 36.0W for 2min	Galactan-rich rhamnogalacturonan I yield of 21.6%	Khodaei, Karboune, and Orsat (2016)
Sugar beet pulp	Enzymatic (Celluclast 1.5L) and nitric acid-assisted extractions followed by combined hydrolysis/fractionation	Enzymatic-assisted extraction (1): substrate loading: 12% (w/v), Celluclast: 20FPU/g, 48°C, at 160rpm, for 48h Nitric acid-assisted (2): substrate loading: 5% (w/v), pH: 1.4, 80°C, at 125rpm for 4h Both processes are followed by hydrolysis (using viscozyme) and fractionation (applying two types of membrane: MWCO of 10 or 5kDa)	Galacturonic acid concentration: (1) 8.9%–9.7% (2) 8.6%–11.8%	Prandi et al. (2018)

AraOS, arabino-oligosaccharides; *GalOS*, galacto-oligosaccharides; *GOS*, glucooligosaccharides; *OGalA*, oligogalacturonides.

24h, respectively), and yields of 13.1 and 10.8g/100g dry lemon peel wastes (for 7.5 and 24h, respectively) (Table 5.2).

However, enzymes are incapable of degrading complex structures of polysaccharides (Liang et al., 2017; Rasmussen & Meyer, 2010). This is the case of glycoside hydrolases (xylanase and cellulase) which presented inefficiency to hydrolyze flaxseed gum reported by Liang et al. (2017). As an attempt to circumvent this limitation, authors applied a chemical treatment, concretely, H_2O_2 oxidative degradation under combined effects of temperature and pressure (Table 5.2). Zhang et al. (2018) obtained pectin-derived oligosaccharides fractions from citrus peel pectin with different molecular weights (lower than 4000Da) and with prebiotic potential by means of a controlled chemical degradation either by trifluoroacetic acid and hydrogen peroxide (Table 5.2). Despite the promising results obtained through chemical treatments, some limitations are identified in terms of environmental safety and controlled degree of polymerization (Babbar et al., 2015; Kim & Rajapakse, 2005).

Hydrothermal treatment is an environmentally friendly method, faster as compared to the enzymatic hydrolysis, and with the capacity for generating satisfactory oligosaccharides yields during the extraction, thus reducing the need for additional depolymerization (Martínez et al., 2010). This extraction technique has also been successfully employed for producing pectin-derived oligosaccharides (Gómez, Gullón, et al., 2016; Martínez et al., 2010). Martínez et al. (2010) proposed an hydrothermal processing (maximum temperature of 160°C, and severity factor of 287min) to solid from water extraction to obtain mixtures of pectin-derived oligosaccharides (25.1kg/100kg of extracted orange peel wastes) from orange peel waste with mass ratios oligogalacturonides/arabinooligosaccharides/galactooligosaccharides of 10:3.8:3.2 (Table 5.2). In turn, Gómez, Gullón, et al. (2016)) applied hydrothermal treatment at a maximum temperature of

160°C and severity factor of 326min aiming at producing pectin-derived oligosaccharides with prebiotic potential from lemon peel wastes (Table 5.2).

In addition to the methods described above, physical pectin degradation by dynamic high pressure microfluidization and microwave-assisted extraction has been attracting wide research interests (Arrutia et al., 2020; Chen et al., 2013; Chen, Liang, et al., 2012; Chen, Ma, et al., 2012). Dynamic high-pressure microfluidization has been suggested by different authors as a promising alternative to produce pectin-derived oligosaccharides, since this physical method allows reducing the molecular weight of polymers (Chen et al., 2013; Holck et al., 2014). Chen et al. (2013) used this technique to produce POS from commercial apple pectin, yielding 32.91% of pectin-derived oligosaccharides (Table 5.2). Chen, Liang, et al. (2012) and Chen, Ma, et al. (2012) had also achieved the degradation of high-methoxyl pectin by using dynamic high pressure microfluidization. It was found that the extent of degradation increased with pressure, probably due to the rupture of glycosidic bonds, and with no effect on the primary structure of pectin. Recent studies suggest microwave-assisted extraction as an emerging technique to produce POS with shorter processing times, higher yields, lower energy consumption, and decreased use of solvents (Arrutia et al., 2020; Zhu et al., 2019). This method was recently applied for extracting pectin from sugar beet pulp (Mao et al., 2019), potato pulp (Arrutia et al., 2020), rape meal, and wheat straw (Budarin et al., 2015). Arrutia et al. (2020) produced pectin-derived oligosaccharides from potato pulp using continuous-flow system for microwave-assisted extraction (Table 5.2). Authors highlighted the potential of this technique to extract hairy pectins as alternative to conventional pectin extraction methods, giving a significant advantage on the preservation hairy regions.

In addition to the conventional and emerging techniques, combined methods are also described for the same purpose. Khodaei et al. (2016) applied microwave-assisted extraction in combination with alkaline extraction for the obtainment of galactan-rich rhamnogalacturonan I from potato pulp by-product. Optimal conditions were 1.5M KOH under microwave power of 36.0W for 2.0min, yielding 21.6% of galactan-rich rhamnogalacturonan I (Table 5.2). Zykwinska et al. (2008) produced pectin and POS from different food industry by-products (chicory roots, citrus peel, cauliflower florets and leaves, endive, and sugar beet pulps) using processing schemes involving ethanol extraction followed by enzymatic extractions with different enzyme mixtures. Other study was conducted by Prandi et al. (2018) using enzymatic and nitric acid-assisted extractions followed by combined hydrolysis/fractionation approaches for extracting pectin oligosaccharides from sugar beet pulp. Important findings of this study are that the fractions from nitric acid-assisted extraction (8.6g/100g dry matter—11.8g/100g dry matter) had higher concentrations of galacturonic acid than those from enzymatic extraction (8.9g/100g dry matter—9.7g/100g dry matter) (Table 5.2).

5.4 BIOLOGICAL PROPERTIES OF POS

Besides the prebiotic activity of POS, these OS are also associated with a number of other health benefits. These includes, (i) antitumor properties (Li et al., 2019; Wicker et al., 2014); (ii) antioxidant activity (Chen, Liang, et al., 2012; Chen, Ma, et al., 2012; Kang et al., 2006); (iii) immunomodulatory effects; (iv) gastro-protective activities (Chen et al., 2016); (v) lowering the serum levels of total cholesterol and triglycerides and the inhibition in the accumulation of body fat, and (vi) protection against cardiovascular diseases (Li et al., 2010). Some of them may be directly linked to their prebiotic activity and associated SCFA production. Additionally, in vivo and in vitro studies have confirmed that acidic POS are not cytotoxic or mutagenic, being suitable for their use in foods for children and babies (Garthoff et al., 2010).

There are a number of reports available on the application of POS in food and pharmaceutical industry; however, their potential in the feed industry is yet to be exploited. Gaggìa, Mattarelli, and Biavati (2010) have reviewed, in detail, the application of prebiotics in animal feeding. Nevertheless, there is only limited information available on the use of POS in animal feeds to promote the health of the animal or acting as therapeutic agents. In vitro studies have shown that POS have a potential to be used as feed additives. However, extensive in vivo studies may be required in different animal models due to the complex structure of the gastrointestinal tract and diverse microbiota (Babbar et al., 2015). Moreover, in vitro methods are absent of synergistic, antagonistic, and/or competitive effects as well as absent of an immune system. Nevertheless, the health effects and biological benefits, attributed to POS and supported by scientific evidences throughout the years, make them a potential functional ingredient. Some of the most relevant in vitro research outcomes are summarized in Table 5.3 and described throughout the next sections. Reference to a much less number to of in vivo studies are also made.

TABLE 5.3
Main Findings of In Vitro Studies With Pectin Oligosaccharides (POS) Extracted From Different Sources.

Source/Type of POS	MW/DP	Production	Main Finding(s)	Reference
Citrus pectin	NR	High pH- and temperature-modified citrus pectin (MCP)	MCP inhibits in vitro and in vivo carbohydrate-mediated angiogenesis by blocking the association of galectin-3 to its receptors	Nangia-Makker et al. (2002)
POS I and POS II	NR	POS I was obtained through controlled hydrolysis from HMP4 and POS II from LMP5	POS I and POS II better candidate prebiotics than the pectins	Olano-Martin, Gibson, and Rastall (2002)
POS I and POS II	NR	POS I was obtained through controlled hydrolysis from HMP and POS II from LMP	This study has shown that derivatives of pectin, have the ability to inhibit Shiga-like toxin, of *Escherichia coli* O157:H7	Olano-Martin, Williams, Gibson, and Rastall (2003)
POS I and POS II	NR	POS I was obtained through controlled hydrolysis from HMP and POS II from LMP	Significant reduction in attached cell numbers was observed after 3days incubation	Olano-Martin, Rimbach, Gibson, and Rastall (2003)
Mixture of POS from orange peel	NR	Thermal treatment to extract orange peel POS and further membrane filtration	Increase in *Eubacterium rectale* population and production of lactic, propionic, and butyric	Manderson et al. (2005)
Sugar beet pulp	≤10kDa	Arabinan was extracted from sugar beet pulp and subjected to enzymatic hydrolysis	Increase *Bifidobacterium*, *Bacteroids*, and *Lactobacilli* numbers and decrease of *Clostridia* numbers. Production of acetate and propionate. Microbial populations and metabolites varied as a function of molecular weight	Al-Tamimi, Palframan, Cooper, Gibson, and Rastall (2006)
Citrus fruits POS	37 and 500kDa	Citrus POS prepared by irradiation	Low-molecular-weight pectin almost maintained the activity of reducing the cholesterol content. Oligomerized pectin increased the inhibitory effect on cancer cell proliferation	Kang et al. (2006)
Mixture of bergamot peel	DP 3–7 (1400–1700kDa)	Enzymatic treatment	Promote bifidogenic microbiota. Increase of *Bifidobacterium*, *Lactobacilli* and *Eubacteria* and decrease of *Clostridia* numbers	Mandalari et al. (2007)
Valencia orange-derived POS	NR	Flash extraction with steam injection, followed by membrane filtration	POS significantly inhibited cell invasion of Caco-2 cells by *Campylobacter*, suggesting that POS could be potentially useful in the control of *Campylobacter jejuni*. as alternatives to antibiotics in animal feed	Ganan et al. (2010)

Source	Molecular weight	Process	Effects	Reference
Potato pulp	10–100 kDa (homogalacturonan); >100 kDa (rhamnogalacturonan I polysaccharides)	Destarched crude potato pulp (CPP) was heat treated and ultrafilterated in different molecular weight fractions	CPP>100 increased significantly the number of *Bifidobacterium* and CPP10–100 and CPP>100 stimulated *Lactobacillus* equally well, in comparison with FOS. These results indicate that CPP10–100 and CPP>100 have beneficial effects on the fecal microbiota composition	Thomassen, Vigsnæs, Licht, Mikkelsen, and Meyer (2011)
Sugar beet pup derived arabino-oligosaccharides (AOS)	2–10 (branched and linear AOS)	Liquid side-stream from the ultrafiltration and diafiltration step in the sequential acid extraction of pectin with nitric acid from sugar beet pulp	Increase of *Bifidobacterium* and *Lactobacilli* population and high production of acetate after fermentation with AOS of fecal samples from ulcerative colitis patients	Vigsnæs, Holck, Meyer, and Licht (2011)
Sugar beet pup derived arabino-oligosaccharides (AOS)	DP 2–14	Liquid side-stream from the ultrafiltration and diafiltration of the acid extraction of pectin from sugar beet pulp and further hydrolyzed	Both LAOS and LFAOS had high selective stimulation of *Bifidobacterium*. Lack of induced growth of *Clostridium difficile*	Holck et al. (2011)
POS mixture from apple pomace apple pomace	≤1 kDa	Apple pomace samples were subjected to simultaneous saccharification and fermentation (SSF) and further ion exchange treatment. The resultant streams were further concentrated/purified by membrane filtration	Major increase of acetic acid at 7–14 h in coincidence with the highest increase of *Bifidobacterium*, *Lactobacillus* and *Atopobium*, *Eubacterium rectale*, and *Clostridium coccoides* numbers	Gullón, Gullón, Sanz, Alonso, and Parajó (2011)
POS from Nopal	NR	Pectin extraction and hydrolysis	Increase of 16% on total SCFA production and 50% of butyrate and propionate. Reduction of putrefactive ammonium production and increase in *Bifidobacterium* population	Guevara-Arauza et al. (2012)
Apple-derived POS	NR	POS were prepared from apple pectin by dynamic high-pressure micro-fluidization (DHPM)	Promoted *Bifidobacterium* and *Lactobacilli* growth and reduced the numbers of *Bacteroids* and *Clostridia*. Production of acetic, lactic, and propionic acid	Chen et al. (2013)
POS derived from orange peel wastes (OPW)	≤1 kDa	OPW were subjected to an aqueous extraction, hydrothermal processing, auto-hydrolysis, and further concentration/purification by membrane filtration	Boost on the growth of *Bifidobacterium* and *Lactobacilli*. POS fermentation resulted in a significant increase of SCFA (acetate>butyrate>propionate)	Gómez et al. (2014)

Continued

TABLE 5.3
Main Findings of In Vitro Studies With Pectin Oligosaccharides (POS) Extracted From Different Sources—cont'd

Source/Type of POS	MW/DP	Production	Main Finding(s)	Reference
Lycium ruthenicum Murr-derived pectin	137kDa	Hot water extraction and precipitation of the polysaccharides	Promotion of macrophage proliferation	Peng, Xu, Yin, Huang, and Du (2014)
Root bark, stem bark, and leaves of *Terminalia macroptera*-derived pectins	0.8–19kDa	Water extraction and enzymatic hydrolysis	The 50°C water extracts of root bark, stem bark, and leaves of *T. macroptera* have immunomodulation properties	Zou et al. (2015)
POS mixtures from sugar beet pulp (SBPOS) and lemon peel wastes (LPOS)	≤1kDa	LPOS were obtained by hydrothermal treatment and purified by a two-step membrane process and SBPOS were submitted to an aqueous extraction followed by enzymatic hydrolysis	The joint populations of *Bifidobacterium* and *lactobacilli* increased from 19% up to 29%, and 34% in cultures with LPOS and SBPOS respectively. *Faecalibacterium* and *Roseburia* also increased their counts with all the substrates (especially with LPOS)	Gómez, Gullón, et al. (2016)
Pectin-curcumin (PEC-CCM) conjugates	35Da	Mixture of a pectin and curcumin solution	PEC-CCM conjugates obtained higher inhibitory activity against cancer cells and reduced cytotoxicity for normal cells	Bai et al. (2017)
Sedum dendroideum-derived polysaccharides	≤100kDa	Ethanol extraction followed by membrane separation	Stimulation of the secretion of the cytokines, TNF-α, IL-1β and IL-10 by THP-1 macrophages, acting as immune-stimulatory agents	de Oliveira, do Nascimento, Iacomini, Cordeiro, and Cipriani (2017)
POS mixtures starting from sugar beet pulp	<400Da to 10kDa	Enzymatic and nitric acid extraction of pectin followed by hydrolysis to tailored POS by combined hydrolysis/fractionation approaches	Low-medium molecular weight, obtained from the enzymatic pretreatment, were in general the most efficient at stimulating the *lactobacilli*, whereas the same low molecular weight fractions, obtained from nitric acid extraction, were much less efficient or even lead to inhibition. In addition, no POS fraction was able to stimulate pathogenic *Escherichia coli* strains	Prandi et al. (2018)
Jackfruit-derived pectins	~52kDa	Acid extraction with ultrasound and microwave assisted extraction	Scavenging of DPPH• and ABTS•+free radicals. The antioxidant activities of the pectin are closely correlated with their monosaccharide composition, methyl-esterification, and intrinsic viscosity	Xu et al. (2018)

Source	MW/DP	Effect	Extraction method	Reference
Pomegranate peel-derived POS	NR	Strong reducing power and good scavenging activities on superoxide anion, hydroxyl, and DPPH radicals	Enzymatic extraction	Zhai et al. (2018)
Pectins and enzymatic-modified pectins (MP) from citrus, sunflower, and artichoke	3–10kDa	Significant increase of *Bifidobacterium*, *Bacteroides/Prevotella*, *Clostridium coccoides/Eubacterium rectale Lactobacillus/Enterococcus* populations. *Clostridium histolyticum* population displayed the lowest changes in all substrates. Acetate was the most abundant SCFA, followed by propionic and butyric acids in all substrate. A positive effect of decreasing MW on fermentation was observed but not on SCFA production	Pectin was extracted by acid and enzymatic hydrolysis, treated enzymatically for MW reduction, and separated by membrane filtration	Ferreira-Lazarte, Kachrimanidou, Villamiel, Rastall, and Moreno (2018)
Ultrasonic-modified sweet potato pectin	100–741kDa	The sonicated pectin exhibited higher cell antiproliferation compared to native pectin	Cold water ultrasound extraction	Ogutu, Mu, Sun, and Zhang (2018)
Orange, lemon, lime, and sugar beet derived pectins	NR	Increase in *Bacteroidales*, genera *Prevotella*, *Faecalibacterium prausnitzii* and family *Enterobacteriaceae*. Low methoxyl citrus pectins were efficient to decrease proportions of *P. copri*, (linked with insulin resistance) and increase the levels of *Coprococcus* (linked to reduced IBS)	Pectins were harsh, mild, and differentially extracted and in some cases with additional chemic or enzymatic de-esterification or amidated	Larsen et al. (2019)
Apple pomace (AP)	DP 2–10	All POS tested (DP 2–10) stimulated bowel colonization with lactic acid bacteria and inhibit the development of infections caused by pathogens. The main metabolites were lactic acid and acetic acid	Mild acid combined with enzymatic hydrolysis (P1) and enzymatic hydrolysis only (P2)	Wilkowska et al. (2019)
Cheongkyool peels (CCE)-derived polysaccharides	62kDa	CCE-I (RGI-type) highly enhanced the production of IL-6, TNF-α, and NO in RAW 264.7 cell lines	Enzymatic hydrolysis, precipitation, and dialysis	Park et al. (2019)

CPP>100, crude potato pulp, fiber released by pectin lyase and polygalacturonase, fraction >100kDa; *CPP10-100*, crude potato pulp, fiber released by pectin lyase and polygalacturonase, fraction 10–100kDa; *DP*, degree of polymerization; *FOS*, fructo-oligosaccharides; *HMP*, high methylated citrus pectin with 66% methylation; *LAOS*, arabino-oligosaccharides with high molecular weight; *LFAOS*, feruloylated arabino-oligosaccharides with high molecular weight; *LMP*, low methylated apple pectin with 8% methylation; *LPOS*, POS mixtures from lemon peel wastes; *MW*, molecular weight; *NR*, not reported; *SBPOS*, POS mixtures from sugar beet pulp.

5.4.1 Prebiotic Effect

The concept of prebiotics was first proposed by Gibson and Roberfroid (1995) in response to the fact that non-digestible food ingredients (almost exclusively until now nondigestible oligosaccharides), after reaching the colon, are selectively fermented by one or a limited number of the colonic bacteria known to have positive effects on gut physiology. Being selectively fed by a preferential substrate, these bacteria get a proliferative advantage over the others (Roberfroid, 1996). As food ingredients, prebiotics have an acceptable odor and are low-calorie, which allows their utilization in antiobesity diets (Chen et al., 2013). POS have been suggested as a new class of prebiotics, which are capable of exerting a number of health-promoting effects (Gullón et al., 2013). The colonic fermentation of POS, as well as from other dietary fiber (DF), generate SCFA, mainly acetate, propionate, and butyrate. On its hand, these SCFA exert a number of health effects, including:

(i) a key role in the prevention and treatment of the metabolic syndrome, bowel disorders, and cancer (Donohoe et al., 2011)

(ii) protection against diet-induced obesity and regulation of the gut hormones (Lin et al., 2012), and

(iii) a positive effect on the treatment of ulcerative colitis, Crohn's disease, and antibiotic-associated diarrhea and obesity (Binder, 2010).

Several studies have shown that POS can be used effectively as a prebiotic by promoting the growth of *Bifidobacterium* and *Eubacterium rectale*, with the subsequent increase in butyrate concentrations (Manderson et al., 2005), increasing *Lactobacilli* numbers and reducing *Clostridia* (Mandalari et al., 2007) and *Bacteroides* (Chen et al., 2013).

Evidences on the link between the prebiotic effect of pectins and POS and their physicochemical characteristics have been previous referred (Gullón et al., 2009). For instance, POS with low degree of methylation have shown higher prebiotic index than high degree of methylation POS (Olano-Martin et al., 2002) and low molecular weight AOS provoked higher increases in *Bifidobacterium* than other AOS (Al-Tamimi et al., 2006). Larsen et al. (2019) provided evidence that modulation of the gut microbiota by pectins depended on their structural features as specific bacterial populations were differentially affected. The authors suggested that the main factors, linked to differences in microbiota composition, are:

(i) Degree of esterification (DE) of polygalacturonic acid;

(ii) Composition of neutral sugars;

(iii) Distribution of RG and HG fractions;

(iv) Degree of branching (DBr);

(v) Modification of pectic backbone, e.g., by amidation.

On the other hand, both pectin and enzymatically-modified pectin from sunflower and artichoke by-products have shown similar ability to promote the in vitro growth of beneficial gut bacteria as *Bifidobacterium* and *Lactobacillus*, in comparison to well-recognized prebiotics, namely inulin and FOS. In addition, no significant effects of the molecular weight of pectin samples on SCFA production were observed. Likewise, the degree of methoxylation did not have any significant impact on the fermentability or SCFA production, regardless of the origin of the pectic compounds (Ferreira-Lazarte et al., 2018). Also POS preparations, containing structurally different oligosaccharides (DPs 2–10 with different ratios of higher order oligosaccharides, as well as DPs 2–6), have stimulated bowel colonization with lactic acid bacteria and inhibit the development of infections caused by pathogens (Wilkowska et al., 2019). Nevertheless, understanding the interplay between the gut commensals and the structural properties of pectins is essential to predict the physiological effects of ingested pectins and to provide ideas for development of pectin-containing DF, targeting beneficial bacteria to facilitate more balanced microbiota profiles. Additionally, further comparative in vitro and in vivo studies with structurally diverse pectins and their derivatives are needed to achieve the detailed knowledge of structure-function relationship of pectins in the gut (Larsen et al., 2019).

5.4.2 Antidiabetic, Anticholesterolemic and Antiobesity Properties

POS have been regarded to have antiobesity activities (Jiang et al., 2016; Šefcíkova & Raček, 2016), via regulating appetite and satiety signals (Pelkman, Navia, Miller, & Pohle, 2007). An earlier report shows that POS can significantly decrease lipid accumulation by affecting lipid metabolism (Story, 1985). This may be associated to the decrease of sterol regulatory element-binding protein and low density lipoprotein (LDL) receptor, suggesting that POS improves lipid metabolism (Palou, Sánchez, García-Carrizo, Palou, & Picó, 2015). Leptin or obesity receptor (Ob-R) can induce cardiac disorders (Wu & Sun, 2017) and is also linked with obesity development. Long-term pectin consumption can remarkably reduce lipid contents and decrease insulin and leptin resistance (Palou et al., 2015). POS and pectin diets have been shown to significantly suppress weight gain in mice (Li et al., 2010) and reduce plasma leptin significantly by more than 60% in an obesity animal model (Adam et al., 2016). Therefore, by binding Ob-R, POS may be a potential antiobesity ingredient.

POS have been also used to treat hypercholesterolemia (Hosobuchi, Rutanassee, Bassin, & Wong, 1999) and have been shown to have an important role on cholesterol lowering (Brouns et al., 2012). Several studies have shown that pectins improved glucose tolerance, hepatic glycogen content, and blood lipid levels in diabetic rats, in addition to significantly reducing insulin resistance, which plays an important role in the resulting antidiabetic effect (Liu, Dong, Yang, & Pan, 2016) and therefore may be considered as potential therapeutic agents for the treatment of diabetes mellitus (Chen et al., 2017). In 2016, Food Standards Australia New Zealand (FSANZ) stated that the intake of 10–14.5g pectin in a meal significantly reduced peak postprandial blood glucose concentration. However, the focus of this review was on lower intakes, hence it was concluded that such high pectin-doses are usually not found in foods eaten in Australia and New Zealand in a single meal (Food Standards Australia New Zealand, 2016). The European Union (EU), which regulates food health claims, only authorizes the use of two pectin-related health claims. The claim on the reduction of blood glucose rise after a meal may be used only for food, which contains 10g of pectins per quantified portion. And the claim on maintenance of blood cholesterol levels may be used only for food which provides a daily intake of 6g of pectins (European Commision, 2012).

5.4.3 Antitumor Property

The antitumor mechanisms of dietary pectin may be the result of their prebiotic activity (Flint, Bayer, Rincon, Lamed, & White, 2008), immune-potentiation (Chen et al., 2006), tumor growth inhibition (Cheng et al., 2011) and antimutagenic potential (Hensel & Meier, 1999). A number of in vitro and in vivo studies on the antitumor activity of native and modified pectin revealed a decrease of adhesion and cell proliferation, as well as the induction of apoptosis and migration (Bush, 2014). POS from flowers of *Lonicera japonica* revealed high potential for pancreatic cancer treatment (Delphi & Sepehri, 2016; Lin et al., 2016) and verified that pectic acid from apple presented apoptotic activity and was able to inhibit breast tumor cells growth in vivo. Ogutu et al. (2018) suggested that sweet potato pectin possesses anticancer activity and to induce the apoptosis of colon cancer (CC) cells and larch arabinogalactan (a kind of POS) has been reported to inhibit p38 phosphorylation in mitogen-activated protein kinases (MAPK) pathways (Lim, 2017), suggesting that POS may prevent the risk or progression of CC by suppressing MAPK signaling pathway. The MAPK signaling pathway plays an important role in most immune responses (Kakavand et al., 2017). According to Ma et al. (2018), downregulation of MAPK signaling pathway can inhibit the proliferation, invasion, and angiogenesis of CC. Moreover, the activation of MAPK signaling pathway will increase antioxidant activities (Choi et al., 2017) properties, which will result in the apoptosis of CC cells (Ajayi, Adedara, & Farombi, 2016).

The majority of the antitumor studies of dietary pectin have focused on CC due to its activity within the gastrointestinal tract and even more with its fermentation in the colon, impact on probiotic bacteria, and production of SCFA. Thus, POS anticarcinogenic mechanisms may be directly correlated with its prebiotic activity.

Although natural pectin is reported to prevent colon cancer as a DF. To enhance its bioavailability and bioactivity, pectin is often modified into bioavailable modified pectin fragments (MPs) with low molecular mass (Zhang, Xu, & Zhang, 2015). However, studies have reported that the bioactivity, namely anticarcinogenic, depends on both the extraction procedure and the cell line. Thus, different extracts might be more efficacious with particular cell lines that represent a type of cancer (Cobs-Rosas, Concha-Olmos, Weinstein-Oppenheimer, & Zúñiga-Hansen, 2015). Leclere, Van Cutsem, and Michiels (2013) showed that pH-modified pectin as well as galactan-rich pectin are capable of inhibiting cell-cell interactions and cancer cell metastasis and heat-modified pectin initiates apoptosis in cancer cells. According to Almeida et al. (2015), modified pectin (derivatization with maleoyl groups) had higher antitumor activity for CC than nonmodified pectin. Cobs-Rosas et al. (2015) compared the antiproliferative effect of acidic and neutral pectins obtained from rapeseed cake using different extraction methods and concluded that the different methods of extraction yield products that might differ on their activity when tested on certain cell lines. The pectic substances exhibited different antiproliferative activity on MCF-7 breast cancer cell line and Caco2 colorectal carcinoma cell line and the authors suggested that even though there are more scientific reports on the antiproliferative activity for acid pectins, there is a promise for neutral pectins as antiproliferative, anticancer, or chemopreventive products with application for the pharmaceutical and nutraceutical industry.

5.4.4 Antioxidant, Antiinflammatory, and Antimicrobial Properties

Oxidation is vital to plenty of organisms that can generate energy to supply biological processes. In normal circumstances, free radicals govern cell growth, and suppress viruses and bacteria. Nevertheless, in large quantities and without regulation, the production of free radicals induced by oxygen cause cell damage, which renders the pathological progressions. The oxidative stress (OxS) is associated with chronic obstructive pulmonary disease, asthma, diabetes, inflammation, cardiovascular diseases, and myocardial infarction (Carneiro et al., 2013; Wang, Hu, Nie, Yu, & Xie, 2016). Pectin oligosaccharides, as soluble DF with various health-promoting functions, have a good potential in controlling OxS and inflammatory situation by affecting antioxidant and antiinflammatory mediated signaling pathways (Tan et al., 2018). POS exert its antioxidant properties by significantly increasing the levels of antioxidant biomarkers while reducing oxidative biomarkers (Koriem, Arbid, & Emam, 2014). The redox system may be regulated by POS, through the normalization of the activity of glutathione reductase (GR) and glutathione peroxidase (GPx) (Khasina, Kolenchenko, Sgrebneva, Kovalev, & Khotimchenko, 2003). Several natural herbs and plants-derived POS have been studied regarding to their antioxidant capacity. Yerba mate (YM) (*Ilex paraguariensis*) polysaccharide demonstrated to possess antioxidant effect in addition to a significant antimicrobial activity against definite fungal and bacterial strains (Kungel et al., 2018). A high-purity polysaccharide from *Ligusticum chuanxiong* showed effective protection against the oxidative action of the intestinal stress (Huang et al., 2017). Pomegranate peel POS extracted by pectinase demonstrated its scavenging influence on hydroxyl radicals, 2,2-diphenyl-1-picrylhydrazyl (DPPH) radicals, and superoxide anion radicals (Zhai et al., 2018). Pectins from jackfruit peel (Xu et al., 2018), pectic polysaccharide from tangerine peel (Chen et al., 2016), Jerusalem artichoke (Liu, Shi, Xu, & Yi, 2016), and a pectin film with acerola extract (Eça, Machado, Hubinger, & Menegalli, 2015) were suggested as natural antioxidants that could be used in the preparation of food.

The pathology of inflammation is a complicated process triggered by microbial pathogens, such as viruses, bacteria, prion, and fungi (de Bezerra et al., 2018; Vitaliti, Pavone, Mahmood, Nunnari, & Falsaperla, 2014). Macrophages account for the first defense line of the human body. Various inflammatory moderators are produced by macrophages that are influenced by inflammatory mediator nitric oxide and lipopolysaccharides (LPS), including cytokines such as tumor necrosis factor alpha (TNF-α) and interleukin-1β (IL-1β) (Lee et al., 2014). POS have been demonstrated to have antiinflammatory properties through a number of different pathways, which may be directly or indirectly influenced by these compounds. A native pectin extracted from sweet pepper fruits was able to control the TNF-α, IL-1β, and interleukin-10 (IL-10) secretion using a spontaneously immortalized monocyte-like cell line—THP1—macrophages (do Nascimento, Winnischofer, Ramirez, Iacomini, & Cordeiro, 2017). Also, de Oliveira et al. (2017) showed that polysaccharides derived from *Sedum dendroideum* stimulate secretion of the cytokines TNF-α, IL-1β and IL-10 by THP1 macrophages, acting as immune-stimulatory agents. On the other hand, they observed a reduction of TNF-α and IL1-β secretion, induced by a pro-inflammatory agent (i.e., LPS), showing antiinflammatory effect. It is known that the structure of pectic polysaccharides (including size, composition, and branching pattern) can vary, depending on their source and the pectin structure has been previously reported to influence its bioactivity (Minzanova et al., 2018). For instance, Popov et al. (2013) showed that low methyl-esterified citrus pectin inhibits local and systemic inflammation, while pectin with a higher degree of esterification can inhibit intestinal inflammation. Therefore, the structure-function relationship has to be taken into account when studying POS bioactivities. POS extracted from the leaves of *Sedum dendroideum* displayed their influence on the secretion of pro-inflammatory and antiinflammatory cytokines by macrophages (de Oliveira et al., 2017). de Bezerra et al. (2018) investigated the antiinflammatory properties of several polysaccharides structures in wines comprising mannan, type II AG and traces of type I and II RG. The pooled fractions and fractions with extracted polysaccharides produced an antiinflammatory effect on the LPS-induced mouse macrophage cells in vitro. These effects were promoted by reducing inflammatory cytokines (TNF-α and IL-1β). YM-derived polysaccharide comprising mainly RG-I demonstrated antiinflammatory effect by its capacity to reduce the tissue expression of inducible nitric oxide synthase (iNOS) and cyclooxygenase-2 (COX-2) (Dartora et al., 2013).

There are evidences that POS increase natural killer bioactivity and the levels of antiinflammatory cytokines (Ye & Lim, 2010) and reduce the levels of proinflammatory cytokines (Tan et al., 2018). In addition, this study showed that the degree of methyl esterification, molecular size, and the characteristics of pectin structure were closely associated with the regulation of cytokine. All the cytokines can be inhibited by preventing the activity of signal transducer and activator of transcription (STAT) pathway in macrophages (Lee, Lee, Shin, Jang, & Lee, 2017). Thus, POS may affect the release of cytokines by regulating STAT signaling pathway (Tan et al., 2018). The STATs are proteins with the dual function of transducing signals from the cell surface to the nucleus and activating transcription of genes. Many cytokines, hormones, and growth factors use STAT signaling pathways to control a remarkable variety of biological responses, including development, differentiation, cell proliferation, and survival (Lee & Gao, 2005). Viral replication and inflammation are also associated with STAT pathway and its inactivation seems to improve antiinflammatory activities (Nunes, Almeida, Barbosa, & Laranjinha, 2017). Therefore, POS may have an important role in the inflammatory process by STATs regulation.

Regarding their antimicrobial properties, POS have been reported to suppress the activity of enteroputrefactive and a number of pathogenic organisms (Manderson et al., 2005), such as *Listeria monocytogenes and Staphylococcus aureus* (Nisar et al., 2018), yeast-like fungus *Candida albicans* (Minzanova et al., 2018) and *Salmonella* (Zofou et al., 2019). In addition, POS have been shown to be relevant in the treatment of several infectious diseases, including *Helicobacter pylori-related illnesses* (Zou et al., 2015) and able to alleviate sepsis (Kungel et al., 2018). However, the application of pectin-derived compounds is determined by its chemical features, including galacturonic acid and methoxyl content and degree of acetylation (Bush, 2014), as their antimicrobial activity was influenced by the extent of pectin esterification. For instance, pectins containing more than 80% of galacturonic acid residues suppress the activity of macrophages and inhibit the delayed-type hypersensitivity reaction. In addition, the branched region of the pectin macromolecule mediates the stimulation of phagocytosis and the increased production of antibodies (Popov et al., 2013). Krivorotova, Staneviciene, Luksa, Serviene, and Sereikaite (2016) demonstrated that the significant antimicrobial activity of nisin-loaded pectin-inulin particles corresponds to a low or zero degree of pectin esterification, while the particles of a more considerable degree of pectin esterification by contrast decreased the antimicrobial activity. Moreover, some studies suggested that conjugating synergistically pectin derived compounds with other functional ingredients, may enhance their benefic power, e.g., by improving stability and solubility (Bai et al., 2017). By creating a synergistic effect between different compounds, these pectin conjugates with cross-linking and antioxidant characteristics possess new applications that have been usefully employed in the pharmaceutical, food, biomedicine, and cosmetic fields (Ahn, Halake, & Lee, 2017).

5.5 CONCLUSIONS

The extraction of neutral and acidic polymers in the form of POS is a promising step towards the manufacture of prebiotics from agricultural by-products, with applications in food, feed, and pharmaceutical industry, will set new directions for future research. POS, either naturally extracted, modified, or conjugated, have a great potential as functional ingredient by exerting a number of bioactive properties. These compounds could be valuable in the development of nutritional and drug therapies to improve several health conditions. It is also important to conduct extensive research on the application of POS as bio-preservatives, due to their antimicrobial properties they may be used as antimicrobial edible films for coatings in the food industry. POS may be a valuable ingredient for animal feed, as their contribution will have direct impact animal products quality.

Over the past years, extensive research has proven the several health beneficial effects of POS; however, more in vivo studies may be required to validate such claims, especially due to the absence of synergistic, antagonistic, competitive effects and immune system when performing in vitro methods.

ACKNOWLEDGMENTS

Pablo G. del Río and Beatriz Gullón would like to express their gratitude to the Ministry of Science, Innovation and Universities of Spain for his FPU research grant (FPU16/04077) and her RYC grant (RYC2018-026177-I), respectively. Thanks to GAIN (Axencia Galega de Innovación) for supporting this study (grant number IN607A2019/01).

REFERENCES

Abang Zaidel, D. N., Ismail, N. H., Mohd Jusoh, Y. M., Hashim, Z., & Wan Azelee, N. I. (2020). Optimization of sweet potato pectin extraction using hydrochloric acid. *IOP Conference Series: Materials Science and Engineering, 736*(2). https://doi.org/10.1088/1757-899X/736/2/022042, 022042.

Adam, C. L., Gratz, S. W., Peinado, D. I., Thomson, L. M., Garden, K. E., Williams, P. A., et al. (2016). Effects of dietary fibre (pectin) and/or increased protein (casein or pea) on satiety, body weight, adiposity and caecal fermentation in high fat diet-induced obese rats. *PLoS One, 11*(5), 1–16. https://doi.org/10.1371/journal.pone.0155871.

Ahn, S., Halake, K., & Lee, J. (2017). Antioxidant and ion-induced gelation functions of pectins enabled by polyphenol conjugation. *International Journal of Biological Macromolecules, 101*, 776–782. https://doi.org/10.1016/j.ijbiomac.2017.03.173.

Ajayi, B. O., Adedara, I. A., & Farombi, E. O. (2016). Benzo (a) pyrene induces oxidative stress, pro-inflammatory cytokines, expression of nuclear factor-kappa B and deregulation of wnt/beta-catenin signaling in colons of BALB/c mice. *Food and Chemical Toxicology*, 95, 42–51. https://doi.org/10.1016/j.fct.2016.06.019.

Almeida, E. A. M. S., Facchi, S. P., Martins, A. F., Nocchi, S., Schuquel, I. T. A., Nakamura, C. V., et al. (2015). Synthesis and characterization of pectin derivative with antitumor property against Caco-2 colon cancer cells. *Carbohydrate Polymers*, 115, 139–145. https://doi.org/10.1016/j.carbpol.2014.08.085.

Al-Tamimi, M. A. H. M., Palframan, R. J., Cooper, J. M., Gibson, G. R., & Rastall, R. A. (2006). *In vitro* fermentation of sugar beet arabinan and arabino-oligosaccharides by the human gut microflora. *Journal of Applied Microbiology*, 100(2), 407–414. https://doi.org/10.1111/j.1365-2672.2005.02780.x.

Ameer, K., Shahbaz, H. M., & Kwon, J. H. (2017). Green extraction methods for polyphenols from plant matrices and their byproducts: A review. *Comprehensive Reviews in Food Science and Food Safety*, 16(2), 295–315. https://doi.org/10.1111/1541-4337.12253.

Arrutia, F., Adam, M., Calvo-Carrascal, M.Á., Mao, Y., & Binner, E. (2020). Development of a continuous-flow system for microwave-assisted extraction of pectin-derived oligosaccharides from food waste. *Chemical Engineering Journal*, 395, 125056. https://doi.org/10.1016/j.cej.2020.125056.

Babbar, N., Dejonghe, W., Gatti, M., Sforza, S., & Elst, K. (2015). Pectic oligosaccharides from agricultural byproducts: Production, characterization and health benefits. *Critical Reviews in Biotechnology*, 36(4), 594–606. https://doi.org/10.3109/07388551.2014.996732.

Bai, F., Diao, J., Wang, Y., Sun, S., Zhang, H., Liu, Y., et al. (2017). A new water-soluble nanomicelle formed through self-assembly of pectin-curcumin conjugates: Preparation, characterization, and anticancer activity evaluation. *Journal of Agricultural and Food Chemistry*, 65(32), 6840–6847. https://doi.org/10.1021/acs.jafc.7b02250.

Baldassarre, S., Babbar, N., Van Roy, S., Dejonghe, W., Maesen, M., Sforza, S., et al. (2018). Continuous production of pectic oligosaccharides from onion skins with an enzyme membrane reactor. *Food Chemistry*, 267, 101–110. https://doi.org/10.1016/j.foodchem.2017.10.055.

Bermúdez-Oria, A., Rodríguez-Gutiérrez, G., Alaiz, M., Vioque, J., Girón-Calle, J., & Fernández-Bolaños, J. (2019). Polyphenols associated to pectic polysaccharides account for most of the antiproliferative and antioxidant activities in olive extracts. *Journal of Functional Foods*, 62(August), 103530. https://doi.org/10.1016/j.jff.2019.103530.

Binder, H. J. (2010). Role of colonic short-chain fatty acid transport in diarrhea. *Annual Review of Physiology*, 72(1), 297–313. https://doi.org/10.1146/annurev-physiol-021909-135817.

Brouns, F., Theuwissen, E., Adam, A., Bell, M., Berger, A., & Mensink, R. P. (2012). Cholesterol-lowering properties of different pectin types in mildly hyper-cholesterolemic men and women. *European Journal of Clinical Nutrition*, 66(5), 591–599. https://doi.org/10.1038/ejcn.2011.208.

Budarin, V. L., Shuttleworth, P. S., De Bruyn, M., Farmer, T. J., Gronnow, M. J., Pfaltzgraff, L., et al. (2015). The potential of microwave technology for the recovery, synthesis and manufacturing of chemicals from bio-wastes. *Catalysis Today*, 239, 80–89. https://doi.org/10.1016/j.cattod.2013.11.058.

Bush, P. (2014). *Pectin: Chemical properties, uses and health benefits (Food science and technology)*. N. S. P. Inc. (ed.).

Carneiro, A. A. J., Ferreira, I. C. F. R., Dueñas, M., Barros, L., Da Silva, R., Gomes, E., et al. (2013). Chemical composition and antioxidant activity of dried powder formulations of *Agaricus blazei* and *Lentinus edodes*. *Food Chemistry*, 138(4), 2168–2173. https://doi.org/10.1016/j.foodchem.2012.12.036.

Chen, R., Jin, C., Tong, Z., Lu, J., Tan, L., Tian, L., et al. (2016). Optimization extraction, characterization and antioxidant activities of pectic polysaccharide from tangerine peels. *Carbohydrate Polymers*, 136, 187–197. https://doi.org/10.1016/j.carbpol.2015.09.036.

Chen, J., Liang, R. H., Liu, W., Li, T., Liu, C. M., Wu, S. S., et al. (2013). Pectic-oligosaccharides prepared by dynamic high-pressure microfluidization and their *in vitro* fermentation properties. *Carbohydrate Polymers*, 91(1), 175–182. https://doi.org/10.1016/j.carbpol.2012.08.021.

Chen, J., Liang, R. H., Liu, W., Liu, C. M., Li, T., Tu, Z. C., et al. (2012). Degradation of high-methoxyl pectin by dynamic high pressure microfluidization and its mechanism. *Food Hydrocolloids*, 28(1), 121–129. https://doi.org/10.1016/j.foodhyd.2011.12.018.

Chen, G. T., Ma, X. M., Liu, S. T., Liao, Y. L., & Zhao, G. Q. (2012). Isolation, purification and antioxidant activities of polysaccharides from *Grifola frondosa*. *Carbohydrate Polymers*, 89(1), 61–66. https://doi.org/10.1016/j.carbpol.2012.02.045.

Chen, C. H., Sheu, M. T., Chen, T. F., Wang, Y. C., Hou, W. C., Liu, D. Z., et al. (2006). Suppression of endotoxin-induced proinflammatory responses by citrus pectin through blocking LPS signaling pathways. *Biochemical Pharmacology*, 72(8), 1001–1009. https://doi.org/10.1016/j.bcp.2006.07.001.

Chen, Q., Zhu, L., Tang, Y., Zhao, Z., Yi, T., & Chen, H. (2017). Preparation-related structural diversity and medical potential in the treatment of diabetes mellitus with ginseng pectins. *Annals of the New York Academy of Sciences*, 1401(1), 75–89. https://doi.org/10.1111/nyas.13424.

Cheng, H., Li, S., Fan, Y., Gao, X., Hao, M., Wang, J., et al. (2011). Comparative studies of the antiproliferative effects of ginseng polysaccharides on HT-29 human colon cancer cells. *Medical Oncology*, 28(1), 175–181. https://doi.org/10.1007/s12032-010-9449-8.

Choi, S. I., Lee, J. H., Kim, J. M., Jung, T. D., Cho, B. Y., Choi, S. H., et al. (2017). *Ulmus macrocarpa* hance extracts attenuated H_2O_2 and UVB-induced skin photo-aging by activating antioxidant enzymes and inhibiting MAPK pathways. *International Journal of Molecular Sciences*, 18(6), 1200. https://doi.org/10.3390/ijms18061200.

Cobs-Rosas, M., Concha-Olmos, J., Weinstein-Oppenheimer, C., & Zúñiga-Hansen, M. E. (2015). Assessment of antiproliferative activity of pectic substances obtained by different extraction methods from rapeseed cake on cancer cell lines. *Carbohydrate Polymers, 117*, 923–932. https://doi.org/10.1016/j.carbpol.2014.10.027.

Concha Olmos, J., & Zúñiga Hansen, M. E. (2012). Enzymatic depolymerization of sugar beet pulp: Production and characterization of pectin and pectic-oligosaccharides as a potential source for functional carbohydrates. *Chemical Engineering Journal, 192*, 29–36. https://doi.org/10.1016/j.cej.2012.03.085.

Dartora, N., De Souza, L. M., Paiva, S. M. M., Scoparo, C. T., Iacomini, M., Gorin, P. A. J., et al. (2013). Rhamnogalacturonan from *Ilex paraguariensis*: A potential adjuvant in sepsis treatment. *Carbohydrate Polymers, 92*(2), 1776–1782. https://doi.org/10.1016/j.carbpol.2012.11.013.

de Bezerra, I. L., Caillot, A. R. C., Palhares, L. C. G. F., Santana-Filho, A. P., Chavante, S. F., & Sassaki, G. L. (2018). Structural characterization of polysaccharides from cabernet franc, cabernet sauvignon and sauvignon blanc wines: Anti-inflammatory activity in LPS stimulated RAW 264.7 cells. *Carbohydrate Polymers, 186*, 91–99. https://doi.org/10.1016/j.carbpol.2017.12.082.

de Oliveira, A. F., do Nascimento, G. E., Iacomini, M., Cordeiro, L. M. C., & Cipriani, T. R. (2017). Chemical structure and anti-inflammatory effect of polysaccharides obtained from infusion of *Sedum dendroideum* leaves. *International Journal of Biological Macromolecules, 105*, 940–946. https://doi.org/10.1016/j.ijbiomac.2017.07.122.

Delphi, L., & Sepehri, H. (2016). Apple pectin: A natural source for cancer suppression in 4T1 breast cancer cells *in vitro* and express p53 in mouse bearing 4T1 cancer tumors, *in vivo. Biomedicine and Pharmacotherapy, 84*, 637–644. https://doi.org/10.1016/j.biopha.2016.09.080.

do Nascimento, G. E., Winnischofer, S. M. B., Ramirez, M. I., Iacomini, M., & Cordeiro, L. M. C. (2017). The influence of sweet pepper pectin structural characteristics on cytokine secretion by THP-1 macrophages. *Food Research International, 102*, 588–594. https://doi.org/10.1016/j.foodres.2017.09.037.

Domínguez-Rodríguez, G., García, M. C., Plaza, M., & Marina, M. L. (2019). Revalorization of *Passiflora* species peels as a sustainable source of antioxidant phenolic compounds. *Science of the Total Environment, 696*, 134030. https://doi.org/10.1016/j.scitotenv.2019.134030.

Donohoe, D. R., Garge, N., Zhang, X., Sun, W., O'Connell, T. M., Bunger, M. K., et al. (2011). The microbiome and butyrate regulate energy metabolism and autophagy in the mammalian colon. *Cell Metabolism, 13*(5), 517–526. https://doi.org/10.1016/j.cmet.2011.02.018.

Eça, K. S., Machado, M. T. C., Hubinger, M. D., & Menegalli, F. C. (2015). Development of active films from pectin and fruit extracts: Light protection, antioxidant capacity, and compounds stability. *Journal of Food Science, 80*(11), C2389–C2396. https://doi.org/10.1111/1750-3841.13074.

European Commission. (2012). *Commission Regulation (EU) No of 16 May 2012 establishing a list of permitted health claims made on foods, other than those referring to the reduction of disease risk and to children's development and health.* Official Journal of the European Union.

Fazio, A., La Torre, C., Dalena, F., & Plastina, P. (2020). Screening of glucan and pectin contents in broad bean (*Vicia faba* L.) pods during maturation. *European Food Research and Technology, 246*(2), 333–347. https://doi.org/10.1007/s00217-019-03347-4.

Ferreira-Lazarte, A., Kachrimanidou, V., Villamiel, M., Rastall, R. A., & Moreno, F. J. (2018). *In vitro* fermentation properties of pectins and enzymatic-modified pectins obtained from different renewable bioresources. *Carbohydrate Polymers, 199*, 482–491. https://doi.org/10.1016/j.carbpol.2018.07.041.

Flint, H. J., Bayer, E. A., Rincon, M. T., Lamed, R., & White, B. A. (2008). Polysaccharide utilization by gut bacteria: Potential for new insights from genomic analysis. *Nature Reviews Microbiology, 6*(2), 121–131. https://doi.org/10.1038/nrmicro1817.

Food Standards Australia New Zealand. (2016). *Systematic review of the evidence for a relationship between phytosterols and blood cholesterol. November,* 44.

Gaggìa, F., Mattarelli, P., & Biavati, B. (2010). Probiotics and prebiotics in animal feeding for safe food production. *International Journal of Food Microbiology, 141*, S15–S28. https://doi.org/10.1016/j.ijfoodmicro.2010.02.031.

Ganan, M., Collins, M., Rastall, R., Hotchkiss, A. T., Chau, H. K., Carrascosa, A. V., et al. (2010). Inhibition by pectic oligosaccharides of the invasion of undifferentiated and differentiated Caco-2 cells by *Campylobacter jejuni. International Journal of Food Microbiology, 137*(2), 181–185. https://doi.org/10.1016/j.ijfoodmicro.2009.12.007.

Garthoff, J. A., Heemskerk, S., Hempenius, R. A., Lina, B. A. R., Krul, C. A. M., Koeman, J. H., et al. (2010). Safety evaluation of pectin-derived acidic oligosaccharides (pAOS): Genotoxicity and sub-chronic studies. *Regulatory Toxicology and Pharmacology, 57*(1), 31–42. https://doi.org/10.1016/j.yrtph.2009.12.004.

Gibson, G. R., & Roberfroid, M. B. (1995). Dietary modulation of the human colonic microbiota: Introducing the concept of prebiotics. *The Journal of Nutrition, 125*(6), 1401–1412. https://doi.org/10.1093/jn/125.6.1401.

Gómez, B., Gullón, B., Remoroza, C., Schols, H. A., Parajó, J. C., & Alonso, J. L. (2014). Purification characterization, and prebiotic properties of pectic oligosaccharides from orange peel. *Journal of Agricultural and Food Chemistry, 62*, 9769–9782. https://doi.org/10.1021/jf503475b.

Gómez, B., Gullón, B., Yáñez, R., Schols, H., & Alonso, J. L. (2016). Prebiotic potential of pectins and pectic oligosaccharides derived from lemon peel wastes and sugar beet pulp: A comparative evaluation. *Journal of Functional Foods, 20*, 108–121. https://doi.org/10.1016/j.jff.2015.10.029.

Gómez, B., Míguez, B., Yáñez, R., & Alonso, J. L. (2017). Extraction of oligosaccharides with prebiotic properties from agro-industrial wastes. In *Vol. 3. Water extraction of bioactive compounds* (pp. 131–161). Elsevier. https://doi.org/10.1016/B978-0-12-809380-1.00005-X. Issue 2.

Gómez, B., Yáñez, R., Parajó, J. C., & Alonso, J. L. (2016). Production of pectin-derived oligosaccharides from lemon peels by extraction, enzymatic hydrolysis and membrane filtration. *Journal of Chemical Technology and Biotechnology*, *91*(1), 234–247. https://doi.org/10.1002/jctb.4569.

Grassino, A. N., Halambek, J., Djaković, S., Rimac Brnčić, S., Dent, M., & Grabarić, Z. (2016). Utilization of tomato peel waste from canning factory as a potential source for pectin production and application as tin corrosion inhibitor. *Food Hydrocolloids*, *52*, 265–274. https://doi.org/10.1016/j.foodhyd.2015.06.020.

Guevara-Arauza, J. C., de Jesús Ornelas-Paz, J., Pimentel-González, D. J., Rosales Mendoza, S., Soria Guerra, R. E., & Paz Maldonado, L. M. T. (2012). Prebiotic effect of mucilage and pectic-derived oligosaccharides from nopal (*Opuntia ficus-indica*). *Food Science and Biotechnology*, *21*(4), 997–1003. https://doi.org/10.1007/s10068-012-0130-1.

Gullón, P., Eibes, G., Lorenzo, J. M., Pérez-Rodríguez, N., Lú-Chau, T. A., & Gullón, B. (2020). Green sustainable process to revalorize purple corn cobs within a biorefinery frame: Co-production of bioactive extracts. *Science of the Total Environment*, *709*, 136236. https://doi.org/10.1016/j.scitotenv.2019.136236.

Gullón, B., Gómez, B., Martínez-Sabajanes, M., Yáñez, R., Parajó, J. C., & Alonso, J. L. (2013). Pectic oligosaccharides: Manufacture and functional properties. *Trends in Food Science and Technology*, *30*(2), 153–161. https://doi.org/10.1016/j.tifs.2013.01.006.

Gullón, B., Gullón, P., Eibes, G., Cara, C., De Torres, A., López-Linares, J. C., et al. (2018). Valorisation of olive agro-industrial by-products as a source of bioactive compounds. *Science of the Total Environment*, *645*, 533–542. https://doi.org/10.1016/j.scitotenv.2018.07.155.

Gullón, P., Gullón, B., Moure, A., Alonso, J. L., Domínguez, H., & Parajó, J. C. (2009). In D. Charalampopoulos, & R. A. Rastall (Eds.), *Manufacture of prebiotics from biomass sources BT—Prebiotics and probiotics science and technology* (pp. 535–589). New York: Springer. https://doi.org/10.1007/978-0-387-79058-9_14.

Gullón, B., Gullón, P., Sanz, Y., Alonso, J. L., & Parajó, J. C. (2011). Prebiotic potential of a refined product containing pectic oligosaccharides. *LWT—Food Science and Technology*, *44*(8), 1687–1696. https://doi.org/10.1016/j.lwt.2011.03.006.

Hensel, A., & Meier, K. (1999). Pectins and xyloglucans exhibit antimutagenic activities against nitroaromatic compounds. *Planta Medica*, *65*(5), 395–399. https://doi.org/10.1055/s-1999-14013.

Holck, J., Hjernø, K., Lorentzen, A., Vigsnæs, L. K., Hemmingsen, L., Licht, T. R., et al. (2011). Tailored enzymatic production of oligosaccharides from sugar beet pectin and evidence of differential effects of a single DP chain length difference on human faecal microbiota composition after *in vitro* fermentation. *Process Biochemistry*, *46*(5), 1039–1049. https://doi.org/10.1016/j.procbio.2011.01.013.

Holck, J., Hotchkiss, A. T., Meyer, A. S., Mikkelsen, J. D., & Rastall, R. A. (2014). Production and bioactivity of pectic oligosaccharides from fruit and vegetable biomass. In *Food oligosaccharides: Production, analysis and bioactivity* (pp. 76–87). https://doi.org/10.1002/9781118817360.ch5. (Vol. 9781118426, Issue October 2017.

Holck, J., Lorentzen, A., Vigsnæs, L. K., Licht, T. R., Mikkelsen, J. D., & Meyer, A. S. (2011). Feruloylated and nonferuloylated arabino-oligosaccharides from sugar beet pectin selectively stimulate the growth of *Bifidobacterium* spp. in human fecal *in vitro* fermentations. *Journal of Agricultural and Food Chemistry*, *59*(12), 6511–6519. https://doi.org/10.1021/jf200996h.

Hosobuchi, C., Rutanassee, L., Bassin, S. L., & Wong, N. D. (1999). Efficacy of acacia, pectin, and guar gum-based fiber supplementation in the control of hypercholesterolemia. *Nutrition Research*, *19*(5), 643–649. https://doi.org/10.1016/S0271-5317(99)00029-9.

Hotchkiss, A. T., Manderson, K., Olano-Martin, E., Grace, W. E., Gibson, G. R., & Rastall, R. A. (2004). Orange peel pectic oligosaccharide prebiotics with food and feed applications. In *228th ACS national meeting*.

Huang, C., Cao, X., Chen, X., Fu, Y., Zhu, Y., Chen, Z., et al. (2017). A pectic polysaccharide from *Ligusticum chuanxiong* promotes intestine antioxidant defense in aged mice. *Carbohydrate Polymers*, *174*, 915–922. https://doi.org/10.1016/j.carbpol.2017.06.122.

Ismail, B. B., Pu, Y., Fan, L., Dandago, M. A., Guo, M., & Liu, D. (2019). Characterizing the phenolic constituents of baobab (*Adansonia digitata*) fruit shell by LC-MS/QTOF and their *in vitro* biological activities. *Science of the Total Environment*, *694*, 133387. https://doi.org/10.1016/j.scitotenv.2019.07.193.

Jiang, T., Gao, X., Wu, C., Tian, F., Lei, Q., Bi, J., et al. (2016). Apple-derived pectin modulates gut microbiota, improves gut barrier function, and attenuates metabolic endotoxemia in rats with diet-induced obesity. *Nutrients*, *8*(3), 126. https://doi.org/10.3390/nu8030126.

Kakavand, H., Rawson, R. V., Pupo, G. M., Yang, J. Y. H., Menzies, A. M., Carlino, M. S., et al. (2017). PD-L1 expression and immune escape in melanoma resistance to MAPK inhibitors. *Clinical Cancer Research*, *23*(20), 6054–6061. https://doi.org/10.1158/1078-0432.CCR-16-1688.

Kang, H. J., Jo, C., Kwon, J. H., Son, J. H., An, B. J., & Byun, M. W. (2006). Antioxidant and cancer cell proliferation inhibition effect of citrus pectin-oligosaccharide prepared by irradiation. *Journal of Medicinal Food*, *9*(3), 313–320. https://doi.org/10.1089/jmf.2006.9.313.

Kazemi, M., Khodaiyan, F., Hosseini, S. S., & Najari, Z. (2019). An integrated valorization of industrial waste of eggplant: Simultaneous recovery of pectin, phenolics and sequential production of pullulan. *Waste Management*, *100*, 101–111. https://doi.org/10.1016/j.wasman.2019.09.013.

Khasina, E. I., Kolenchenko, E. A., Sgrebneva, M. N., Kovalev, V. V., & Khotimchenko, Y. S. (2003). Antioxidant activities of a low etherified pectin from the seagrass *Zostera marina*. *Russian Journal of Marine Biology*, *29*(4), 259–261. https://doi.org/10.1023/A:1025493128327.

Khodaei, N., Karboune, S., & Orsat, V. (2016). Microwave-assisted alkaline extraction of galactan-rich rhamnogalacturonan I from potato cell wall by-product. *Food Chemistry*, *190*, 495–505. https://doi.org/10.1016/j.foodchem.2015.05.082.

Kim, S. K., & Rajapakse, N. (2005). Enzymatic production and biological activities of chitosan oligosaccharides

(COS): A review. *Carbohydrate Polymers, 62*(4), 357–368. https://doi.org/10.1016/j.carbpol.2005.08.012.

Koriem, K. M. M., Arbid, M. S., & Emam, K. R. (2014). Therapeutic effect of pectin on octylphenol induced kidney dysfunction, oxidative stress and apoptosis in rats. *Environmental Toxicology and Pharmacology, 38*(1), 14–23. https://doi.org/10.1016/j.etap.2014.04.029.

Krivorotova, T., Staneviciene, R., Luksa, J., Serviene, E., & Sereikaite, J. (2016). Preparation and characterization of nisin-loaded pectin-inulin particles as antimicrobials. *LWT—Food Science and Technology, 72*, 518–524. https://doi.org/10.1016/j.lwt.2016.05.022.

Kungel, P. T. A. N., Correa, V. G., Corrêa, R. C. G., Peralta, R. A., Soković, M., Calhelha, R. C., et al. (2018). Antioxidant and antimicrobial activities of a purified polysaccharide from yerba mate (*Ilex paraguariensis*). *International Journal of Biological Macromolecules, 114*, 1161–1167. https://doi.org/10.1016/j.ijbiomac.2018.04.020.

Larsen, N., De Souza, C. B., Krych, L., Cahú, T. B., Wiese, M., Kot, W., et al. (2019). Potential of pectins to beneficially modulate the gut microbiota depends on their structural properties. *Frontiers in Microbiology, 10*, 1–13. https://doi.org/10.3389/fmicb.2019.00223.

Leclere, L., Van Cutsem, P., & Michiels, C. (2013). Anti-cancer activities of pH- or heat-modified pectin. *Frontiers in Pharmacology, 4*, 1–8. https://doi.org/10.3389/fphar.2013.00128.

Lee, S. O., & Gao, A. C. (2005). STAT3 and transactivation of steroid hormone receptors. *Vitamins and Hormones, 70*, 333–357. Academic Press https://doi.org/10.1016/S0083-6729(05)70011-X.

Lee, S. B., Lee, W. S., Shin, J. S., Jang, D. S., & Lee, K. T. (2017). Xanthotoxin suppresses LPS-induced expression of iNOS, COX-2, TNF-α, and IL-6 via AP-1, NF-κB, and JAK-STAT inactivation in RAW 264.7 macrophages. *International Immunopharmacology, 49*, 21–29. https://doi.org/10.1016/j.intimp.2017.05.021.

Lee, K. P., Sudjarwo, G. W., Kim, J. S., Dirgantara, S., Maeng, W. J., & Hong, H. (2014). The anti-inflammatory effect of Indonesian *Areca catechu* leaf extract *in vitro* and *in vivo*. *Nutrition Research and Practice, 8*(3), 267–271. https://doi.org/10.4162/nrp.2014.8.3.267.

Li, T., Li, S., Du, L., Wang, N., Guo, M., Zhang, J., et al. (2010). Effects of haw pectic oligosaccharide on lipid metabolism and oxidative stress in experimental hyperlipidemia mice induced by high-fat diet. *Food Chemistry, 121*, 1010–1013. https://doi.org/10.1016/j.foodchem.2010.01.039.

Li, J., Li, S., Zheng, Y., Zhang, H., Chen, J., Yan, L., et al. (2019). Fast preparation of rhamnogalacturonan I enriched low molecular weight pectic polysaccharide by ultrasonically accelerated metal-free Fenton reaction. *Food Hydrocolloids, 95*, 551–561. https://doi.org/10.1016/j.foodhyd.2018.05.025.

Liang, S., Liao, W., Ma, X., Li, X., & Wang, Y. (2017). H_2O_2 oxidative preparation, characterization and antiradical activity of a novel oligosaccharide derived from flaxseed gum. *Food Chemistry, 230*, 135–144. https://doi.org/10.1016/j.foodchem.2017.03.029.

Lim, S.-H. (2017). Larch arabinogalactan attenuates myocardial injury by inhibiting apoptotic cascades in a rat model of ischemia–reperfusion. *Journal of Medicinal Food, 20*(7), 691–699. https://doi.org/10.1089/jmf.2016.3886.

Lin, H. V., Frassetto, A., Kowalik, E. J., Nawrocki, A. R., Lu, M. M., Kosinski, J. R., et al. (2012). Butyrate and propionate protect against diet-induced obesity and regulate gut hormones via free fatty acid receptor 3-independent mechanisms. *PLoS One, 7*(4), 1–9. https://doi.org/10.1371/journal.pone.0035240.

Lin, L., Wang, P., Du, Z., Wang, W., Cong, Q., Zheng, C., et al. (2016). Structural elucidation of a pectin from flowers of *Lonicera japonica* and its antipancreatic cancer activity. *International Journal of Biological Macromolecules, 88*, 130–137. https://doi.org/10.1016/j.ijbiomac.2016.03.025.

Liu, Y., Dong, M., Yang, Z., & Pan, S. (2016). Anti-diabetic effect of citrus pectin in diabetic rats and potential mechanism via PI3K/Akt signaling pathway. *International Journal of Biological Macromolecules, 89*, 484–488. https://doi.org/10.1016/j.ijbiomac.2016.05.015.

Liu, S., Shi, X., Xu, L., & Yi, Y. (2016). Optimization of pectin extraction and antioxidant activities from Jerusalem artichoke. *Chinese Journal of Oceanology and Limnology, 34*(2), 372–381. https://doi.org/10.1007/s00343-015-4314-4.

Luo, J., Ma, Y., & Xu, Y. (2020). Valorization of apple pomace using a two-step slightly acidic processing strategy. *Renewable Energy, 152*, 793–798. https://doi.org/10.1016/j.renene.2020.01.120.

Ma, J., Su, H., Yu, B., Guo, T., Gong, Z., Qi, J., et al. (2018). CXCL12 gene silencing down-regulates metastatic potential via blockage of MAPK/PI3K/AP-1 signaling pathway in colon cancer. *Clinical and Translational Oncology, 20*(8), 1035–1045. https://doi.org/10.1007/s12094-017-1821-0.

Mandalari, G., Nueno Palop, C., Tuohy, K., Gibson, G. R., Bennett, R. N., Waldron, K. W., et al. (2007). *In vitro* evaluation of the prebiotic activity of a pectic oligosaccharide-rich extract enzymatically derived from bergamot peel. *Applied Microbiology and Biotechnology, 73*(5), 1173–1179. https://doi.org/10.1007/s00253-006-0561-9.

Manderson, K., Pinart, M., Tuohy, K. M., Grace, W. E., Hotchkiss, A. T., Widmer, W., et al. (2005). *In vitro* determination of prebiotic properties of oligosaccharides derived from an orange juice manufacturing byproduct stream. *Applied and Environmental Microbiology, 71*(12), 8383–8389. https://doi.org/10.1128/AEM.71.12.8383-8389.2005.

Mao, Y., Lei, R., Ryan, J., Arrutia Rodriguez, F., Rastall, B., Chatzifragkou, A., et al. (2019). Understanding the influence of processing conditions on the extraction of rhamnogalacturonan-I "hairy" pectin from sugar beet pulp. *Food Chemistry, 2*, 100026. https://doi.org/10.1016/j.fochx.2019.100026.

Maran, J. P., & Priya, B. (2015). Ultrasound-assisted extraction of pectin from sisal waste. *Carbohydrate Polymers, 115*, 732–738. https://doi.org/10.1016/j.carbpol.2014.07.058.

Martínez, M., Gullón, B., Schols, H. A., Alonso, J. L., & Parajó, J. C. (2009). Assessment of the production of oligomeric compounds from sugar beet pulp. *Industrial* and *Engineering Chemistry Research*, *48*(10), 4681–4687. https://doi.org/10.1021/ie8017753.

Martínez, M., Gullón, B., Yáñez, R., Alonso, J. L., & Parajó, J. C. (2009). Direct enzymatic production of oligosaccharide mixtures from sugar beet pulp: Experimental evaluation and mathematical modeling. *Journal of Agricultural and Food Chemistry*, *57*(12), 5510–5517. https://doi.org/10.1021/jf900654g.

Martínez, M., Yáñez, R., Alonsó, J. L., & Parajó, J. C. (2010). Chemical production of pectic oligosaccharides from orange peel wastes. *Industrial and Engineering Chemistry Research*, *49*(18), 8470–8476. https://doi.org/10.1021/ie101066m.

Martínez, M., Yáñez, R., Alonso, J. L., & Parajó, J. C. (2012). Pectic oligosaccharides production from orange peel waste by enzymatic hydrolysis. *International Journal of Food Science and Technology*, *47*(4), 747–754. https://doi.org/10.1111/j.1365-2621.2011.02903.x.

Míguez, B., Gómez, B., Gullón, P., Gullón, B., & Alonso, J. L. (2016). Pectic oligosaccharides and other emerging prebiotics. In *Vol. 15. Probiotics and prebiotics in human nutrition and health*. https://doi.org/10.5772/62830.

Minzanova, S. T., Mironov, V. F., Arkhipova, D. M., Khabibullina, A. V., Mironova, L. G., Zakirova, Y. M., et al. (2018). Biological activity and pharmacological application of pectic polysaccharides: A review. *Polymers*, *10*(12), 1–31. https://doi.org/10.3390/polym10121407.

Muñoz, A. L., Rodríguez-Gutiérrez, G., Rubio-Senent, F., & Fernández-Bolaños, J. (2011). Production, characterization and isolation of neutral and pectic oligosaccharides with low molecular weights from olive by-products thermally treated. *Food Hydrocolloids*, *28*. https://doi.org/10.1016/j.foodhyd.2011.11.008.

Nangia-Makker, P., Hogan, V., Honjo, Y., Baccarini, S., Tait, L., Bresalier, R., et al. (2002). Inhibition of human cancer cell growth and metastasis in nude mice by oral intake of modified citrus pectin. *Journal of the National Cancer Institute*, *94*(24), 1854–1862. https://doi.org/10.1093/jnci/94.24.1854.

Ni'mah, L., Makhyarini, I., & Normalina. (2020). *Musa acuminata* L. (Banana) peel wastes as edible coating based on pectin with addition of *Cinnamomum burmannii* extract. *Asian Journal of Chemistry*, *32*(3), 703–705. https://doi.org/10.14233/ajchem.2020.22392.

Nisar, T., Wang, Z. C., Yang, X., Tian, Y., Iqbal, M., & Guo, Y. (2018). Characterization of citrus pectin films integrated with clove bud essential oil: Physical, thermal, barrier, antioxidant and antibacterial properties. *International Journal of Biological Macromolecules*, *106*, 670–680. https://doi.org/10.1016/j.ijbiomac.2017.08.068.

Nunes, C., Almeida, L., Barbosa, R. M., & Laranjinha, J. (2017). Luteolin suppresses the JAK/STAT pathway in a cellular model of intestinal inflammation. *Food and Function*, *8*(1), 387–396. https://doi.org/10.1039/c6fo01529h.

Ogutu, F. O., Mu, T.-H., Sun, H., & Zhang, M. (2018). Ultrasonic modified sweet potato pectin enduces apoptosis like cell death in colon cancer (HT-29) cell line. *Nutrition and Cancer*, *70*(1), 136–145. https://doi.org/10.1080/01635581.2018.1406123.

Olano-Martin, E., Gibson, G. R., & Rastall, R. A. (2002). Comparison of the *in vitro* bifidogenic properties of pectins and pectic-oligosaccharides. *Journal of Applied Microbiology*, *93*(3), 505–511. https://doi.org/10.1046/j.1365-2672.2002.01719.x.

Olano-Martin, E., Rimbach, G. H., Gibson, G. R., & Rastall, R. A. (2003). Pectin and pectic-oligosaccharides induce apoptosis in *in vitro* human colonic adenocarcinoma cells. *Anticancer Research*, *23*(1(A)), 341–346.

Olano-Martin, E., Williams, M. R., Gibson, G. R., & Rastall, R. A. (2003). Pectins and pectic-oligosaccharides inhibit *Escherichia coli* O157:H7 Shiga toxin as directed towards the human colonic cell line HT29. *FEMS Microbiology Letters*, *218*(1), 101–105. https://doi.org/10.1016/S0378-1097(02)01119-9.

Palou, M., Sánchez, J., García-Carrizo, F., Palou, A., & Picó, C. (2015). Pectin supplementation in rats mitigates age-related impairment in insulin and leptin sensitivity independently of reducing food intake. *Molecular Nutrition & Food Research*, *59*(10), 2022–2033. https://doi.org/10.1002/mnfr.201500292.

Panesar, P. S., Bali, V., Kumari, S., Babbar, N., & Oberoi, H. S. (2014). In S. K. Brar, G. S. Dhillon, & C. R. Soccol (Eds.), *Prebiotics BT—Biotransformation of waste biomass into high value biochemicals* (pp. 237–259). New York: Springer. https://doi.org/10.1007/978-1-4614-8005-1_10.

Park, H. R., Lee, S. J., Im, S. B., Shin, M. S., Choi, H. J., Park, H. Y., et al. (2019). Signaling pathway and structural features of macrophage-activating pectic polysaccharide from Korean citrus, *Cheongkyool* peels. *International Journal of Biological Macromolecules*, *137*, 657–665. https://doi.org/10.1016/j.ijbiomac.2019.07.012.

Pasandide, B., Khodaiyan, F., Mousavi, Z., & Hosseini, S. S. (2018). Pectin extraction from citron peel: Optimization by Box–Behnken response surface design. *Food Science and Biotechnology*, *27*(4), 997–1005. https://doi.org/10.1007/s10068-018-0365-6.

Pelkman, C. L., Navia, J. L., Miller, A. E., & Pohle, R. J. (2007). Novel calcium-gelled, alginate-pectin beverage reduced energy intake in nondieting overweight and obese women: Interactions with dietary restraint status. *American Journal of Clinical Nutrition*, *86*(6), 1595–1602. https://doi.org/10.1093/ajcn/86.6.1595.

Peng, Q., Xu, Q., Yin, H., Huang, L., & Du, Y. (2014). Characterization of an immunologically active pectin from the fruits of *Lycium ruthenicum*. *International Journal of Biological Macromolecules*, *64*, 69–75. https://doi.org/10.1016/j.ijbiomac.2013.11.030.

Popov, S. V., Markov, P. A., Popova, G. Y., Nikitina, I. R., Efimova, L., & Ovodov, Y. S. (2013). Anti-inflammatory activity of low and high methoxylated citrus pectins. *Biomedicine and Preventive Nutrition*, *3*(1), 59–63. https://doi.org/10.1016/j.bionut.2012.10.008.

Prandi, B., Baldassarre, S., Babbar, N., Bancalari, E., Vandezande, P., Hermans, D., et al. (2018). Pectin oligosaccharides from sugar beet pulp: Molecular characterization and potential prebiotic activity. *Food and Function*, *9*(3), 1557–1569. https://doi.org/10.1039/c7fo01182b.

Rahmani, Z., Khodaiyan, F., Kazemi, M., & Sharifan, A. (2020). Optimization of microwave-assisted extraction and structural characterization of pectin from sweet lemon peel. *International Journal of Biological Macromolecules, 147*, 1107–1115. https://doi.org/10.1016/j.ijbiomac.2019.10.079.

Rasmussen, L. E., & Meyer, A. S. (2010). Endogeneous β-D-xylosidase and α L arabinofuranosidase activity in flax seed mucilage. *Biotechnology Letters, 32*(12), 1883–1891. https://doi.org/10.1007/s10529-010-0367-9.

Rico, X., Gullón, B., Alonso, J. L., & Yáñez, R. (2020). Recovery of high value-added compounds from pineapple, melon, watermelon and pumpkin processing by-products: An overview. *Food Research International, 132*, 109086. https://doi.org/10.1016/j.foodres.2020.109086.

Roberfroid, M. B. (1996). Functional effects of food components and the gastrointestinal system: Chicory fructooligosaccharides. *Nutrition Reviews, 54*(11), S38–S42. https://doi.org/10.1111/j.1753-4887.1996.tb03817.x.

Sabater, C., Ferreira-Lazarte, A., Montilla, A., & Corzo, N. (2019). Enzymatic production and characterization of pectic oligosaccharides derived from citrus and apple pectins: A GC-MS study using random forests and association rule learning. *Journal of Agricultural and Food Chemistry, 67*(26), 7435–7447. https://doi.org/10.1021/acs.jafc.9b00930.

Šefčíková, Z., & Raček, L. (2016). Effect of pectin feeding on obesity development and duodenal alkaline phosphatase activity in Sprague-Dawley rats fed with high-fat/high-energy diet. *Acta Physiologica Hungarica, 103*(2), 183–190. https://doi.org/10.1556/036.103.2016.2.5.

Showpanish, K., Sonhom, N., Pilasombut, K., Woraprayote, W., Prachom, N., Buathong, R., et al. (2020). Effect of extracted pectin from fruit wastes on growth of *Pediococcus pentosaceus* RSU-Nh1 and *Lactobacillus plantarum* RSU-SO2. *International Journal of Agricultural Technology, 16*(2), 403–420.

Story, J. A. (1985). Dietary fiber and lipid metabolism. *Proceedings of the Society for Experimental Biology and Medicine, 180*(3), 447–452. https://doi.org/10.3181/00379727-180-42201.

Sun, J., Deering, R. W., Peng, Z., Najia, L., Khoo, C., Cohen, P. S., et al. (2019). Pectic oligosaccharides from cranberry prevent quiescence and persistence in the uropathogenic *Escherichia coli* CFT073. *Scientific Reports, 9*(1), 1–9. https://doi.org/10.1038/s41598-019-56005-w.

Talekar, S., Patti, A. F., Vijayraghavan, R., & Arora, A. (2018). An integrated green biorefinery approach towards simultaneous recovery of pectin and polyphenols coupled with bioethanol production from waste pomegranate peels. *Bioresource Technology, 266*, 322–334. https://doi.org/10.1016/j.biortech.2018.06.072.

Talma, S. V., Regis, S. A., Ferreira, P. R., Mellinger-Silva, C., & de Resende, E. D. (2019). Characterization of pericarp fractions of yellow passion fruit: Density, yield of flour, color, pectin content and degree of esterification. *Food Science and Technology, 39*, 683–689. https://doi.org/10.1590/fst.30818.

Tan, H., Chen, W., Liu, Q., Yang, G., & Li, K. (2018). Pectin oligosaccharides ameliorate colon cancer by regulating oxidative stress- and inflammation-activated signaling pathways. *Frontiers in Immunology, 9*, 1–13. https://doi.org/10.3389/fimmu.2018.01504.

Tan, J., Hua, X., Liu, J., Wang, M., Liu, Y., Yang, R., et al. (2020). Extraction of sunflower head pectin with superfine grinding pretreatment. *Food Chemistry, 320*(March), 126631. https://doi.org/10.1016/j.foodchem.2020.126631.

Thomassen, L. V., Vigsnæs, L. K., Licht, T. R., Mikkelsen, J. D., & Meyer, A. S. (2011). Maximal release of highly bifidogenic soluble dietary fibers from industrial potato pulp by minimal enzymatic treatment. *Applied Microbiology and Biotechnology, 90*(3), 873–884. https://doi.org/10.1007/s00253-011-3092-y.

Tovar, A. K., Godínez, L. A., Espejel, F., Ramírez-Zamora, R. M., & Robles, I. (2019). Optimization of the integral valorization process for orange peel waste using a design of experiments approach: Production of high-quality pectin and activated carbon. *Waste Management, 85*, 202–213. https://doi.org/10.1016/j.wasman.2018.12.029.

Vigsnæs, L. K., Holck, J., Meyer, A. S., & Licht, T. R. (2011). *In vitro* fermentation of sugar beet arabino-oligosaccharides by fecal microbiota obtained from patients with ulcerative colitis to selectively stimulate the growth of *Bifidobacterium* spp. and *Lactobacillus* spp. *Applied and Environmental Microbiology, 77*(23), 8336–8344. https://doi.org/10.1128/AEM.05895-11.

Vitaliti, G., Pavone, P., Mahmood, F., Nunnari, G., & Falsaperla, R. (2014). Targeting inflammation as a therapeutic strategy for drug-resistant epilepsies: An update of new immune-modulating approaches. *Human Vaccines and Immunotherapeutics, 10*(4), 868–875. https://doi.org/10.4161/hv.28400.

Wang, J., Hu, S., Nie, S., Yu, Q., & Xie, M. (2016). Reviews on mechanisms of *in vitro* antioxidant activity of polysaccharides. *Oxidative Medicine and Cellular Longevity, 2016*. https://doi.org/10.1155/2016/5692852.

Wang, W., Wu, X., Chantapakul, T., Wang, D., Zhang, S., Ma, X., et al. (2017). Acoustic cavitation assisted extraction of pectin from waste grapefruit peels: A green two-stage approach and its general mechanism. *Food Research International, 102*, 101–110. https://doi.org/10.1016/j.foodres.2017.09.087.

Wicker, L., Kim, Y., Kim, M. J., Thirkield, B., Lin, Z., & Jung, J. (2014). Pectin as a bioactive polysaccharide—Extracting tailored function from less. *Food Hydrocolloids, 42*(P2), 251–259. https://doi.org/10.1016/j.foodhyd.2014.01.002.

Wilkowska, A., Nowak, A., Antczak-Chrobot, A., Motyl, I., Czyzowska, A., & Paliwoda, A. (2019). Structurally different pectic oligosaccharides produced from apple pomace and their biological activity *in vitro*. *Food, 8*(9), 365. https://doi.org/10.3390/foods8090365.

Wu, L., & Sun, D. (2017). Leptin receptor gene polymorphism and the risk of cardiovascular disease: A systemic review and meta-analysis. *International Journal of Environmental Research and Public Health, 14*(4), 375. https://doi.org/10.3390/ijerph14040375.

Xu, S. Y., Liu, J. P., Huang, X., Du, L. P., Shi, F. L., Dong, R., et al. (2018). Ultrasonic-microwave assisted extraction, characterization and biological activity of pectin from jack-fruit peel. *LWT—Food Science and Technology*, *90*(January), 577–582. https://doi.org/10.1016/j.lwt.2018.01.007.

Ye, M. B., & Lim, B. O. (2010). Dietary pectin regulates the levels of inflammatory cytokines and immunoglobulins in interleukin-10 knockout mice. *Journal of Agricultural and Food Chemistry*, *58*(21), 11281–11286. https://doi.org/10.1021/jf103262s.

Zhai, X., Zhu, C., Li, Y., Zhang, Y., Duan, Z., & Yang, X. (2018). Optimization for pectinase-assisted extraction of poly-saccharides from pomegranate peel with chemical com-position and antioxidant activity. *International Journal of Biological Macromolecules*, *109*, 244–253. https://doi.org/10.1016/j.ijbiomac.2017.12.064.

Zhang, S., Hu, H., Wang, L., Liu, F., & Pan, S. (2018). Preparation and prebiotic potential of pectin oligosaccharides ob-tained from citrus peel pectin. *Food Chemistry*, *244*, 232–237. https://doi.org/10.1016/j.foodchem.2017.10.071.

Zhang, W., Xu, P., & Zhang, H. (2015). Pectin in cancer ther-apy: A review. *Trends in Food Science & Technology*, *44*, 258–271. https://doi.org/10.1016/j.tifs.2015.04.001.

Zhu, R., Wang, C., Zhang, L., Wang, Y., Chen, G., Fan, J., et al. (2019). Pectin oligosaccharides from fruit of *Actinidia arguta*: Structure-activity relationship of prebiotic and antiglycation potentials. *Carbohydrate Polymers*, *217*, 90–97. https://doi.org/10.1016/j.carbpol.2019.04.032.

Zhu, R., Zhang, X., Wang, Y., Zhang, L., Wang, C., Hu, F., et al. (2019). Pectin oligosaccharides from hawthorn (*Crataegus pinnatifida* Bunge. Var. *major*): Molecular characterization and potential antiglycation activities. *Food Chemistry*, *286*, 129–135. https://doi.org/10.1016/j.foodchem.2019.01.215.

Zofou, D., Shu, G. L., Foba-Tendo, J., Tabouguia, M. O., & Assob, J. C. N. (2019). *In vitro* and *in vivo* anti-salmonella evaluation of pectin extracts and hydroly-sates from "cas mango" (*Spondias dulcis*). *Evidence-Based Complementary and Alternative Medicine*, *2019*. https://doi.org/10.1155/2019/3578402.

Zou, Y. F., Barsett, H., Ho, G. T. T., Inngjerdingen, K. T., Diallo, D., Michaelsen, T. E., et al. (2015). Immunomodulating pectins from root bark, stem bark, and leaves of the Malian medicinal tree *Terminalia macroptera*, structure activity re-lations. *Carbohydrate Research*, *403*, 167–173. https://doi.org/10.1016/j.carres.2014.05.004.

Zykwinska, A., Boiffard, M. H., Kontkanen, H., Buchert, J., Thibault, J. F., & Bonnin, E. (2008). Extraction of green labeled pectins and pectic oligosaccharides from plant by-products. *Journal of Agricultural and Food Chemistry*, *56*(19), 8926–8935. https://doi.org/10.1021/jf801705a.

CHAPTER 6

Biopreservation: Foodborne Virus Contamination and Control in Minimally Processed Food

DORIS SOBRAL MARQUES SOUZA[A,B,*] • VILAINE CORRÊA DA SILVA[A,*] • MARÍLIA MIOTTO[C] • JULIANO DE DEA LINDNER[B] • DAVID RODRÍGUEZ-LÁZARO[D] • GISLAINE FONGARO[A]

[a]Laboratory of Applied Virology, Department of Microbiology, Immunology and Parasitology, Federal University of Santa Catarina (UFSC), Florianópolis, SC, Brazil, [b]Food Technology and Bioprocess Research Group, Department of Food Science and Technology, UFSC, Florianópolis, SC, Brazil, [c]Food Microbiology Laboratory, UFSC, Florianópolis, SC, Brazil, [d]Microbiology Division, Faculty of Sciences, University of Burgos, Burgos, Spain

6.1 INTRODUCTION

Unprocessed foods can be defined as edible food parts of vegetables or algae, animals, fungi, or water. Similarly, minimally processed foods (MPFs) can be considered as those foods that have been altered with the purpose of preserving and extending their self-life, without substantial change in the nutritional content and without the addition of substances such as salt, sugar, or other food ingredients. Many fresh fruits, vegetables whole grains, nuts, grains and cereals, meats, chilled, or frozen seafood, and milk fall into this category (Monteiro et al., 2016). Cleaning and removing inedible parts, refrigeration, freezing, pasteurization, fermentation, and vacuum-packaging are examples of minimal processing procedures in the food industry. The advantages of minimal food processing include the preservation of the nutritional value and organoleptic characteristics of foods, and an easy preparation prior to consumption. However, the treatments applied may not be enough to eliminate foodborne pathogens that may contaminate these foods (Webb, Barker, Goodburn, & Peck, 2019).

The greater availability of molecular tools and methods of viral concentration from food has demonstrated that the enteric viruses are responsible for an important percentage of foodborne illness (Rodríguez-Lázaro, Fongaro, Marques Souza, & Hernández, 2021), representing a relevant food safety risk that

occurred particularly for MPFs (FAO/WHO, 2008; Rodríguez-Lázaro et al., 2012; Suffredini et al., 2017), and some foodborne viral outbreaks are reported annually worldwide (Dufour, 2015). The risk increases as these food products are often eaten raw or only lightly cooked. Water contaminated with human sewage and the use of manure in crops are the main causes of foodborne viral contamination of vegetables and bivalve molluscs, mainly by irrigation, lixiviation, and contamination of shellfish growing areas. The use of animal manure that may content zoonotic viral pathogens represents an additional public health concern. Furthermore, MPFs can be contaminated by incorrect food production practices by infected food handlers or by contaminated utensils.

Enteric viruses have fecal-oral route and are excreted in diarrheic feces in large concentrations (up to 10^{11} viral particles per g) (Gerba, 2000). They have low infective doses (1–100) and are much more stable in the environment than enteric bacteria, usually used as fecal contamination indicators (for example *Escherichia coli*, or *Salmonella* spp.) (Noble, Lee, & Schiff, 2004). Enteric viruses are nonenveloped and therefore more resistant in the environment. Most of the main enteric viruses have a single and positive-stranded RNA genome (ssRNA), such as the human calicivirus—norovirus and sapovirus, hepatitis A and E viruses, but other viral pathogens possess segmented double-stranded

---

*These authors contributed equally to the work

Sustainable Production Technology in Food. https://doi.org/10.1016/B978-0-12-821233-2.00009-5
Copyright © 2021 Elsevier Inc. All rights reserved.

RNA (dsRNA) (rotavirus), or DNA (dsDNA) (mastadenoviruses, usually called human adenoviruses) (Bosch, Abad, & Pintó, 2005). Enteric viruses group of viruses generally replicate in the human intestine causing mainly gastroenteritis, but enteric hepatitis, meningitis, and respiratory and urinary diseases are also reported (Bosch et al., 2005). In addition to the enteric viruses, other emergent viral pathogens are excreted by urine and stools and contaminate water and food posing an inherent public health concern.

Enteric viral pathogens may persist infectious after treatments applied in wastewater treatment plants (Da Silva et al., 2007), and can tolerate pH, refrigeration, salinity, drying, UV radiation, the usually used disinfectants in the food industry, and other stresses present in the food related environment or in wastewater treatment plants (Table 6.1). They can stay stable outside of their host for extended periods in the environment (Rzezutka & Cook, 2004).

6.2 MAIN FOODBORNE VIRUSES ASSOCIATED TO THE CONSUMPTION OF MPFS

6.2.1 Human Pathogens

Human noroviruses (NoVs) are the most frequently reported pathogenic agent involved in foodborne diseases. Hepatitis A virus (HAV) is also an important foodborne virus and is the second most frequently reported (NSSP, 2017). According to the Food and Agriculture Organization (FAO), NoVs and HAV were the third and seventh most reported foodborne agents in alerts and border rejections by microbiological causes between 2012 and 2019 in the European Union (Fig. 6.1) (FAO, 2020a).

6.2.1.1 Caliciviridae *family*

The family *Caliciviridae* has two genera involved with foodborne diseases, *Norovirus* and *Sapovirus*, both causing gastroenteritis illness.

(a) Noroviruses

Human NoVs are a major cause of acute gastroenteritis, accounting for almost 18% of outbreaks and around 200,000 deaths worldwide (Cardemil & Hall, 2020). They represent an important threat to public health causing gastroenteritis diseases mainly in winter. NoVs have a single-stranded RNA genome, with approximately 7.5 Kb in length (Chhabra et al., 2019). NoV genome is organized in three open reading frames (ORFs): ORF1, encodes to a polyprotein for six nonstructural proteins; ORF2 encodes the major structural protein (VP1), containing shell (S) and protruding (P) domains. The P2 subdomain is highly variable, having a major of neutralizing epitopes and interacting regions with histo-blood groups antigens (HBGAs); ORF3, encodes a minor structural protein (VP2). The murine NoVs only possess ORFs 1 and 2 (Chhabra et al., 2019). NoVs are distributed among 10 genogroups (GI–GX) with several genotypes (denominated by Arabic numbers) and variants, based on phylogenetic analysis of VP1 and RdRp (the RNA-dependent RNA polymerase) (Chhabra et al., 2019). Five genogroups cause diseases in humans, GI, GII, GIV, GVIII, and GIX (Chhabra et al., 2019), and GII is the most prevalent in waterborne and

TABLE 6.1
Main regulations regarding microbiological limits for assessing the microbiological quality of bivalve molluscs.

Producer	Sample	Bacterial parameter	Status	Posttreatment
European Union	Shellfish (100 g of flesh)	80th percentile of samples <230 MPN of *E. coli*	Class A	No posttreatment required
		<4600 MPN of *E. coli*	Class B	Heat-treated or purified
		46,000 MPN of *E. coli*	Class C	Debugged for 2 months
United States of America	Water (100 ml)	Geometric means <70 of Total Coliforms <14 MPN or MF of *E. coli*	Approved	No posttreatment required
		Total Coliforms <700 <88 MPN or MF of *E. coli*	Restricted	Depuration or relaying in an approved area
		Total coliforms >700 <88 MPN or MF of *E. coli*	Prohibited	Harvest not permitted

MF, membrane filter; *MPN*, most probable number.

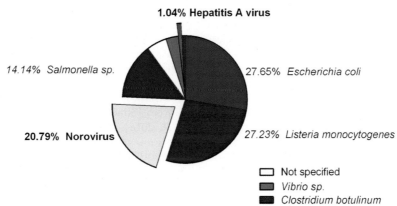

FIG. 6.1 Microbiological causes for alerts and border rejections of shellfish in the EU, between 2012 and 2019, reported by GLOBEFISH, FAO (FAO, 2020a).

foodborne outbreaks (in primary or along the food chain). NoV GII.4 strains are the most prevalent reported among noroviruses outbreaks (person-to-person and food), however, GI is more frequently found in episodes of illness after bivalve mollusc consumption (Le Guyader, Atmar, & Le Pendu, 2012).

Although recently the NoV replication has been reported using stem cell-derived nontransformed human enteroid monolayer cultures (Ettayebi et al., 2016), wild type foodborne food NoVs are not adapted to the routine culture replication. Only Murine NoV-1 (MNV-1) can be easily propagated in cell lines models, and is used as a model surrogate for human NoVs in inactivation studies (Baert et al., 2008).

NoVs are resistant to environmental stresses, including disinfectants in the concentrations usually used in water treatment plants (Cook, Knight, & Richards, 2016). NoVs are very stable in water (Ngazoa, Fliss, & Jean, 2008), including the marine environment, and may persist in the oyster's organs for 29 days (Nappier, Graczyk, Tamang, & Schwab, 2010), by attaching to their cells, in the digestive gland, because of the similarity of their ligands with the HBGA (Le Guyader et al., 2012). This characteristic explains why they remain in the oysters, even after depuration processes.

(b) Sapoviruses

Sapovirus (SAV) prototype species is *Sapporo* virus and it was first reported in 1976 in human diarrheic feces by electron microscopy (Madeley & Cosgrove, 1976). They are small, about 30–38 nm in diameter, and based on the Manchester virus strain, the full-length genomic sequence (7.4 kb) showed that the genome is organized into three ORFs (Oka, Wang, Katayama, & Saifb, 2015). The phylogenetic analysis of the VP1 amino acids sequence exhibits a high heterogeneity,

and these differences are the base of genogroups grouping. From 19 genogroups, only GI, GII, GIV, and GVI infect humans (Oka et al., 2015). SaV GIII, a porcine enteric calicivirus, is the only species adapted to growing in cell lines (Chang et al., 2004). SaV contamination in vegetables and shellfish is well documented (Gallimore et al., 2005; Hansman, Oka, Okamoto, & Nishida, 2007), and can cause gastroenteritis in different ages, by outbreaks or sporadic cases.

6.2.1.2 Picornaviridae *family*

All members of the family *Picornaviridae* have the same genome organization, with a 5′ untranslated region (UTR) having an internal ribosome entry site (IRES), ORF that encodes a single precursor polypeptide, and a poly-A tail in 3′ region (Elrick, Pekosz, & Duggal, 2021). Among this family, are several species that are associated with waterborne and foodborne outbreaks.

(a) The hepatitis A virus

The *Hepatovirus* genus is organized into nine species, however only the *Hepatovirus* A species has genotypes (IIIA and IIIB) that cause human diseases. Like other picornaviruses, Hepatitis A virus (HAV) particles are small, 27–32 nm in diameter. The genome (7.5 kb) has only one ORF that encodes a single polyprotein. During replication, this polyprotein gives rise to three precursors (P1, P2, and P3) that will generate all constituents of its viral particle (Knipe & Howley, 2013).

HAV infects epithelial cells of the small intestine and hepatocytes of some mammals, including humans. It is replicated in the liver and excreted by feces in high concentrations (10^9 infectious viruses per ml of diarrheic stool) (Knipe & Howley, 2013). The incubation period is very long (about 30 days), and viral particles are excreted for several weeks, which facilitates their spreading (Knipe & Howley, 2013). The disease can curse an unapparent infection (mainly in children) even in asymptomatic forms, however, it can progress to fulminant hepatitis in adults (< 1% of cases).

Although the primary HAV transmission source is via contaminated water or foods, the person-to-person transmission has increased, with a relevant contribution to the growth of hepatitis A worldwide (CDC, 2020). HAV is considered the most common cause of acute viral hepatitis in many countries (Pinto, de Oliveira, & González, 2017). According to the World Health Organization (WHO), approximately 1.5 million clinical cases of hepatitis A occur worldwide annually, but the infection rate is probably up to 10 times higher (WHO, 2017). The CDC (2020) reported that hepatitis A incidence increased 850% from 2014 to 2018 in the United States.

This virus is very stable in environmental matrices, tolerant to acid pH (pH 1.0 for 1 h), and elevated temperature (60 °C for 10 min) (Scholz, Heinricy, & Flehmig, 1989; Wang et al., 2015). Its persistence represents risks of foodborne diseases because HAV may stay infectious in food, after lightly cooking or pH variation (Croci et al., 1999). Moreover, HAV can persist in oysters for up to 6 weeks after contamination, even under simulated depuration conditions (Kingsley & Richards, 2003). They found that HAV is preferentially accumulated in basal cells of the stomach of oysters and that these viruses are also found contaminating oysters' hemocytes, tolerating the acidic conditions into the phagolysosomal vesicles of these cells (Kingsley, 2013; Provost et al., 2011). It facilitates the permanence of these viruses in the oysters making difficult to remove them during debugging.

(b) *Enterovirus* genus

Enterovirus genus consists in 15 species, infect 7 to humans (*Enterovirus* A–D, and *Rhinovirus* A–C). *Enterovirus* and *Rhinovirus* genera are diverse in antigenic properties and tropism (Simmonds et al., 2020), but all the species of this genera have primary replication in the gastrointestinal tract regardless of the disease caused (Knipe & Howley, 2013). Some species are reported as human waterborne and foodborne pathogens, causing mainly gastroenteritis. However, more serious diseases that affect the central nervous system may also occur. Poliovirus (types 1–3), belonging to *Enterovirus* C are the most studied species in this genus. It was the cause of the poliomyelitis pandemic, a disease that has been known since the second millennium BC, in Egypt (Knipe & Howley, 2013). The resolution adopted by the World Health Assembly in 1988 to eradicate poliomyelitis has reduced the countries where wild poliovirus is still transmitted (WHO, 2021).

Enterovirus genus was included in the Drinking Water Contaminant Candidate List and in the Unregulated Contaminant Monitoring Rule (UCMR) as viral pathogens that may pose risks for drinking water (EPA, 2016). They are very stable in the environment and can contaminate food. Studies reported Enteroviruses in bivalve molluscs (Le Guyader et al., 2008; Vinatea, Sincero, Simões, & Barardi, 2006) and crops (Prez et al., 2018).

(c) *Aichi Virus* species

Another picornavirus (*Kobuvirus* genus) that caused foodborne illness in humans is the *Aichi virus* A-1 (AiV-A1). It is associated to gastroenteritis outbreaks. AiV-A1 can contaminate oysters (Terio et al., 2018) and other foods (Rivadulla & Romalde, 2020), and epidemiological studies in several EU countries, Tunisia, and Japan reported AiV seroprevalence up to 80 and 99% (Rivadulla & Romalde, 2020).

6.2.1.3 *Human Mastadenovirus species*

Human Mastadenovirus belongs to the family *Adenoviridae*, genus *Mastadenovirus*. They are dsDNA viruses with 70–90 nm in diameter and have seven species that infect humans (A–G) producing several diseases (gastroenteritis, conjunctivitis, acute respiratory outbreaks, and others). Human Mastadenovirus species are reported as water and food contaminants (Marti & Barardi, 2016; Mena & Gerba, 2008; Souza et al., 2012), and are considered the second leading cause of childhood gastroenteritis (International Committee on Taxonomy of Viruses-ICTV, 2019; Knipe & Howley, 2013). Even species that do not cause enteric diseases are replicated in intestine cells (Knipe & Howley, 2013).

Mastadenovirus are very resistant to environmental conditions, mainly UV radiation (Eischeid, Meyer, & Linden, 2009). Its high prevalence and persistence in the environment contributed to their inclusion in the Drinking Water Contaminant Candidate List (CCL) by the US Environmental Protection Agency (USEPA, 2016) as a viral contaminant monitoring parameter.

6.2.2 Zonotic Enteric Viruses

Zoonotic viruses are those that replicate in a host-reservoir animal and spill over infecting humans. Among enteric viruses, the hepatitis E virus and rotaviruses are reported as zoonotic viruses. Although other species belonging to the enteric viruses group are pointed to as potential risks of zoonotic transmission route (Bank-Wolf, König, & Thiel, 2010; Villabruna, Koopmans, & de Graaf, 2019), up until now, there is no evidence that it occurs.

6.2.2.1 *Hepatitis E virus*

HEV is a member of the *Orthohepevirus A* species, belonging to the family *Hepeviridae*, *Orthohepevirus* genus (Purdy et al., 2017). It is found worldwide but is more common in low and middle-income countries. In these places, genotypes 1 and 2 are found in humans, causing epidemic hepatitis, through contaminated water via the fecal-oral route (Knipe & Howley, 2013). The epidemic hepatitis E occurs in some parts of Asia, Africa, and America. In industrialized countries, zoonotic transmission is more frequent (Purdy et al., 2017), associated to genotypes 3 and 4 (de Oliveira-Filho et al., 2019; Knipe & Howley, 2013). Domestic pigs are an important host of these genotypes. Zoonotic hepatitis E is usually transmitted by the consumption of meat from viremic animals (Knipe & Howley, 2013). All these four genotypes are spread by stools (from human or animal) and when they reach the shellfish areas, HEV may contaminate the coastal areas and be bioaccumulated by bivalve molluscs (Grodzki et al., 2014). Beyond the oral route, hepatitis E also occurs by vertical transmission and blood transfusion (Knipe & Howley, 2013). It is estimated 20 million HEV infections worldwide annually, with approximately 44,000 deaths in 2015. In pregnant women, hepatitis E is serious and can lead to fatal diseases (WHO, 2020).

6.2.2.2 *Rotavirus*

Rotavirus genus (RV) is organized into 10 species or groups based on their serological evidence, which are named with alphabet letters (Rotavirus A–J). These species cause diseases in different vertebrate species, including humans (International Committee on Taxonomy of Viruses, 2020; Knipe & Howley, 2013). Human diseases are caused by species A, B, and C, but they are also found infecting herds of animals (bovines, swine, goats, and sheep). RV-A is the most important species involved in human diseases and with zoonotic episodes (Matthijnssens et al., 2011). This species is the main cause of seasonal diarrhea in children (International Committee on Taxonomy of Viruses, 2020; Vos et al., 2016). RVB is associated with epidemic diarrhea in Asia and RVC with sporadic cases in children (Knipe & Howley, 2013).

The viral genome is a segmented (11 segments) double-stranded RNA and facilitates reassortment events among the RV strains, during RV infections. Each segment encodes to one specific protein (structural protein VP1–VP7 and nonstructural NSP1–NSP6) (International Committee on Taxonomy of Viruses, 2020; Matthijnssens et al., 2011).

RV species are more resistant to heat than the majority of enteric viruses (Araud et al., 2016), resistant to some organic solvents and nonionic detergents, and support acidic pH (Knipe & Howley, 2013). Several studies reported RV-A contamination in fresh vegetables, fruits, and bivalve molluscs (De Keuckelaere et al., 2015; Ito et al., 2019).

6.2.3 Other Human Viral Pathogens That May Contaminate Water and Food

6.2.3.1 *Polyomavirus*

The family *Polyomaviridae* is composed of nonenveloped and icosahedral viruses. It is organized in four genera (*Alpha-* to *Deltapolyomavirus*) that infect mammals, birds, fish, and humans. Only three genera cause human diseases: *Alphapolyomavirus*, *Betapolyomavirus*, and *Deltapolyomavirus*, that can infect mammals, birds, and fish (Knipe & Howley, 2013; Moens et al., 2017). Studies reported human Polyomaviruses (PyV) (mainly strains BK and JC) contaminating water and food (Abreu et al., 2020; Souza et al., 2012). These strains are, respectively, associated with nephropathy and progressive multifocal leukoencephalopathy (Moens et al., 2017).

PyV virions are about 40 nm in diameter and have a circular ds 5-Kbp dsDNA genome (Moens et al., 2017). It has already been detected in the gastrointestinal tract, with elimination in urine and feces, suggesting fecal-urine oral transmission (Bofill-Mas, Formiga-Cruz, Clemente-Casares, Calafell, & Girones, 2001). It can also be transmitted through the respiratory tract, being found in tonsillar tissue and respiratory aspirates, and also through drink and food (Bofill-Mas et al., 2013).

6.2.3.2 *Coronavirus*

The first coronavirus (CoV) was described in the 1930s, in poultry, and several strains were discovered until nowadays. They cause respiratory, liver, gastrointestinal, and neurological diseases, able to infect humans, and other vertebrates (Cui, Li, & Shi, 2019; Knipe & Howley, 2013). Coronaviruses are enveloped viruses, with a genome with positive sense ssRNA (Knipe & Howley, 2013). Three species are epidemiologically more relevantly associated with human diseases causing infection and obstruction of the lungs (Knipe & Howley, 2013; Zhao et al., 2020): SARS-CoV-1, responsible for the SARS epidemic in 2002 (Cheng, Lau, Woo, & Yuen, 2007); MERS-CoV, cause of the Middle East respiratory syndrome in 2012 (Mubarak, Alturaiki, & Hemida, 2019) and SARs-CoV-2 that is the cause of the COVID-19 pandemic (Ahmed et al., 2020). Although SARs-CoV-2 transmission is mainly by aerosols, via the respiratory tract, other transmission routes are not yet completely ruled out. Studies reported that this virus has a high resistance in the environment (Lodder & de Roda Husman, 2020) and, also, since it is excreted by feces it has the potential risk of water and food contamination (Wu et al., 2020). Recent studies pointed to SARs-CoV-2 detection in the sewage worldwide, even before the COVID-19 first report in China, 2019 (Ahmed et al., 2020; Fongaro et al., 2020; La Rosa et al., 2021). According to the WHO's January 17, 2021 case information report, CoV has reached about 93 million reported cases and more than 2 million deaths worldwide (WHO, 2021). However, there is, to date, no documented evidence that food is a likely source or route of transmission of Sars-CoV-2 (EFSA, 2020).

6.3 VIRAL CONTAMINATIONS IN CROPS AND SHELLFISH PRODUCTION AREAS

6.3.1 Fresh Vegetables and Rinsed Fruits

Fresh vegetables and rinsed fruits are vulnerable to viral contamination from crops to sale sites (Fig. 6.2). Animal manure is a valuable nutrient source for crops, however, studies reported that it can have zoonotic pathogens, including viruses (Farzan et al., 2011; Fongaro et al., 2014; Miller et al., 2015; Souza et al., 2020).

The zoonotic HEV and RV-A are found in some animal herds, such as domestic pigs. HEV and RVA are resistant to stresses in the environment and may persist infectious in soil for extended periods contaminating vegetables and fruits (Bosch, Pintó, & Abad, 2006; Parashar, Khalkar, & Arankalle, 2011). Other studies reported that swine and poultry manure can contain a high concentration of zoonotic viruses (Doro, Farkas, Martella, & Banyai, 2015; Dufour, 2015). The use of manure in crops make viral zoonotic contamination the most found in vegetables (fresh and fresh-cut leafy vegetables and herbs) and rinsed fruits contaminated in the primary production sites. At other stages of the food-chain, nonzoonotic human viruses are more frequently found in food contamination by infectious food handlers and fomites (Rzezutka & Cook, 2004).

6.3.2 Bivalve Molluscs

According to the *Food and Agriculture Organization* (FAO, 2020b), the world production of molluscs was 17.5 million tons in 2018. Many countries have great potential in the production of bivalve molluscs because they have an extensive coastal strip. However, several

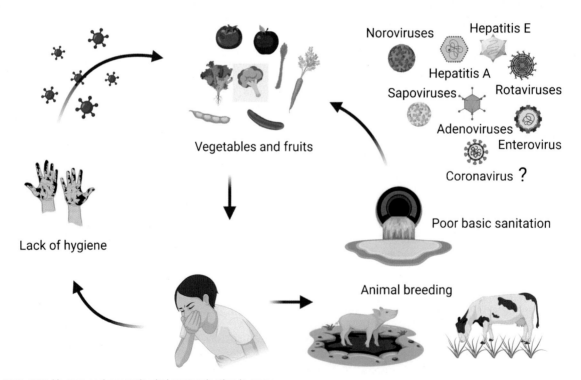

FIG. 6.2 Human and zoonotic viral contamination in crops.

of them have shortcomings in collecting processes and sewage treatments, which affect the sanitary quality of the animals produced in these areas. This lack of sanitary controls greatly affects bivalve mollusc production. Because they are filter-feeders their microbiological condition is closely related to the growing area condition (Hassard et al., 2017; Souza et al., 2012). Bivalve molluscs accumulate several contaminants inside their organs (Souza et al., 2012; Souza, Dominot, Moresco, & Barardi, 2018). Some species filtered up to 10 liters per hour and studies reported that human viral pathogens concentration is much higher in these animals than in the growing water (Butt, Aldridge, & Sanders, 2004).

6.3.2.1 Foodborne outbreaks linked to oysters

Analyzing data on foodborne outbreaks reported by the European Food Safety Authority (EFSA, 2016), the virus-related outbreaks were the 36.8% of the 2015 total cases. Around 27% was associated with crustacean or bivalve molluscs consumption. Oysters are the shellfish mostly involved in foodborne outbreaks, associated to contaminations in the growing areas (Kingsley & Richards, 2003; Le Guyader et al., 2012). They are typically eaten raw or lightly cooked and harbor several viral contaminants. Moreover, oyster meat is a complex matrix containing many substances that would protect enteric viruses from heat inactivation. Bertrand et al. (2012) demonstrated that food matrices cause a protective effect to the viral inactivation during cooking processes. Studies pointed to the failure of some heat-treatments for complete inactivation of enteric virus contaminations in bivalve molluscs species (Croci et al., 1999; Croci, De Medici, Di Pasquale, & Toti, 2005; Polo, Álvarez, Vilariño, Longa, & Romalde, 2014; Souza et al., 2018).

The sanitary controls worldwide for bivalve molluscs cultivation and marketing are usually based on fecal bacteria monitoring in the seawater and in the mollusc meat (Food Standard Agency, 2020; NSSP, 2017). According to bacteria indicator levels, they establish shellfish harvesting permission with or without postharvest treatments (no pretreatment; or depuration; or relaying; or heating treatment) before shellfish commercialization. However, studies reported that enteric bacteria inactivation in seawater occurs until 5 days, while enteric virus persistence in the marine environment (water and shellfish) may remain for up to 130 days (NSSP, 2017).

The Europe Union follow the standards recommended by EU SQAP (*The European Union Shellfish Quality Assurance Program*), which classifies mollusc farming waters into three classes: A, B, and C (Table 6.1),

according to the level of *E. coli* in 100 g of shellfish flesh and intravalvular liquid (Food Standard Agency, 2020). In the United States, the National Shellfish Sanitation Program was prepared by the FDA (Food and Drug Administration) for the control of the growth, processing, and transportation of bivalve molluscs (NSSP, 2017). They established that the criteria used for the control of growing areas are based on the bacterial parameters in surface waters—total and thermotolerant coliforms (*E. coli*) (Table 6.1).

Bacteria monitoring in shellfish (in surface water or animal meat) is based on culture protocols, while enteric viruses are monitored using molecular techniques (RT-PCR and real-time RT-PCR), and consequently the viral investigation on environmental samples, such as water and food, is complex. Usually, the viruses are diluted in large volumes, requiring concentration techniques, and these samples may contain substances that inhibit the detection and quantification of the viruses by molecular techniques. In addition, few enteric viruses are adapted to cell culture assays, and genome detection by molecular methods cannot determine the viral infectivity on environmental samples (Barardi et al., 2012).

6.4 FOODBORNE VIRUSES CONTROL IN MINIMALLY PROCESSED FOOD

The resistance of the waterborne and foodborne viruses in the environment is a great challenge for the food industry. The capability of these viruses to support extreme acidification and the increase of temperature, as demonstrated in Table 6.2, coupled with their persistence in bivalve molluscs makes them difficult to control in minimally processed foods.

6.4.1 Vegetables and Fruits

Fruits and vegetables are among the main types of food that can be contaminated by bacteria and viruses (Fung, Wang, & Menon, 2018). This contamination occurs through water, soil, dust, and postharvest processes (Mritunjay & Kumar, 2015). Microbiological contamination can occur in several stages: production, distribution, processing, packaging, and its preparation (Harding et al., 2017). To reduce fruit and vegetable contamination after harvesting, some management procedures are performed, mainly chlorination of washing water, refrigerated transport, and storage. Postharvest treatments are an efficient way to sanitize vegetables and fruits. One of these is through chemical products, the most common being sodium hypochlorite, often used in concentrations of 200–500 ppm, in the rinsing

TABLE 6.2
Enteric virus resistance to environmental stresses.

Virus	Matrix	Stress condition	Persistence (infectious)	Reference
Astrovirus	Fomites	Drying	7 to > 60 days	Abad et al. (2001)
Poliovirus, Hepatitis A virus and Mastadenovirus	Water	Drinking water	up to 41 days up to 56 days up to 92 days	Enriquez (1995)
Poliovirus	Fresh vegetables	Celery and spinach	up to 76 days up to 55 days	Ward and Irving (1987)
Enteroviruses (Poliovirus and Coxsackievirus)	Dairy products	Pasteurized milk and boiled milk	15 to > 30 days	Rzezutka and Cook (2004)

process (Bhilwadikar, Pounraj, Manivannan, Rastogi, & Negi, 2019). This chemical does not produce toxic by-products and can reduce pesticide residues in fruits and vegetables (Bhilwadikar et al., 2019).

Jeong, Park, and Ha (2018) evaluated the effectiveness of sodium hypochlorite and peroxyacetic acid on the inactivation of MNV-1 in cabbage and green onion. For peroxyacetic acid treatment, the concentration of 300 ppm for green onion or 500 ppm for cabbage were suitable for inactivation of MNV-1 and does not adversely affected the food quality. The sodium hypochlorite treatment at 400 ppm decreased MNV-1 by more than 1 $\log_{10}$ in both samples, when the chlorine concentration increased, however, it produced an unpleasant chlorinated-odor in cabbage.

Another method commonly used as a disinfectant in the food industry is electrolyzed water, which contains chlorine and ozone in its composition, making this treatment an effective oxidizing agent, capable of destroying pathogenic microorganisms and reducing pesticide residues. According to Rahman, Khan, and Oh (2016), electrolyzed water showed virucidal activity in different sectors including fruits, vegetables, seafood, eggs and poultry. However, with very low pH (2.7) it could affect the organoleptic properties of some foods, which limits its use. Furthermore, the inactivation of pathogens through irradiation is effective and has advantages such as: not heating the product and not leaving toxic residues, especially in aqueous solutions (Bhilwadikar et al., 2019).

The food sanitization alternatives above mentioned must address the particularities of each virus, since they may behave differently in the presence of these agents, especially in foods. Thus, the efficacy shown against bacteria is not always the same against viruses.

6.4.2 Bivalve Molluscs

Depuration uses the natural ability of bivalve molluscs to clean themselves while filtering water for food. This process may occur in tanks or the sea (relaying). Depuration in tanks was initially developed to reduce and control fecal bacteria pathogens. The animals remain in the tanks for periods of 24–48 h, so that they can be decontaminated (Lee, Lovatelli, & Ababouch, 2008; McLeod, Polo, Le Saux, & Le Guyader, 2017). These depurate systems may be open, semiopen, or closed (with recirculation of water), according to water flow in the tank. In some cases, to ensure the sanitary quality of the water in the tank, physical or chemical processes are applied in this water (Lees, Younger, & Dore, 2010). The main and best known are chlorination, ozone, and ultraviolet (UV). Chlorination has a high disinfectant capacity and low cost, but generates organoleptic changes, inhibits the mollusc filtration activity, and even generates cancer-causing by-products (Lees et al., 2010). Although ozone does not alter the characteristics of bivalves, it has a high cost, and its efficiency is influenced by changes in pH and temperature (Lees et al., 2010). Another disadvantage is that it can interrupt or reduce animal activity (Lee et al., 2008). UV does not affect the organoleptic or filtration activity of molluscs, however, its performance is affected by turbidity of water and Mastadenovirus are more resistant to UV disinfection than the other viral pathogens (Eischeid et al., 2009).

The oysters' depuration carried out in tanks is not enough for the NoV removal. These viruses (especially from genogroup I) can persist attached to the cells of the oyster's digestive tissues, preventing their removal within the time limit established by the depuration protocols (Le Guyader et al., 2012). Relaying is another protocol for shellfish depuration that allows bivalve molluscs decontamination by transferring the animals to sites with clean water (based on indicator bacteria parameters). Relaying allows self-purification in the natural environment for 2 months, longer purification time would facilitate noroviruses removal (McLeod et al., 2017). However, an efficient relaying process is dependent on strict control of microorganisms (fecal bacteria and enteric viruses) in the relaying area. Oyster species are adapted to grow in bays and coves (usually near seaside towns) it may not be easy to find relaying places, with less anthropogenic impact, and especially in developing countries.

The natural compounds are promising in the disinfection process. Some of them have antimicrobial activity without any harmful effect on human health (Su, Howell, & D'Souza, 2010). Curcuma (also called turmeric or saffron) has numerous biological properties of interest for application in food, with proven antioxidant, antifungal, and antimicrobial activity. Wu et al. (2015) used photodynamic action mediated by curcumin, with a wavelength of 470 nm, energy density 3.6 J/cm^2, and a concentration of 10 µM curcumin, demonstrating that MNV-1, a surrogate of NoV, was significantly inhibited from 1–3 log PFU/mL in oysters.

However, tests are not yet available performed in live oysters, applying natural products for viral decontamination in tanks (Fig. 6.3).

6.5 THE CHALLENGE OF EMERGENT VIRAL PATHOGENS IN FOOD

Foods have a high complexity of substances and virus concentration protocols are required which are capable of reducing as much as the matrix presence and retain the viral particles in small volumes. Organic and inorganic substances may produce inhibitors effects to the molecular techniques used in viral detection and quantification in food. PCR and real-time PCR are the most common techniques usually in viral food analysis worldwide. They have a high sensibility, but are impacted by inhibitory substances, such as polysaccharides, some proteins (hemoglobin, collagen, and others), humic acids, and others that are frequently present in food matrices (Schrader, Schielke, Ellerbroek, & Johne, 2012).

Although protocols for detection of enteric viruses in vegetables, fruits, and shellfish are currently available (CEFAS, 2016; ISO/TS 15216-1:2017, 2017), emerging viral pathogens that represent potential risks of

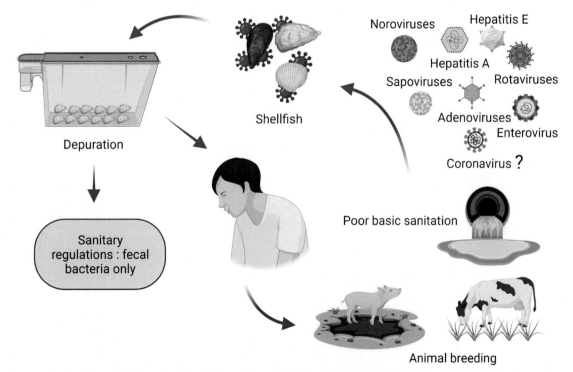

FIG. 6.3 Methods used to bivalve molluscs depuration and disinfection and new potential based on natural compounds.

transmission by minimally processed foods are continually discovered. The COVID-19 pandemic raised questions about the capability of the current protocols for correct detection of nonenteric viruses, such as SARs-CoV-2. The great variability of food matrix compounds coupled with the frequency which new viral strains appear, require continuous studies about better protocols of elution/concentration, nucleic acid extraction, and viral detection in food.

ACKNOWLEDGMENTS

The authors thanks CAPES/PNPD for the postdoc fellowship (N 88887.473179/2020-00) of Doris Sobral Marques Souza and CAPES/Master for the master fellowship of Vilaine Correa da Silva.

REFERENCES

Abad, F. X., Villena, C., Guix, S., Caballero, S., Pintó, R. M., & Bosch, A. (2001). Potential role of fomites in the vehicular transmission of human astroviruses. *Applied and Environmental Microbiology, 67*(9), 3904–3907.

Abreu, I. N., Cortinhas, J. M., Dos Santos, M. B., Queiroz, M. A. F., Da Silva, A. N. M. R., Cayres-Vallinoto, I. M. V., et al. (2020). Detection of human polyomavirus 2 (HPyV2) in oyster samples in northern Brazil. *Virology Journal, 17*(1), 1–6.

Ahmed, W., Angel, N., Edson, J., Bibby, K., Bivins, A., O'Brien, J. W., et al. (2020). First confirmed detection of SARS-CoV-2 in untreated wastewater in Australia: A proof of concept for the wastewater surveillance of COVID-19 in the community. *The Science of the Total Environment,* 138764. [Internet]. Available from: https://linkinghub.elsevier.com/retrieve/pii/S0048969720322816.

Araud, E., DiCaprio, E., Ma, Y., Lou, F., Gao, Y., Kingsley, D., et al. (2016). Thermal inactivation of enteric viruses and bioaccumulation of enteric foodborne viruses in live oysters (*Crassostrea virginica*). *Applied and Environmental Microbiology, 82*(7), 2086–2099.

Baert, L., Wobus, C. E., Van Coillie, E., Thackray, L. B., Debevere, J., & Uyttendaele, M. (2008). Detection of murine norovirus 1 by using plaque assay, transfection assay, and real-time reverse transcription-PCR before and after heat exposure. *Applied and Environmental Microbiology, 74*(2), 543–546.

Bank-Wolf, B. R., König, M., & Thiel, H. J. (2010). Zoonotic aspects of infections with noroviruses and sapoviruses. *Veterinary Microbiology, 140*(3–4), 204–212.

Barardi, C. R., Viancelli, A., Rigotto, C., Correa, A. A., Moresco, V., Souza, D. S., et al. (2012). Monitoring viruses in environmental samples. *International Journal of Environmental Science and Engineering Research, 3*(3), 1–9.

Bertrand, I., Schijven, J. F., Sánchez, G., Wyn-Jones, P., Ottoson, J., Morin, T., et al. (2012). The impact of temperature on the inactivation of enteric viruses in food and water: A review. *Journal of Applied Microbiology, 112*(6), 1059–1074.

Bhilwadikar, T., Pounraj, S., Manivannan, S., Rastogi, N. K., & Negi, P. S. (2019). Decontamination of microorganisms and pesticides from fresh fruits and vegetables: A comprehensive review from common household processes to modern techniques. *Comprehensive Reviews in Food Science and Food Safety, 18*(4), 1003–1038. [Internet]. Available from: https://onlinelibrary.wiley.com/doi/abs/10.1111/1541-4337.12453.

Bofill-Mas, S., Formiga-Cruz, M., Clemente-Casares, P., Calafell, F., & Girones, R. (2001). Potential transmission of human polyomaviruses through the gastrointestinal tract after exposure to virions or viral DNA. *Journal of Virology, 75*(21), 10290–10299.

Bofill-Mas, S., Rusiñol, M., Fernandez-Cassi, X., Carratalà, A., Hundesa, A., & Girones, R. (2013). Quantification of human and animal viruses to differentiate the origin of the fecal contamination present in environmental samples. *BioMed Research International,* 2013.

Bosch, A., Abad, F. X., & Pintó, R. M. (2005). Human pathogenic viruses in the marine environment. In *Oceans and health: Pathogens in the marine environment* (pp. 109–131). Springer.

Bosch, A., Pintó, R. M., & Abad, F. X. (2006). Survival and transport of enteric viruses in the environment. In *Viruses in foods* (pp. 151–187). Springer.

Butt, A. A., Aldridge, K. E., & Sanders, C. V. (2004). Infections related to the ingestion of seafood Part I: Viral and bacterial infections. *The Lancet Infectious Diseases, 4*(4), 201–212. [Internet]. Available from: http://www.thelancet.com/pdfs/journals/laninf/PIIS1473-3099(04)00969-7.pdf.

Cardemil, C. V., & Hall, A. J. (2020). Travel-related infectious diseases. Norovirus. In *CDC yellow book* (p. 78). Centre for Disease Control and Prevention. [Internet]. (Chapter 4). Available from: https://wwwnc.cdc.gov/travel/yellowbook/2020/travel-related-infectious-diseases/norovirus.

CDC. (2020). *Viral hepatitis surveillance report 2018—Hepatitis A. Vol. 247* (pp. 2011–2018). CDC. [Internet]. Available from: https://www.cdc.gov/hepatitis/statistics/2018surveillance/HepA.htm.

CEFAS. (2016). *Quantitative detection of norovirus and hepatitis A virus in bivalve molluscan shellfish.* Weymouth: CEFAS.

Chang, K. O., Sosnovtsev, S. V., Belliot, G., Kim, Y., Saif, L. J., & Green, K. Y. (2004). Bile acids are essential for porcine enteric calicivirus replication in association with down-regulation of signal transducer and activator of transcription 1. *Proceedings of the National Academy of Sciences of the United States of America, 101*(23), 8733–8738.

Cheng, V. C. C., Lau, S. K. P., Woo, P. C. Y., & Yuen, K. Y. (2007). Severe acute respiratory syndrome coronavirus as an agent of emerging and reemerging infection. *Clinical Microbiology Reviews, 20*(4), 660–694. [Internet]. Available from: https://cmr.asm.org/content/20/4/660.

Chhabra, P., de Graaf, M., Parra, G. I., Chan, M. C. W., Green, K., Martella, V., et al. (2019). Updated classification of norovirus genogroups and genotypes. *The Journal of General Virology, 100*(10), 1393–1406.

Cook, N., Knight, A., & Richards, G. P. (2016). Persistence and elimination of human norovirus in food and on food contact surfaces: A critical review. *Journal of Food Protection*, *79*(7), 1273–1294.

Croci, L., Ciccozzi, M., De Medici, D., Di Pasquale, S., Fiore, A., Mele, A., et al. (1999). Inactivation of hepatitis A virus in heat-treated mussels. *Journal of Applied Microbiology*, *87*(6), 884–888.

Croci, L., De Medici, D., Di Pasquale, S., & Toti, L. (2005). Resistance of hepatitis A virus in mussels subjected to different domestic cookings. *International Journal of Food Microbiology*, *105*(2), 139–144.

Cui, J., Li, F., & Shi, Z. L. (2019). Origin and evolution of pathogenic coronaviruses. *Nature Reviews. Microbiology*, *17*(3), 181–192.

Da Silva, A. K., Le Saux, J. C., Parnaudeau, S., Pommepuy, M., Elimelech, M., & Le Guyader, F. S. (2007). Evaluation of removal of noroviruses during wastewater treatment, using real-time reverse transcription-PCR: Different behaviors of genogroups I and II. *Applied and Environmental Microbiology*, *73*(24), 7891–7897.

De Keuckelaere, A., Jacxsens, L., Amoah, P., Medema, G., Mcclure, P., Jaykus, L. A., et al. (2015). Zero risk does not exist: Lessons learned from microbial risk assessment related to use of water and safety of fresh produce. *Comprehensive Reviews in Food Science and Food Safety*, *14*(4), 387–410.

de Oliveira-Filho, E. F., Dos Santos, D. R. L., Durães-Carvalho, R., da Silva, A., de Lima, G. B., Filho, A. F. B. B., et al. (2019). Evolutionary study of potentially zoonotic hepatitis E virus genotype 3 from swine in northeast Brazil. *Memórias do Instituto Oswaldo Cruz*, *114*(17), 1–5.

Doro, R., Farkas, S. L., Martella, V., & Banyai, K. (2015). Zoonotic transmission of rotavirus: Surveillance and control. *Expert Review of Anti-Infective Therapy*, *13*(11), 1337–1350.

Dufour, A. (2015). *Animal waste, water quality and human health. Vol. 12*. Water Intelligence Online.

EFSA. (2016). *European food safety authority—Annual accounts 2015*. [Internet]. Parma. Available from: www.efsa.europa.eu.

EFSA. (2020). Coronavirus: No evidence that food is a source or transmission route. *EFSA Journal*, *6*(9), 726. [Internet]. Available from: https://www.efsa.europa.eu/it/news/mammalnet-live-download-simple-app-and-help-us-collect-data-wild-mammals-europe.

Eischeid, A. C., Meyer, J. N., & Linden, K. G. (2009). UV disinfection of adenoviruses: Molecular indications of DNA damage efficiency. *Applied and Environmental Microbiology*, *75*(1), 23–28.

Elrick, M. J., Pekosz, A., & Duggal, P. (2021). Enterovirus D68 molecular and cellular biology, and pathogenesis. *The Journal of Biological Chemistry*, *269*, 100317. [Internet]. Available from: https://doi.org/10.1016/j.phrs.2020.104743.

Enriquez, C. (1995). Survival of the enteric adenoviruses 40 and 41 in tap, sea, and waste water. *Water Research*, *29*(11), 2548–2553. [Internet]. Available from: https://linkinghub.elsevier.com/retrieve/pii/0043135495000702.

EPA. (2016). *Drinking water contaminant candidate list 4-final* (pp. 1–36).

Ettayebi, K., Crawford, S. E., Murakami, K., Broughman, J. R., Karandikar, U., Tenge, V. R., et al. (2016). Replication of human noroviruses in stem cell-derived human enteroids. *Science (80-)*, *353*(6306), 1387–1393. [Internet]. Available from: https://www.sciencemag.org/lookup/doi/10.1126/science.aaf5211.

FAO. (2020a). *Alerts & border rejections by causes (2012-2019)—Microbiological causes _ GLOBEFISH—Information and Analysis on World Fish Trade _ Food and Agriculture Organization of the United Nations _ GLOBEFISH _ Food and Agriculture Organization of the Unite.pdf* (Internet). Food and Agriculture Organization of the United Nations. [cited 2021 Jan 12]. Available from: http://www.fao.org/in-action/globefish/border-rejections/border-rejections-eu/en/.

FAO. (2020b). The state of world fisheries and aquaculture 2020 (Internet). *Nature and resources: Vol. 35* (pp. 4–13). Rome: FAO. Available from: http://www.fao.org/documents/card/en/c/ca9229en.

FAO/WHO. (2008). Viruses in food: Scientific advice to support risk management activities: Meeting report. Rome *Microbiological risk assessment series*. Report No.: 13.

Farzan, A., Parrington, L., Coklin, T., Cook, A., Pintar, K., Pollari, F., et al. (2011). Detection and characterization of giardia duodenalis and cryptosporidium spp. on swine farms in Ontario, Canada. *Foodborne Pathogens and Disease*, *8*(11), 1207–1213.

Fongaro, G., Hermes Stoco, P., Sobral Marques Souza, D., Grisard, E. C., Magri, M. E., Rogovski, P., et al. (2020). *SARS-CoV-2 in human sewage in Santa Catalina, Brazil, November 2019*. (January).

Fongaro, G., Viancelli, A., Magri, M. E., Elmahdy, E. M., Biesus, L. L., Kich, J. D., et al. (2014). Utility of specific biomarkers to assess safety of swine manure for biofertilizing purposes. *The Science of the Total Environment*, *479–480*(1), 277–283. [Internet]. Available from: https://doi.org/10.1016/j.scitotenv.2014.02.004.

Food Standard Agency. (2020). *Shellfish classification how production areas are classified, the method of treatment and how to apply for classification* (Internet). FSA. [cited 2020 Sep 28]. Available from: https://www.food.gov.uk/business-guidance/shellfish-classification.

Fung, F., Wang, H.-S., & Menon, S. (2018). Food safety in the 21st century. *Biomedical Journal*, *41*(2), 88–95. [Internet]. Available from: https://linkinghub.elsevier.com/retrieve/pii/S2319417017304055.

Gallimore, C. I., Pipkin, C., Shrimpton, H., Green, A. D., Pickford, Y., McCartney, C., et al. (2005). Detection of multiple enteric virus strains within a foodborne outbreak of gastroenteritis: An indication of the source of contamination. *Epidemiology and Infection*, *133*(1), 41–47.

Gerba, C. P. (2000). Assessment of enteric pathogen shedding by bathers during recreational activity and its impact on water quality. *Quantitative Microbiology*, *2*, 55–68.

Grodzki, M., Schaeffer, J., Piquet, J. C., Le Saux, J. C., Chevé, J., Ollivier, J., et al. (2014). Bioaccumulation efficiency, tissue distribution, and environmental occurrence of hepatitis E virus in bivalve shellfish from France. *Applied and Environmental Microbiology, 80*(14), 4269–4276.

Hansman, G. S., Oka, T., Okamoto, R., & Nishida, T. (2007). Human sapovirus in clams, Japan. *Emerging Infectious Diseases, 13*(4), 5–7.

Harding, M. W., Butler, N., Dmytriw, W., Rajput, S., Burke, D. A., & Howard, R. J. (2017). Characterization of microorganisms from fresh produce in Alberta, Canada reveals novel food-spoilage fungi. *Research Journal of Microbiology, 12*(1), 20–32. [Internet]. Available from: http://www.scialert.net/abstract/?doi=jm.2017.20.32.

Hassard, F., Sharp, J. H., Taft, H., LeVay, L., Harris, J. P., McDonald, J. E., et al. (2017). Critical review on the public health impact of norovirus contamination in shellfish and the environment: A UK perspective. *Food and Environmental Virology, 9*(2), 123–141.

International Committee on Taxonomy of Viruses. (2020). *Reoviridae* (Internet). ICTV. [cited 2020 Oct 26]. Available from: https://talk.ictvonline.org/.

International Committee on Taxonomy of Viruses-ICTV. (2019). *Adenoviridae. Vol. 2009* (pp. 1–21). ICTV 9th Report (2011). [Internet] [cited 2021 Jan 21]. Available from: https://talk.ictvonline.org/.

ISO/TS 15216-1:2017. (2017). *ISO 15216-1:2017 Microbiology of the food chain—Horizontal method for determination of hepatitis A virus and norovirus using real-time RT-PCR—Part 1: Method for quantification.* [Internet]. Available from: https://www.iso.org/standard/65681.html.

Ito, E., Pu, J., Miura, T., Kazama, S., Nishiyama, M., Ito, H., et al. (2019). Weekly variation of rotavirus a concentrations in sewage and oysters in Japan, 2014–2016. *Pathogens, 8*(3), 2014–2016.

Jeong, M.-I., Park, S. Y., & Ha, S.-D. (2018). Effects of sodium hypochlorite and peroxyacetic acid on the inactivation of murine norovirus-1 in Chinese cabbage and green onion. *LWT, 96*, 663–670. [Internet]. Available from: https://linkinghub.elsevier.com/retrieve/pii/S0023643818305309.

Kingsley, D. H. (2013). High pressure processing and its application to the challenge of virus-contaminated foods. *Food and Environmental Virology, 5*(1), 1–12.

Kingsley, D. H., & Richards, G. P. (2003). Persistence of hepatitis A virus in oysters. *Journal of Food Protection, 66*(2), 331–334.

Knipe, D. M., & Howley, P. M. (Eds.). (2013). *Fields virology* (6th ed.). Philadelphia: Lippincott Williams & Wilkins, a Wolters Kluwer Business.

La Rosa, G., Mancini, P., Bonanno Ferraro, G., Veneri, C., Iaconelli, M., Bonadonna, L., et al. (2021). SARS-CoV-2 has been circulating in northern Italy since December 2019: Evidence from environmental monitoring. *The Science of the Total Environment, 750*, 141711. [Internet]. Available from: https://linkinghub.elsevier.com/retrieve/pii/S0048969720352402.

Le Guyader, F. S., Atmar, R. L., & Le Pendu, J. (2012). Transmission of viruses through shellfish: When specific ligands come into play. *Current Opinion in Virology, 2*(1), 103–110.

Le Guyader, F. S., Le Saux, J. C., Ambert-Balay, K., Krol, J., Serais, O., Parnaudeau, S., et al. (2008). Aichi virus, norovirus, astrovirus, enterovirus, and rotavirus involved in clinical cases from a French oyster-related gastroenteritis outbreak. *Journal of Clinical Microbiology, 46*(12), 4011–4017.

Lee, R., Lovatelli, A., & Ababouch, L. (2008). *Bivalve depuration: Fundamental and practical aspects. Vol. 511.* [Internet]. FAO Fisheries Technical Paper. Rome. Available from: https://www.m-culture.go.th/mculture_th/download/king9/Glossary_about_HM_King_Bhumibol_Adulyadej's_Funeral.pdf.

Lees, D., Younger, A., & Dore, B. (2010). *Depuration and relaying* (pp. 145–181). Safe Management of Shellfish and Harvest Waters.

Lodder, W., & de Roda Husman, A. M. (2020). SARS-CoV-2 in wastewater: Potential health risk, but also data source. *The Lancet Gastroenterology & Hepatology, 5*(6), 533–534.

Madeley, C. R., & Cosgrove, B. P. (1976). Caliciviruses in man. *Lancet, 307*(7952), 199–200. [Internet]. Available from: https://linkinghub.elsevier.com/retrieve/pii/S014067367691309X.

Marti, E., & Barardi, C. R. M. (2016). Detection of human adenoviruses in organic fresh produce using molecular and cell culture-based methods. *International Journal of Food Microbiology, 230*, 40–44.

Matthijnssens, J., Ciarlet, M., McDonald, S. M., Attoui, H., Bányai, K., Brister, J. R., et al. (2011). Uniformity of rotavirus strain nomenclature proposed by the Rotavirus Classification Working Group (RCWG). *Archives of Virology, 156*(8), 1397–1413.

McLeod, C., Polo, D., Le Saux, J. C., & Le Guyader, F. S. (2017). Depuration and relaying: A review on potential removal of norovirus from oysters. *Comprehensive Reviews in Food Science and Food Safety, 16*(4), 692–706.

Mena, K., & Gerba, C. (2008). Waterborne adenovirus. *Reviews of Environmental Contamination and Toxicology, 198*, 133–167.

Miller, L. A., Colby, K., Manning, S. E., Hoenig, D., McEvoy, E., Montgomery, S., et al. (2015). Ascariasis in humans and pigs on small-scale farms, Maine, USA, 2010–2013. *Emerging Infectious Diseases, 21*(2), 332–334. [Internet]. Available from: http://wwwnc.cdc.gov/eid/article/21/2/14-0048_article.htm.

Moens, U., Calvignac-Spencer, S., Lauber, C., Ramqvist, T., Feltkamp, M. C. W., Daugherty, M. D., et al. (2017). ICTV virus taxonomy profile: Polyomaviridae. *The Journal of General Virology, 98*(6), 1159–1160.

Monteiro, C. A., Cannon, G., Levy, R. B., Moubarac, J.-C., Jaime, P., Martins, A. P., et al. (2016). NOVA. The star shines bright (Food classification. Public health). *World Nutrition, 7*(1–3), 28–38. with Ricardo C, Calixto G, Machado P, Martins C, Martinez E, Baraldi L, Garzillo J SI.

Mritunjay, S. K., & Kumar, V. (2015). Fresh farm produce as a source of pathogens: A review. *Research Journal of Environmental Toxicology, 9*(2), 59–70. [Internet]. Available from: http://www.scialert.net/abstract/?doi=rjet.2015.59.70.

Mubarak, A., Alturaiki, W., & Hemida, M. G. (2019). Middle East respiratory syndrome coronavirus (MERS-CoV): Infection, immunological response, and vaccine development. *Journal of Immunology Research, 2019*, 1–11. [Internet]. Available from: https://www.hindawi.com/journals/jir/2019/6491738/.

Nappier, S. P., Graczyk, T. K., Tamang, L., & Schwab, K. J. (2010). Co-localized *Crassostrea virginica* and *Crassostrea ariakensis* oysters differ in bioaccumulation, retention and depuration of microbial indicators and human enteropathogens. *Journal of Applied Microbiology, 108*(2), 736–744.

Ngazoa, E. S., Fliss, I., & Jean, J. (2008). Quantitative study of persistence of human norovirus genome in water using TaqMan real-time RT-PCR. *Journal of Applied Microbiology, 104*(3), 707–715. [Internet]. Available from: http://doi.wiley.com/10.1111/j.1365-2672.2007.03597.x.

Noble, R. T., Lee, I. M., & Schiff, K. C. (2004). Inactivation of indicator micro-organisms from various sources of faecal contamination in seawater and freshwater. *Journal of Applied Microbiology, 96*(3), 464–472.

NSSP. (2017). *National Shellfish Sanitation Program (NSSP) guide for the control of molluscan shellfish 2017 revision.* US FDA.

Oka, T., Wang, Q., Katayama, K., & Saifb, L. J. (2015). Comprehensive review of human sapoviruses. *Clinical Microbiology Reviews, 28*(1), 32–53.

Parashar, D., Khalkar, P., & Arankalle, V. A. (2011). Survival of hepatitis A and E viruses in soil samples optimization of RNA extraction procedure treatment of soil samples with potassium acetate survival of HAV and HEV in the soil samples survival of HAV and HEV in the spiked soil samples. *Clinical Microbiology and Infection, 17*, E1–E4.

Pinto, M. A., de Oliveira, J. M., & González, J. (2017). Hepatitis A and E in South America: New challenges toward prevention and control. In *Human virology in Latin America* (pp. 119–138). Cham: Springer International Publishing. [Internet]. Available from: http://link.springer.com/10.1007/978-3-319-54567-7_7.

Polo, D., Álvarez, C., Vilariño, M. L., Longa, Á., & Romalde, J. L. (2014). Depuration kinetics of hepatitis A virus in clams. *Food Microbiology, 39*, 103–107.

Prez, V. E., Martínez, L. C., Victoria, M., Giordano, M. O., Masachessi, G., Ré, V. E., et al. (2018). Tracking enteric viruses in green vegetables from central Argentina: Potential association with viral contamination of irrigation waters. *The Science of the Total Environment, 637–638*, 665–671.

Provost, K., Dancho, B. A., Ozbay, G., Anderson, R. S., Richards, G. P., & Kingsley, D. H. (2011). Hemocytes are sites of enteric virus persistence within oysters. *Applied and Environmental Microbiology, 77*(23), 8360–8369.

Purdy, M. A., Harrison, T. J., Jameel, S., Meng, X.-J., Okamoto, H., Van der Poel, W. H. M., et al. (2017). ICTV virus taxonomy profile: Hepeviridae. *The Journal of General Virology, 98*, 2645–2646. [Internet]. [cited 2021 Jan 21]. Available from: https://talk.ictvonline.org/ictv-reports/.

Rahman, S., Khan, I., & Oh, D.-H. (2016). Electrolyzed water as a novel sanitizer in the food industry: Current trends and future perspectives. *Comprehensive Reviews in Food Science and Food Safety, 15*(3), 471–490. [Internet]. Available from: http://doi.wiley.com/10.1111/1541-4337.12200.

Rivadulla, E., & Romalde, J. L. (2020). A comprehensive review on human Aichi virus. *Virologica Sinica, 35*(5), 501–516.

Rodríguez-Lázaro, D., Cook, N., Ruggeri, F. M., Sellwood, J., Nasser, A., Nascimento, M. S. J., et al. (2012). Virus hazards from food, water and other contaminated environments. *FEMS Microbiology Reviews, 36*(4), 786–814.

Rodríguez-Lázaro, D., Fongaro, G., Marques Souza, D. S., & Hernández, M. (2021). 3.10—Molecular detection of viruses in foods: From PCR to high-throughput sequencing and beyond. In *Reference module in food science* Elsevier. [Internet]. MBT-RM in FS. Available from: http://www.sciencedirect.com/science/article/pii/B9780081005965228394.

Rzezutka, A., & Cook, N. (2004). Survival of human enteric viruses in the environment and food. *FEMS Microbiology Reviews, 28*(4), 441–453.

Scholz, E., Heinricy, U., & Flehmig, B. (1989). Acid stability of hepatitis A virus. *The Journal of General Virology, 70*(9), 2481–2485.

Schrader, C., Schielke, A., Ellerbroek, L., & Johne, R. (2012). PCR inhibitors—Occurrence, properties and removal. *Journal of Applied Microbiology, 13*(5), 1014–1026.

Simmonds, P., Gorbalenya, A. E., Harvala, H., Hovi, T., Knowles, N. J., Lindberg, A. M., et al. (2020). Recommendations for the nomenclature of enteroviruses and rhinoviruses. *Archives of Virology, 165*(3), 793–797.

Souza, D. S. M., Dominot, A. F.Á., Moresco, V., & Barardi, C. R. M. (2018). Presence of enteric viruses, bioaccumulation and stability in *Anomalocardia brasiliana* clams (Gmelin, 1791). *International Journal of Food Microbiology, 266*(August 2017), 363–371. [Internet]. Available from: https://doi.org/10.1016/j.ijfoodmicro.2017.08.004.

Souza, D. S. M., Ramos, A. P. D., Nunes, F. F., Moresco, V., Taniguchi, S., Guiguet Leal, D. A., et al. (2012). Evaluation of tropical water sources and mollusks in southern Brazil using microbiological, biochemical, and chemical parameters. *Ecotoxicology and Environmental Safety, 76*(1), 153–161.

Souza, D. S. M., Tápparo, D. C., Rogovski, P., Cadamuro, R. D., de Souza, E. B., da Silva, R., et al. (2020). Hepatitis E virus in manure and its removal by psychrophilic anaerobic biodigestion in intensive production Farms, Santa Catarina, Brazil, 2018–2019. *Microorganisms, 8*(12), 2045. [Internet]. Available from: https://www.mdpi.com/2076-2607/8/12/2045.

Su, X., Howell, A. B., & D'Souza, D. H. (2010). Antiviral effects of cranberry juice and cranberry proanthocyanidins on foodborne viral surrogates—A time dependence study in vitro. *Food Microbiology, 27*(8), 985–991. [Internet]. Available from: https://linkinghub.elsevier.com/retrieve/pii/S0740002010001334.

Suffredini, E., Proroga, Y. T. R., Di Pasquale, S., Di Maro, O., Losardo, M., Cozzi, L., et al. (2017). Occurrence and trend of hepatitis A virus in bivalve molluscs production areas following a contamination event. *Food and Environmental Virology, 9*(4), 423–433.

Terio, V., Bottaro, M., Di Pinto, A., Fusco, G., Barresi, T., Tantillo, G., et al. (2018). Occurrence of Aichi virus in retail shellfish in Italy. *Food Microbiology, 74*, 120–124.

USEPA. (2016). *Drinking water contaminant candidate list 4—final* (pp. 47–62).

Villabruna, N., Koopmans, M. P. G., & de Graaf, M. (2019). Animals as reservoir for human norovirus. *Viruses, 11*(5), 478.

Vinatea, C. E. B., Sincero, T. C. M., Simões, C. M. O., & Barardi, C. R. M. (2006). Detection of poliovirus type 2 in oysters by using cell culture and RT-PCR. *Brazilian Journal of Microbiology, 37*(1), 64–69.

Vos, T., Allen, C., Arora, M., Barber, R. M., Brown, A., Carter, A., et al. (2016). Global, regional, and national incidence, prevalence, and years lived with disability for 310 diseases and injuries, 1990–2015: A systematic analysis for the Global Burden of Disease Study 2015. *Lancet, 388*(10053), 1545–1602.

Wang, X., Ren, J., Gao, Q., Hu, Z., Sun, Y., Li, X., et al. (2015). Hepatitis A virus and the origins of picornaviruses. *Nature, 517*(7532), 85–88.

Ward, B. K., & Irving, L. G. (1987). Virus survival on vegetables spray-irrigated with wastewater. *Water Research, 21*(1), 57–63. [Internet]. Available from: https://linkinghub.elsevier.com/retrieve/pii/0043135487900996.

Webb, M. D., Barker, G. C., Goodburn, K. E., & Peck, M. W. (2019). Risk presented to minimally processed chilled foods by psychrotrophic *Bacillus cereus*. *Trends in Food Science and Technology, 93*, 94–105. [Internet]. Available from: https://linkinghub.elsevier.com/retrieve/pii/S0924224418308148.

WHO. (2017). *Global hepatitis report, 2017*. [Internet]. Geneva. Available from: who.int.

WHO. (2021). *Poliomyelitis (polio)* (Internet) (pp. 1–8). World Health Organization. [cited 2021 Jan 25]. Available from: https://www.who.int/health-topics/poliomyelitis#tab=tab_1.

World Health Organization. (2020). *Hepatitis E* (Internet) (pp. 1–7). WHO. [cited 2020 Oct 26]. Available from: https://www.who.int/en/news-room/fact-sheets/detail/hepatitis-e.

Wu, Y., Guo, C., Tang, L., Hong, Z., Zhou, J., Dong, X., et al. (2020). Prolonged presence of SARS-CoV-2 viral RNA in faecal samples. *The Lancet Gastroenterology & Hepatology, 5*(5), 434–435. [Internet]. Available from: https://linkinghub.elsevier.com/retrieve/pii/S2468125320300832.

Wu, J., Hou, W., Cao, B., Zuo, T., Xue, C., Leung, A. W., et al. (2015). Virucidal efficacy of treatment with photodynamically activated curcumin on murine norovirus bio-accumulated in oysters. *Photodiagnosis and Photodynamic Therapy, 12*(3), 385–392. [Internet]. Available from: https://linkinghub.elsevier.com/retrieve/pii/S1572100015000757.

Zhao, J., Yuan, Q., Wang, H., Liu, W., Liao, X., Su, Y., et al. (2020). Antibody responses to SARS-CoV-2 in patients with novel coronavirus disease 2019. *Clinical Infectious Diseases, 71*(16), 2027–2034. [Internet]. Available from: https://academic.oup.com/cid/article/71/16/2027/5812996.

FURTHER READING

Phan, C., & Hollinger, F. B. (2013). Hepatitis A: Natural history, immunopathogenesis, and outcome. *Clinics in Liver Disease, 2*(6), 231–234.

World Health Organization. (2021). *COVID-19 weekly epidemiological update 22* (Internet) (pp. 1–3). World Health Organization. Available from: https://www.who.int/docs/default-source/coronavirus/situation-reports/weekly_epidemiological_update_22.pdf.

CHAPTER 7

Ohmic Heating—An Emergent Technology in Innovative Food Processing

RUI M. RODRIGUES • ANTÓNIO A. VICENTE • ANTÓNIO J. TEIXEIRA • RICARDO N. PEREIRA

CEB - Centre of Biological Engineering, University of Minho, Braga, Portugal

7.1 INTRODUCTION

OH technology working principles are based on known George Ohm law and James Prescott Joule findings. An electrical current when forced to flow through a given food product will be directly proportional to the applied potential difference and inversely proportional to the resistance of the food material (Ohm's Law). At the same time, when electricity flow across the a given semi-conductor material can result in the generation of heat (Joule effect).

First appearance of this technology dates from XIX century with was followed by several patents and studies about electric pasteurization, most of them focused on milk. Table 7.1 gives a short historical overview about the very first documents with reference to ohmic heating technology, regarding electric fields on inactivation of microorganisms and electric pasteurization. In 1919, Anderson and Finklestein published an interesting report about application of OH for milk pasteurization in a private dairy company, process that at that time become known by "Electro-pure process of treatment milk"(Anderson & Finkelstein, 1919). Electro-pure process reveled to be very effective on providing a high temperature in a very short-time treatment (HTST), and was considered "as one of the outstanding achievements of dairy science of the period" (de Alwis & Fryer, 1990; Fetterman, 1928). But Electropure process also brought technical and operational issues that hindered commercial success of the technology, most of them related with lack of process control, the cost of electricity by that time and problems related with chemical contamination of eletrodes material (de Alwis & Fryer, 1990; Getchell, 1935). This technology lived then a kind of long "valley of death" period with attempts to improve the process until being revived in the 1980s. Together with technological development, which included development of new and versatile power supply (such as IGBT), possibility of using inert materials contacting the foods, as well as the reduced price of electricity, brought the opportunity to revive the electricity-based processing technologies. It is then interesting to realize that OH has been able to struggle over the centuries, much because of key findings that have been thriving in different research lines. There is now evidence that interaction of electric fields of moderate intensity with microorganisms and biological structures should not be overlooked. Depending on microorganism nonthermal effects of the application of electric field can contribute to enhanced inactivation. Electric effects can be also used to accelerate inactivation rate of certain enzymes, but also modulate their activity depending on the electric frequency applied for example. The influence of OH on structure, interaction, and functional proprieties of some globular proteins have thoroughly investigated during last decade; the synergy of combining fast internal heating and electric fields application on a single treatment open novel possibilities regarding protein denaturation kinetics, protein-protein interactions, binding with bioactive molecules, as well as different outcomes regarding immunoreactivity and allergenicity potential. From biotechnological point of view OH technology opens novel threads of applications related thermal extraction procedures, bringing the ability of designing treatments, in which well controlled internal temperature and electric fields applied allow damage of cellular material (eventually due to electroporation effects) and selective extraction of compounds of interest (El Darra et al., 2013; Pereira, Rodrigues, Genisheva, et al., 2016; Pereira et al., 2020b; Rocha et al., 2018). For all these reasons, and given the current body of knowledge, currently this technology is much more than thermal and exclusive for commercial pasteurization and

Copyright © 2021 Elsevier Inc. All rights reserved.

TABLE 7.1

Historical Overview About Some of Very First Reported Documents and Key Moments Regarding the Use of Electric Fields on Food Technology.

Document Type	Title	Reference
Patent	Apparatus for electrically treating liquids	(Jones, 1897)
Journal manuscript	Influence of electricity on micro-organisms	(Stone, 1909)
Journal manuscript	Electrical treatment of milk for infant feeding	(Beattie, 1914)
Report	Report on the Electrical Treatment of Milk to the City of Liverpool	(Beattie, 1915)
Journal manuscript	A Study of the Electro-pure process of treating milk	(Anderson & Finkelstein, 1919)
Patent	Method and apparatus for pasteurizing milk	(Anglim, 1921)
Journal manuscript	The electric current (apart from the heat generated). A bacteriological agent in the sterilization of milk and other fluids	(Beattie & Lewis, 1925)
Journal manuscript	Electrical conductivity method of processing milk	(Fetterman, 1928)
Journal manuscript	Electric pasteurization of milk	(Getchell, 1935)
Patent	Milk pasteurization method and apparatus	(Mittelmann, 1949)
Patent	Continuous production of cheese curd	(Richardson & Cornwell, 1968)
Journal manuscript	Growth kinetics of *Lactobacillus acidophilus* under ohmic heating	(Cho, Yousef, & Sastry, 1996)
Journal manuscript	Blanching by electroconductive heating	(Mizrahi, Kopelman, & Perlman, 2007)
Journal manuscript	Pulsed ohmic heating—A novel technique for minimization of electrochemical reactions during processing	(Samaranayake, Sastry, & Zhang, 2005)
Journal manuscript	The effect of electric field on important food processing enzymes: Comparison of inactivation kinetics under conventional and ohmic heating	(Castro, Macedo, Teixei, & Vicente, 2006)
Journal manuscript	Moderate electric fields can inactivate *Escherichia coli* at room temperature	(Machado, Pereira, Martins, Teixeira, & Vicente, 2010)
Journal manuscript	Extraction of polyphenols from red grape pomace assisted by pulsed ohmic heating	(El Darra, Nabil, Vorobiev, Louka, & Maroun, 2013)
Journal manuscript	Effects of ohmic heating on extraction of food-grade phytochemicals from colored potato	(Pereira, Rodrigues, Genisheva, et al., 2016)
Journal manuscript	Technology in-situ activity of α-amylase in the presence of controlled-frequency moderate electric fields	(Samaranayake & Sastry, 2018)
Journal manuscript	Effects of ohmic heating on the immunoreactivity of β-lactoglobulin—A relationship towards structural aspects	(Pereira et al., 2020a)
Journal manuscript	Using ohmic heating effect on grape skins as a pretreatment for anthocyanins extraction	(Pereira et al., 2020b)
Journal manuscript	"Influence of moderate electric fields in β-lactoglobulin thermal unfolding and interactions"	(Rodrigues, Avelar, Vicente, Petersen, & Pereira, 2020)

sterilization processes. Electric field processing can be probably a more comprehensive designation for this technology once encompasses the occurrence of electrical and thermal effects. It is important to point out that joule effect or OH is just a secondary effect of the application of an electric field of a given intensity on a semi-conductive material—i.e., depending on electrical conductivity of food material and intensity of the electric field applied, it is possible to have passage of an electrical current a within material but without occurring significant internal heat dissipation.

This chapter overviews the basics of electric field processing, highlighting OH technology by addressing its fundamental aspects but in particular by point out novel findings and innovative applications.

7.2 OHMIC HEATING: THE BASICS

7.2.1 Electrical Conductivity

The passage of electric current, inherently necessary for the OH process to occur, results in molecular agitation of the charged species within the food and thus heat generation (Sastry, 2004). The internal heat generation rate can then be described by Eq. (7.1):

$$q = E^2 \sigma \qquad (7.1)$$

where q is the energy generation rate (W/m^3), E is the EF strength (V/m) and σ is the electric conductivity. Most of the biological-based matrices (e.g., foodstuff) contain charged species such as salts and acids, thus presenting at some extent of electrical conductivity. The electric conductivity can also be adjusted by controlling the water content, adding salt or acids and, and, by this reason, OH process can be implemented in a wide range of foodstuffs (Goullieux & Pain, 2014). The electrical conductivity of a material is temperature dependent and its relation is given by Eq. (7.2):

$$\sigma = \sigma_0 \bullet (1 + mT) \qquad (7.2)$$

where σ_0 is the electric conductivity at a given temperature, m is the temperature coefficient and T is the temperature variation. By this relationship, the electrical conductivity will increase linearly with the temperature Some exceptions may be found in foods where the electrical conductivity may vary with (a) the position, in case of nonhomogeneous foods (e.g., foods containing particles); (b) with temperature in the case of disruption of cellular material—i.e., breaking of the cell wall can result on the release of intercellular content and an effective increase of the electrical conductivity (Knirsch, dos Santos, Vicente, & Penna, 2010); and (c) through

physical changes induced by the cooking, once as food structure can altered by gelation for example. In an OH system electrical conductivity can be monitored and determined from the process variables such as electrical resistance, intensity, as well as geometry of the heating system using the following equation (Palaniappan & Sastry, 1991)

$$\sigma_T = \frac{1}{R_T} \bullet \frac{L}{A} \qquad (7.3)$$

where σ is the electrical conductivity (S/m) at given temperature, L is the distance between electrodes (m), A is the cross-section surface area of the electrodes (m^2), and R is electrical resistance of the sample at given temperature. Electrical resistance can be calculated from voltage applied and electrical current measurement.

Fig. 7.1 shows examples of food products that show a linear increase of electrical conductivity against temperature increase. Higher or lower levels of electrical conductivity will be dependent on the amounts of dissolved ionic species, such as salts and acids, or presence of less conductive molecules such as a sugar, fats, and alcohols. OH efficiency and uniformity can then be defined by the conductivity of the food product, either on its magnitude as on its distribution through the product.

Generally, the conductivity range of foods is comprehended between 0.01 and 10 S/m. Table 7.2 resumes typical values for electrical conductivity measured at room temperature, which are important to define the power levels to be applied for the application of OH technology. Generally, voltage, and current needed to produce sufficient heat below and above these limits, can

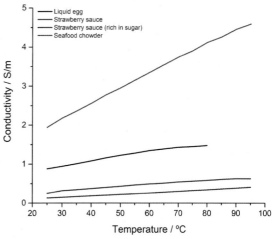

FIG. 7.1 Electrical conductivity of several food products as function of temperature during OH application.

TABLE 7.2
Electrical Conductivity of Several Food
Products Measured at 20°C.

Food Product	σ (S/m) at 20°C
Liquid egg	0.88
Milk	0.39
Soup	1.31
Apple	0.07
Mango	0.12
Chicken breast	0.61
Beef (lean)	0.42

result in constraints for practical applications. However, during development of a given food formulation a fine adjustment of highly conductive and nonconductive molecules can be performed in order to get an optimal electrical conductivity for an efficient OH application.

7.2.2 Electric Field

The movement of electric charges can occur through direct current (DC) or alternating current (AC). In DC charges movement is unidirectional while in AC the charges periodically reverse direction. Though OH will occur in both forms of current, AC is exclusively used to deliver electric power in OH technology. AC is defined not only by the current and voltage, but also by the wave shape produced (e.g., sinusoidal or wave) and frequency of the polarity alternation. A fast periodically change of polarity on the electrodes (i.e., high frequency AC) effectively results on the reduction of the charge build-up on the electrodes surface, thus preventing the occurrence of the electrochemical effects and electrode corrosion. In the case of low frequencies AC, the polarity reversal time may not be enough to eliminate electrochemical reactions and simultaneous cathodic (reduction) or anodic (oxidation) half-reactions do occur. With the increase of the electrical frequency, the time available for the charged species to accumulate on the electrodes surface is reduced. Eventually the alternation frequency will surpass the time needed for the charge build-up and effectively eliminate the electrochemical reactions and electrode erosion. By this reason the use of high frequencies has been systematically implemented on commercial OH devices. The use of electrical frequencies in the range of kHz can reduces to a greater extent electrodes corrosion, and electrochemical, and minimize leakage of metals contaminants to the medium (Pataro et al., 2014).

The passage of electric current through a food product implies that an electric potential must be applied. The applied potential at a defined point is expressed in volts (V) and the resultant electric field ($\vec{E}$) is defined by the magnitude and orientation of the electric force. This parameter is a function of the voltage applied, direction of the current and distance between electrodes, and can be regulated by controlling the voltage output and/or the geometry of the system. The EF used during OH usually ranges between 1 and 1000 V/cm, being this considered a Moderate Electric Field (MEF) application (Machado et al., 2010), in a way to distinguish from other electric field processing technologies, such as Pulsed Electric Field (PEF), where range is often between 1 and 40 kV/cm. In fact, in literature the terms OH and MEF appear often associated, with the informal differentiation of attributing the mention of OH to processes where thermal action is predominant, while MEF to those where electric effects are predominant or the main focus. The EF is then a decisive factor and its regulation can be used to control operational parameters such as the heating rate and operational temperature, besides playing a critical role on secondary phenomena during the OH process, such as electrochemical reaction or nonthermal effects (discussed further on this section).

In the end, the effective control of the OH process relies on the electric conductivity of the product, the electric field applied and its characteristics (i.e., type of current, wave shape and electric frequency). Furthermore, nonthermal effects resultant from the presence of the EF are directly affected by its intensity, direction, and frequency. Inactivation studies in microorganisms revealed a synergetic effect between the thermal action of OH and the EF intensity (Rocha et al., 2018). These effects are positively correlated with EF intensity applied and resulted on reduced temperature load needed to achieve an intended inactivation. Such effects arise from a mild electroporation-type mechanism during OH, capable to cause cellular damage, permeation of vegetable tissues and even inactivate microorganism without an associate thermal effect (Lebovka, Praporscic, Ghnimi, & Vorobiev, 2005; Machado et al., 2010; Pereira, Rodrigues, Genisheva, et al., 2016; Sensoy & Sastry, 2004; Sung-Won & Ki-Myung, 2002). Additionally to the cell permeabilization effects, enzyme activity is also affected by the EF, resulting on an increased inactivation kinetics of some enzymes (Castro et al., 2006). In fact, recent studies concluded that EF effect can be extended to the molecular level, inducing change in food constituents such as enzymes and other proteins, and thus causing changes on their structure and activity (Bekard & Dunstan, 2014; Rodrigues, Avelar, Vicente, et al., 2020; Samaranayake & Sastry, 2018). Both inactivation and molecular effect inherent to the presence of an external EF are also dependent of the frequency used. Several of the mentioned studies of the EF action in microorganisms, enzymes, and other proteins concluded that different electrical frequencies also result in different outputs, generally being more effective in causing inactivation or structural changes at low frequencies (i.e. < 500 Hz). However, these effects should not be generalized, and a case-by-case analysis strategy needs to be adopted; there are still many variables that can influence this responses, such as noncontrolled internal heating, composition of matrices where treatments is applied, physical and chemical environment (e.g., pH, ionic strength, solute concentration), among others.

7.3 MAIN APPLICATIONS

OH processing is a versatile technology, being possible to adapt it to several geometries according to the products and process specifications, as well as operating in batch or continuous mode. The main advantage of OH compared with conventional thermal processing technologies relies on the internal heat generation, resulting in a fast and volumetric heating of the products with extremely high energetic efficiencies (i.e., > 90%). OH does not rely in traditional heating transfer mechanism (i.e., conduction, convection, and radiation), in this sense mechanical stirring or scraping is often minimized or eliminated, allowing operations at low shear. In addition, being a direct heating method without the presence of hot surfaces, thermal load can be drastically reduced, reducing overprocessing, and minimizing the occurrence of fouling. These factors contribute not only to the maintenance of the products quality but also in operational advantages such as longer operation periods, reduced cleaning cycles, and lower maintaining costs.

Nevertheless, there are still some drawbacks associated with OH technology, as follow; i) food products need to be in contact with electrodes which imply proper equipment design and limit some applications; ii) high or low conductive products may impose operational restrictions and the need of compositional adjustments—food products containing large fat fractions may restrict the passage of the electric current and impair the OH process; iii) as any other emerging technology, the economic cost of their

implementation are relatively high to the more established alternatives. However, it is important to highlight that with recent technological developments, particularly on power sources, and with the increase of suppliers and manufacturing at industrial scale, the OH inherent costs lean towards becoming very competitive. Furthermore, the high versatility, compact size, and low dependence of external devices (e.g., boilers to produce steam or extensive piping to transport hot and cold streams), makes of OH n linearly scalable technology with an easy and straightforward implementation on the production lines. In addition to its high energetic efficiency, OH discard the used of boilers and fossil energy, reducing water consumption, and allowing the use of renewable sources to produce electrical power. OH offer an opportunity to reduced environmental footprint being considered a green food processing technology (Pereira & Vicente, 2010; Rodrigues et al., 2019). By all these reasons OH applications have steadily increased and diversified through the last centuries, finding a vast array of target product and processed in research and in industrial applications.

7.3.1 Food Processing

OH technology was developed with the focus in pasteurization as an alternative to conventional heat exchange processes. Naturally, it is in pasteurization that until today OH found most of its applications. Its first and still one of the most predominate applications is in the dairy industry, as it presents operational advantages due its fast and homogeneous heating, reduced fouling and better preservation of nutritional and organoleptic properties (Cappato et al., 2017; Pereira, Martins, & Vicente, 2008; Pereira, Teixeira, et al., 2018). Pasteurization of egg products, such as hole liquid egg, egg white or yolk were also successfully implemented in industrial scale (Rodrigues et al., 2019). These products are particularly sensitive to thermal load and temperature abuse (due to protein coagulation issues). OH processing with its high thermal control and absence of hot surfaces brings an important advantage on the processing of such products. Fouling is dramatically reduced, product quality, and functionality is maintained or improved and the higher inactivation rates resultant from the presence of EF can improve the product's safety, shelf life, and contribute to reduce traditional temperature-time binomials (Alamprese, Cigarini, & Brutti, 2019; Icier & Bozkurt, 2011; Varghese, Pandey, Radhakrishna, & Bawa, 2014). Soymilk thermal coagulation for the production of tofu is another case of a successful industrial application of OH by Yanagiya company. This innovation allowed producing tofu in shorter times, higher yields, as well as improve texture, and the taste. Several other products were successfully processed by OH at commercial scale ranging from vegetables, soups, stews, sauces, precooked meals, juices, and fruit purees (Sakr & Liu, 2014; Varghese et al., 2014). Despites the mentioned processes and products, there is still few information on OH processing industrial facilities or processed products on the market. However, there is already a diversified offer regarding OH industrial equipment (see Table 7.3) mostly of them focused in food processing, which supports an increasing demand and implementation of the technology.

The most reports on OH potential and associated advantages are still related with lab scale or pilot applications focused on academic research or demonstration research projects. In these fields, the possible application and mentioned advantages of OH technology is extensive and brings novel and interesting perspectives to several processes concerning food technology, but also other fields of sciences such as biotechnology and chemical synthesis, among others. A promising example is the processing of baby foods, where OH demonstrated better performance facing conventional processing techniques. OH resulted on the preservation of organoleptic properties, nutritional content and prevented the formation of harmful compounds resultant form thermal degradation such as furan, compounds that result from fatty acids oxidation and Maillard reactions (Hradecky, Kludska, Belkova, Wagner, & Hajslova, 2017; Mesías, Wagner, George, & Morales, 2016; Roux et al., 2016). Several other food processing operations (i.e., thawing, blanching, cooking, dehydration) have been successfully performed with OH technology bringing notorious advantages. Meat and fish products can be processed by OH using thawing or cooking methods and bring several advantages such as reduction of processing time, improvement of organoleptic and functional properties (color, texture, moisture retention) and contribute to enhance microorganism inactivation in products such as hamburger patties, meatballs, whole meat of turkey, beef and shrimp or surimi (Engchuan, Jittanit, & Garnjanagoonchorn, 2014; Lascorz, Torella, Lyng, & Arroyo, 2016; Tadpitchayangkoon, Park, & Yongsawatdigul, 2012; Zell, Lyng, Cronin, & Morgan, 2010). Vegetable processing such as peeling or production of pastes and concentrates have been implemented by OH, again resulting in shorter processing times, reduced resources (e.g., water, energy) and improving the product's quality (Torkian Boldaji, Borghei, Beheshti, &

Hosseini, 2015; Wongsa-Ngasri & Sastry, 2015). Many other examples of OH utilization in food processing can be found (e.g., vegetable blanching, rice or pasta cooking, bread and cake baking) proving the operational flexibility and high potential of this technology (Deleu, Luyts, Haesendonck, Brijs, & Delcour, 2019; Gally, Rouaud, Jury, & Le-Bail, 2016; Guida, Ferrari, Pataro, Di Maro, & Parente, 2013; Kanjanapongkul, 2017; Lyng, McKenna, & Arroyo, 2018).

7.3.2 Food Biotechnology

As previously mentioned, OH was primarily developed aiming to be explored as thermal processing technology to achieve inactivation of microorganisms and endogenous enzymes in food products (e.g., blanching, pasteurization, or sterilization). Nonetheless unique operational advantages along with nonthermal effects associated with the presence of an external EF, placed the OH technology on a unique position to be applied in several branches of biotechnology.

The electroporation of cells by the transmembrane potential induced by the EF is a well described effect and well documented in OH applications (Lebovka et al., 2005; Sensoy & Sastry, 2004; Silva et al., 2017; Sung-Won & Ki-Myung, 2002). Electroporation depends on EF characteristics such as intensity, duration and application form, and synergy with external factors like thermal load and characteristics of the materials being processed (Silva et al., 2017). The electroporation caused by OH/MEF is considered mild, compared with the effects caused by other techniques such as PEF that use EF of much higher intensities (typically > 1 kV/cm). However, large cells and with relatively low resistant cell walls—e.g., vegetable cells in foodstuff, are most susceptible to be permeated under lower EF intensities. By these reasons, the electroporation mechanism caused by EF on the MEF range (mostly during OH) have been often reported. OH demonstrated to successfully permeate and increase diffusion of water and solutes across tissues in products such as beetroot, apples, and strawberries. As a result, processes such as juice extraction, tissue impregnation or dehydration were improved by the permeabilization effect, along with other advantages such as higher enzymatic and microbial inactivation or improvements of physical properties of the products (Allali, Marchal, & Vorobiev, 2009, 2010; Lima, Henkitt, & Sastry, 2001; Moreno et al., 2013; Praporscic, Ghnimi, & Vorobiev, 2005; Simpson et al., 2015). Permeabilization and increase diffusion caused by OH is naturally attractive in extraction processes, along with the associated thermal effects that contribute to tissue softening, increasing

TABLE 7.3
Examples of OH Equipment's Manufacturers and Main Identified Applications.

Constructor	Specifications (Power, Voltage, Electrical Frequency, and Operational Mode)	Applications (Specified by the Constructor)
Alfa Laval	60–300 kW 400 V 20 kHz continuous	Fruits and derivatives, vegetables, prepared foods, cheese, liquid eggs, ready-to-eat dishes, sauces, and juices
Raztek	N.A. 4–12 kV 50 Hz Continuous	Liquid egg and egg white
Emmepiemme	N.A. 30–600 kW 25 kHz continuous	Fruits and derivatives, vegetables, dairy products, egg products, algae, syrups, sauces, and ready-to-eat dishes
INDAG High Power Heating System (IPS)	N.A.	Liquid products including chunks and highly viscous foodstuffs
APV Baker	75–300 kW N.A. 50 Hz Continuous and batch	Fruit preparations and dressings with high fruit content (also sterile, for pH-neutral products)
Yanagiya	1.5 kW From 100 V up N.A. Batch and continuous N.A.	Tofu production
Kasag	N.A.	Caramel, vanilla, or chocolate sauces; convenience food with vegetables and meat; soups; ketchup; salsa sauces; and all foodstuffs which can be pumped
JBT Corporate	N.A. N.A. High frequency continuous	Liquid, semi liquid, concentrated, and high viscosity products containing fibers, small cells and featuring high viscosity such as puree and soups Fruit preparations and fruit jam with dices Soup and sauces Including products containing large size particles, 50–70 mm
CFT group	60–600 kW Up to 5000 V 20–30 kHz Continuous	Fruit and vegetables (whole, diced & purees, juices) fruits, pulps, juices, smoothies soups, ready meals, sauces smooth sauces like cheese-based sauces and white sauces jams with particles
C-tech innovation	Up to 250 kW N.A. N.A. Batch and continuous	Food and beverage Fine chemical industries
Simaco	N.A. N.A. 25 kHz Continuous	N.A.
Agro-process	50 kW 1000–5000 V 17 kHz Continuous	Liquid, semi liquid products, such as fruit juices, jams

solutes solubility and mass transfer. Hence, OH was applied to assist or promote extraction processes of a large variety of compounds in different matrices. The extraction of phenolic compounds from several vegetable matrices such black rice, potato, tomato by-products, mint and tea leaves and other plants were achieved or improved by OH (Coelho, Pereira, Sebastião, Teixeira, & Pintado, 2019; Loypimai, Moongngarm, Chottanom, & Moontree, 2015; Pereira, Rodrigues, Genisheva, et al., 2016; Sensoy & Sastry, 2004). Combination of thermal and mild EF intensity was successfully applied for disruption of vegetable cells of purple potato and subsequent extraction of water-soluble phytochemicals (see Fig. 7.2; short-time treatments (from seconds to min time scale) at high temperatures (from 90 to 100 °C), at EF < 20 V/cm allowed enhanced extraction of phenolic compounds (Pereira, Rodrigues, Genisheva, et al., 2016).

Polysaccharides extraction such as pectin from fruit by-products or carrageenan from seaweed was also promoted by the presence of an EF, resulting in advantages such as higher extraction yields, extraction at lower temperatures and improved functional properties of the recovered products (de Oliveira, Giordani, Poliana, Cladera-Olivera, & Marczak, 2015; Saberian, Hamidi-Esfahani, Gavlighi, Banakar, & Barzegar, 2018; Salengke, Waris, & Mochtar, 2016). Oil extraction and recovery from herbs, seeds, and cereal brans, is another application for the OH process, were in all the reported cases OH contributed with operational advantages as higher yields, reduced extraction time, and lower energy consumption (Aamir & Jittanit, 2017; Gavahian

& Farahnaky, 2018; Lakkakula, Lima, & Walker, 2004; Seidi Damyeh & Niakousari, 2016). Regarding OH/MEF effects in extraction from unicellular microorganisms, the reports are still scarce. Nonetheless, OH demonstrated to promote the leakage of cellular content (e.g., proteins and nucleic acid) from yeast cells (Sung-Won & Ki-Myung, 2002). In microalgae, EF effects (MEF range) were reported to assist or promote extraction of lipids, carotenoids and genetic material, by promoting electroporation and cellular disruption (Bahi, Tsaloglou, Mowlem, & Morgan, 2011; Daghrir, Igounet, Brar, & Drogui, 2014; De Carvalho Neto et al., 2014; Jaeschke, Menengol, Rech, Mercali, & Marczak, 2016).

It is challenging to differentiate the true nature of the OH/MEF effects in cellular matrices and phenomena such as permeabilization and extraction. The application of an external EF will inherently result on the molecular agitation and dissipation of internal heat, even if that it is at limited extent or immensurable. However, through the reported studies about the influence of EF in cells, in which there is an attempt to dissociate thermal from nonthermal effects, it seems that electrical variables play a decisive role on the observed effects. Particularly, the permeabilization effects seems to be directly proportional to the EF strength applied and more effective as the electric frequency is reduced (Gavahian, Chu, & Sastry, 2018). The optimizing process parameters (i.e., temperature, EF intensity, frequency, and treatment time) can then be tuned to optimize OH performance in extraction applications envisaging better yields, extraction selectivity, and improved functionality of the recovered products.

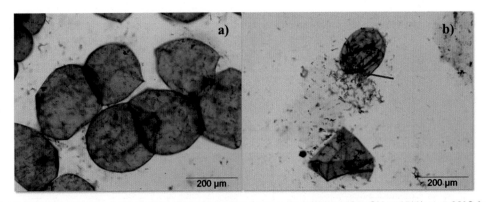

FIG. 7.2 Microscopic pictures of vegetable cells from purple potato before (A) and after OH at 15 V/cm at 90°C for 5 min at 25 kHz (B). *Red arrow (grey in print version)* points out a disrupted cell and cellular debris.(Adapted from Rodrigues, R. M., et al. (2019). 6—Ohmic heating for preservation, transformation, and extraction. In: F. Chemat & B. T. Eugene Vorobiev (Eds.), *Green food processing techniques—Preservation, transformation and extraction* (pp. 159–191). Academic Press.)

The extent of EF related effects during OH/MEF in living cells can be controlled though the EF parameters, with reported effects ranging from mild and dependent on synergic effect of the thermal action up to the unique action of nonthermal effects well succeed on inactivation of microorganisms, (Machado et al., 2010). If the temperatures are kept at sublethal level and the EF strength is maintained relatively low, OH/MEF can be used to regulate the metabolic activity of microorganisms. The use of low intensity EF on the culture of a recombinant *S. cerevisiae* demonstrated to enhance the early stages of fermentation kinetics and achieve higher biomass yield. However, the reported effects were dependent of the EF intensity, with higher values resulting in plasmid instability and reduction of metabolites production (Castro, Oliveira, Domingues, Teixeira, & Vicente, 2012). A similar approach was tested in several works with in *Lactobacillus acidophilus*, also resulting on the lag phase reduction but without improvements on the biomass yield. The observed effects were dependent on the EF intensity but also on the frequency used since only low frequencies ($\geq 60\,Hz$) demonstrated observable effects on cell permeabilization (Cho, Yousef, & Sastry, 1996; Loghavi, Sastry, & Yousef, 2007, 2008; Loghavi, Sastry, & Yousef, 2009). The use of EF technologies was also tested on the harvest of microalgae biomass by causing coagulation and flocculation of cells. Relatively to this process the presence of an EF resulted on the permeabilization and release of metabolites of interest opening new perspectives for a simultaneous biomass recovery and extraction of metabolites or even induce the release of compounds of interest without compromising the culture viability (Daghrir et al., 2014; De Carvalho Neto et al., 2014; Geada et al., 2018).

Additionally, OH can be used as a method to quickly and precisely control fermentation's medium temperature, helping to optimize the fermentation conditions, and improve the process performance. These advantages were attested in bread dough proofing during OH by its fast and homogeneous heating (Gally et al., 2017). OH and MEF can then assist fermentation processes through thermal and nonthermal effects and thus improve industrial fermentations' productivity (Gavahian & Tiwari, 2020).

7.3.3 Biomolecules Functionality
Recently, the effects of OH in biomolecules has been the focus of numerous studies, either as a quality control of the processed products or by targeting more fundamental aspects of the technology aiming at explore the true influence of nonthermal effects. The use of this technology demonstrated to have an impact on stability and functionality of several bio-based compounds, thus opening the perspectives to use this process as a tool to induce and control biomolecule behavior. OH treatment of polysaccharide (i.e., chitosan and starch) solutions was evaluated in film forming solutions and subsequent film properties. The OH treatment of such solutions resulted in structural changes on the formed films such as higher crystallinity, more uniform structure and improved barrier and mechanical properties (Coelho et al., 2017; Souza et al., 2009, 2010).

The processing of protein rich products by OH have demonstrated significant advantages in maintaining and even improving their functional properties. Numerous studies involving OH/MEF application in proteins unveiled EF effects in molecule structure, unfolding pathways, interactions and functional properties (i.e., aggregation, gelation, emulsification, among others) (Pereira et al., 2017; Rodrigues, Avelar, Machado, Pereira, & Vicente, 2020). Sastry and its co-workers have been developing an interesting work on the effects of EF in enzyme activity. Their studies suggest that the reversal of the EF field results in molecular motion with different amplitudes, depending on the frequency and the intensity of the EF applied, causing conformational changes on the proteins and this impacting their activity (Samaranayake & Sastry, 2016, 2018). The EF effects related with molecular motion (linked with EF intensity and frequency) and protein conformation changes was verified in lysozyme and bovine serum albumin when exposed to low intensity EF and without thermal elevation (Bekard & Dunstan, 2014). In both situations protein structure and conformation was more affected with the increase of the EF strength applied at low frequencies. The influence of OH and MEF effects in synergy with thermal effects have been studied in whey protein fraction from bovine milk. These studies highlight that EF can significantly disturb the protein's unfolding pathways. Upon thermal denaturation with simultaneous presence of an EF, structural effects were observed on a large frequency range (i.e., $50\,Hz$ to $1\,MHz$), but more extensive at low frequencies, which were positively correlated with the EF intensity (Rodrigues, Avelar, Vicente, et al., 2020; Rodrigues, Vicente, Petersen, & Pereira, 2019). These changes at molecular level (i.e., changes in secondary and tertiary structure, surface hydrophobicity or free sulfhydryl content) had a significant impact in protein aggregation, network formation and gelation, resulting in protein systems with distinctive functional properties (Rodrigues, Fasolin, Avelar, et al., 2020). The presence of EF of relatively low intensities ($< 10\,V/cm$) during OH at high temperature (from $90\,°C$ to $98\,°C$) at acidic conditions can result in development of fibrillar structures that hold potential for development of nano and micro encapsulating or associating systems (Pereira, Rodrigues, Ramos, et al., 2016). OH has been used as pretreatment to favor development of protein based networks that can be produced at room temperature, such as the case of iron-rich globular protein hydrogels or cold gel-like emulsions (de Figueiredo Furtado, Pereira, Vicente, & Cunha, 2018; Pereira et al., 2017). Fig. 7.3 illustrates transmission electronic microscopy images of fibrillar structures of beta-lactoglobulin and their use to produce a network of fibrils.

Overall, the use of OH has revealed the potential to influence several processes with appreciable effects from nano to macro scale—including tissue and cellular material disruption to molecular level. The unique thermal action and nonthermal effects resultant from the presence of an EF can influence cell permeability, diffusion and extraction of compounds, microbial growth, and metabolite production, as well as the properties of macromolecules and their functionality. The extent of such effects are dependent on operational parameters such as heating rate, EF strength, or electric frequency, which leads to the perspective that a fine-tune control of these parameters will contribute to selectivity of the induced EF effects. For all these reasons, OH technology can be considered as a promising tool develop and control new functionalities in food biotechnology that still need to be further investigated.

7.4 NOVEL PERSPECTIVES

OH in its thermal processing technological aspects can directly replace conventional thermal treatment plants. The previously discussed advantages in terms of operational conditions (i.e., energetic efficiency, operational time, maintenance costs) and product quality

(i.e., improved/preserved nutritional and organoleptic properties) provide to OH technology important competitive advantages. Furthermore, the additional non-thermal effects that contribute to an increased enzyme and microbial inactivation contribute also decisively to the improvement of food safety and/or reduce the thermal load and thus enhance food quality. Fig. 7.4 resumes some of the main applications of OH processing in bioengineering and biotechnology fields. The main drawbacks that prevent OH implementation are the lack of knowledge and equipment's costs. Nevertheless, during the last years and effort have been made for a better elucidation, and several reports and applications have steadily increased together with the number of manufacturers of commercial OH equipment. The conjugation of these two factors may well represent the reaching of an important momentum that can push forward the widespread implementation of OH in the food processing industry.

OH and MEF as an innovative extraction techniques may be tuned to harness advantages from both thermal and nonthermal effects, depending on the process and substrate specifications. Increased cell permeability can be induced at low temperatures using low frequencies, releasing extracts form fresh materials, preserving bioactivity and functionality of target compounds (e.g., low protein denaturation or reduced oxidation). In opposition, high-temperature short-time processes at high EF intensities can be achieved with relatively low energetic input and reduced use of solvent. These conditions, that naturally combine thermal and electrical effects can help to disrupt tissue, cells, and protective structures, stabilize materials allowing recovery of more inaccessible compounds. The conjugation of the operational advantages of OH applied to extraction processes and the growing environmental concerns to reduce waste, valorize underrated resources, reduce

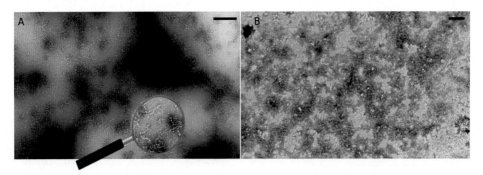

FIG. 7.3 Beta-lactoglobulin aggregates resembling a fibrillar structure produced during OH (A), further used to produce a protein based network through addition of Fe^{2+} (B). The scale bars in the images are of 500 nm.

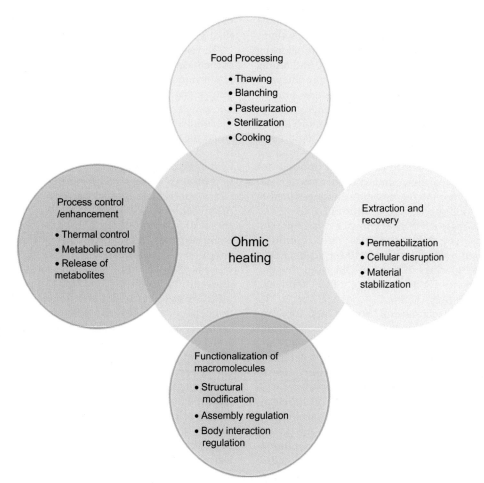

FIG. 7.4 Diagram of the OH applications in different biotechnological fields.

energetic inputs and avoid the use of chemical solvents, place OH on the front line of emerging technologies that can revolutionize traditional food and bioprocess industries.

The use of OH technology in controlling microbial growth and as a tool in fermentation industry is largely unexplored. The available results show some promising perspectives on the use of OH/MEF to stimulate specific phase of the fermentative process, but more detailed studies are required to address and better elucidate the technological potential in this application. Apart from the nonthermal effect in microbial cultures, OH definitely holds the potential to assist the fermentation as process control tool. It can be used to efficiently control the fermentation's media temperature, even in low water content and particulate substrates, as well as to stabilize (trough pasteurization or sterilization) the same media prior or post fermentation. In this sense, further

efforts must be developed into studding the fundamental factors involved in OH assisted fermentation as well as in technical solutions to allow its application from the lab scale to industrial implementation.

One of the most promising perspectives of OH/ MEF applications are the improvement and development of new functionalities through the modification of macromolecules. A growing number or reports deal with OH/MEF effect in polysaccharides, proteins, and enzymes. The further characterization of these phenomena and the development of strategies to control them envision a direct applicability in several branches of biotechnology but also in biomedical fields. The enhancement and control of processes such as structural modifications/denaturation, interactions and complex formation, aggregation, network formation, among other, settle the basis to control functionality and fabrication of innovative biotechnological solutions.

The modification of biological macromolecules can be used in formation of nano and micro-delivery systems to be used in food and medical applications, allowing improving stability of the bioactives/drugs, and tuning their transport and release profiles for any intended functionality. The fabrication of biomaterials is manly focused in macromolecules such as proteins and polysaccharides and the modification of these molecules by OH/MEF action can also bring important implication on the materials performances. A particular example are the hydrogels, described as potential materials for carry drugs or bioactives, production of structure systems to supports cellular grows and tissue regeneration, but also in food applications as structuring agents and smart platforms to develop functional foods. Important innovations have been developed into controlling protein-based hydrogels under OH/MEF action, opening novel perspectives to fine tune their functionality and performance (Rodrigues et al., 2015; Rodrigues, Fasolin, Avelar, et al., 2020).

The particular relationship between food composition and structure with health and wellbeing is gaining particular focus in food science and engineering (Roos et al., 2016). The research attention is turning into the digestibility profile, bioavailability/accessibility, and interaction of the food components with the gastro-intestinal system (Golding, 2019; Sensoy, 2014). It is known that food processing, causing phenomena such as protein denaturation and aggregation can impact the digestibility profiles and even immunoreactive reactions (Pereira, Rodrigues, et al., 2018; Sensoy, 2014). In this sense the role of new processing technologies, and particularly OH, must be evaluated to seek possible negative implications or reduce naturally occurring issues with the consumption of certain foods. Any food product that has undergone OH should be evaluated and not be assumed to be identical to a conventional thermally treated food (Jaeger et al., 2016). Recently a study on the effect of pasteurization protocols, performed under OH and conventional heating methods has been performed in bovine beta-lactoglobulin from milk whey fraction. It has been shown that immunoreactivity of this protein was linked with structural modifications and protein aggregation, which were very dependent on the processing heating kinetics, but also seemed to be influenced by other parameters such as EF intensity and frequency in a lesser extent (Pereira et al., 2020a). OH change the pathways of protein denaturation and aggregation and these events can change the behavior of these structures during gastrointestinal digestion and interaction with intestinal mucosa, thus bringing new outcomes regarding allergenic responses (Pereira, Rodrigues, et al., 2018).

The vast implication of food structuring, physicochemical modification of food components and interaction among food component and the gastrointestinal track makes this field a promising and vastly unexplored branch of science. It is expectable that in the following years, a large body of research will be focused on this area and naturally the OH implication in food matrices will be further explored being aligned with these developments.

Lastly, OH potential has experienced few occasional applications in niche areas that may turn into promising research and industrial applications in chemical and biological engineering areas. Regarding environmental technologies, OH has demonstrated potential in wastewater treatment processes such protein coagulation of proteins from industrial residues and sludge disinfection from human waste, taking advantage of the fast and homogeneous heating nondependent of heat transfer (Kanjanapongkul, Tia, Wongsa-Ngasri, & Yoovidhya, 2009; Yin, Hoffmann, & Jiang, 2018). The homogeneous and precise heating, along with the EF effects in the mobility of charged species has also demonstrated advantages in organic synthesis. Several works reported advantages related with the application of OH in reaction times, yields, selectivity, or use of catalyzers when compared with other heating method (do Cardoso et al., 2015; Pinto et al., 2013; Silva, Santos, & Silva, 2017). Despites the lack of information about OH industrial applications besides food processing, C-Tech innovations, one of the OH equipment suppliers currently on the market, advertises the use of its plants in fine chemicals industries with advantages previously mentioned.

7.5 CONCLUSIONS

With its origins in pasteurization, OH has come a long way and today several applications in different fields can be enumerated. Its implementation in industrial scale is already ongoing in the food processing industry with aseptic lines already producing commercially available product and with several manufactures on the market producing equipment at an industrial scale. This implementation is motivated by the competitive advantages on technical and operation aspects as well as in product's quality. Current research presents evidences of the occurrence of nonthermal effects during OH processing. These effects are related with increase inactivation of enzymes and microorganism, permeabilization of cells and tissue and modification of macromolecules, thus opening new perspectives on the use of the technology along with field of sciences. Furthermore, the possibility of controlling such effect with the variation of operational parameter such as the EF strength or the electric frequency can turn the use of OH as a tool for process control and optimization.

Particularly on the critical moment we are going through as a society, with increasing environmental concerns and general awareness and greater apprehension to factors as sustainability or population's health and wellbeing, it is imperative to find novel solutions to face these problems. OH has an extremely high energetic efficiency and is less dependent on nonrenewable resources as electricity can be generated by renewable means. Lower product degradation, maintenance of the nutritional content and improve safety and stability also contribute to a general improvement on the use of food resources and added value of food products. OH processing fit in the concept bio-resources valorization and circular economy, contributing to a more effective extraction and recovery of valuable compound form several residues and byproducts. The most recent perspectives of achieve modification of macromolecules though OH, aiming at developing new functional foods as well as biomedical and pharmaceutical applications that can contribute to an improved nutrition a general wellbeing of the population. On the edge of such possibilities, it is necessary to have an increased understanding of the influence that OH and its effects will have in complex food matrices, their interaction with the human body and their impact in health. Taking into account all the revised applications and reported operational advantages of OH it is expected that the academic research as well as industrial application in food and biotechnological processing will keep increasing at an accelerated pace. Limiting aspects of the OH process still needs to overcome and many of the forthcoming research needs to be done through a case-by-case assessment due to dynamic and complex behavior of different products and process. Furthermore, technical aspects must be evaluated through a global perspective along with a careful economic analysis to evaluate a possible advantage of OH process implementation.

These efforts must be conducted through collaborative joint actions between research institutions and stakeholders in order to evaluate the full potential and effects of these innovations on the production process, environmental impact, and consumer's benefits.

REFERENCES

Aamir, M., & Jittanit, W. (2017). Ohmic heating treatment for Gac aril oil extraction: Effects on extraction efficiency, physical properties and some bioactive compounds. *Innovative Food Science & Emerging Technologies, 41*, 224–234.

Alamprese, C., Cigarini, M., & Brutti, A. (2019). Effects of ohmic heating on technological properties of whole egg. *Innovative Food Science & Emerging Technologies, 58*(June), 102244.

Allali, H., Marchal, L., & Vorobiev, E. (2009). Effect of blanching by ohmic heating on the osmotic dehydration behavior of apple cubes. *Drying Technology, 27*(6), 739–746.

Allali, H., Marchal, L., & Vorobiev, E. (2010). Blanching of strawberries by ohmic heating: Effects on the kinetics of mass transfer during osmotic dehydration. *Food and Bioprocess Technology, 3*(3), 406–414.

Anderson, A. K., & Finkelstein, R. (1919). A study of the electro-pure process of treating milk. *Journal of Dairy Science, 2*(5), 374–406. http://www.sciencedirect.com/science/article/pii/S0022030219943384. November 9, 2020.

Anglim, H. A. (1921). *Method and apparatus for pasteurizing milk*. Patent US1468871A.

Bahi, M. M., Tsaloglou, M.-N., Mowlem, M., & Morgan, H. (2011). Electroporation and lysis of marine microalga *Karenia brevis* for RNA extraction and amplification. *Journal of the Royal Society Interface, 8*(57), 601–608.

Beattie, J. M. (1914). Electrical treatment of milk for infant feeding. *The British Journal of State Medicine, 24*, 97–113.

Beattie, J. M. (1915). *Report on the electrical treatment of milk to the city of liverpool.*

Beattie, J. M., & Lewis, F. C. (1925). The electric current (apart from the heat generated). A bacteriologibal agent in the sterilization of milk and other fluids. *Journal of Hygiene, 24*(2), 123–137. https://doi.org/10.1017/S0022172400008640. November 11, 2020.

Bekard, I., & Dunstan, D. E. (2014). Electric field induced changes in protein conformation. *Soft Matter, 10*(3), 431–437.

Cappato, L. P., Ferreira, M. V. S. S., Guimaraes, J. T., Portela, J. B., Costa, A. L. R. R., Freitas, M. Q., … Cruz, A. G. (2017). Ohmic heating in dairy processing: Relevant aspects for safety and quality. *Trends in Food Science and Technology, 62*, 104–112.

Castro, I., Macedo, B., Teixeira, J. A., & Vicente, A. A. (2006). The effect of electric field on important food-processing enzymes: Comparison of inactivation kinetics under conventional and ohmic heating. *Journal of Food Science, 69*(9), C696–C701.

Castro, I., Oliveira, C. C. M., Domingues, L., Teixeira, J. A., & Vicente, A. A. (2012). The effect of the electric field on lag phase, β-galactosidase production and plasmid stability of a recombinant *Saccharomyces cerevisiae* strain growing on lactose. *Food and Bioprocess Technology, 5*(8), 3014–3020.

Cho, H.-Y., Yousef, A. E., & Sastry, S. K. (1996). Growth kinetics of lactobacillus acidophilus under ohmic heating. *Biotechnology and Bioengineering, 49*(3), 334–340.

Coelho, C. S., Cerqueira, M. A., Pereira, R. N., Pastrana, L. M., Freitas-Silva, O., Vicente, A. A., … Teixeira, J. C. (2017). Effect of moderate electric fields in the properties of starch and chitosan films reinforced with microcrystalline cellulose. *Carbohydrate Polymers, 174*, 1181–1191.

Coelho, M., Pereira, R. N., Sebastião, R., Teixeira, J. A., & Pintado, M. (2019). Extraction of tomato by-products' bioactive compounds using ohmic technology. *Food and Bioproducts Processing, 117*, 329–339.

Daghrir, R., Igounet, L., Brar, S.-K., & Drogui, P. (2014). Novel electrochemical method for the recovery of lipids from microalgae for biodiesel production. *Journal of the Taiwan Institute of Chemical Engineers, 45*(1), 153–162.

de Alwis, A. A. P., & Fryer, P. J. (1990). The use of direct resistance heating in the food industry. *Journal of Food Engineering, 11*(1), 3–27. http://www.sciencedirect.com/science/article/pii/0260877490900368.

De Carvalho Neto, R. G., Nascimento, J. G. S., Costa, M. C., Lopes, A. C., Neto, E. F. A., Filho, C. R. M., & dos Santos, A. B. (2014). Microalgae harvesting and cell disruption: A preliminary evaluation of the technology electroflotation by alternating current. *Water Science and Technology, 70*(2), 315–320.

de Figueiredo Furtado, G., Pereira, R. N., Vicente, A. A., & Cunha, R. L. (2018). Cold gel-like emulsions of lactoferrin subjected to ohmic heating. *Food Research International, 103*, 371–379.

de Oliveira, C. F., Giordani, D., Poliana, D. G., Cladera-Olivera, F., & Marczak, L. D. F. (2015). Extraction of pectin from passion fruit peel using moderate electric field and conventional heating extraction methods. *Innovative Food Science & Emerging Technologies, 29*, 201–208.

Deleu, L. J., Luyts, A., Haesendonck, I. V., Brijs, K., & Delcour, J. A. (2019). Ohmic versus conventional heating for studying molecular changes during pound cake baking. *Journal of Cereal Science, 89*(January), 102708.

do Cardoso, M. F. C., Gomes, A. T. P. C, Silva, V. L. M., Silva, A. M. S., Neves, M. G. P. M. S., da Silva, F. C., … Cavaleiro, J. A. S. (2015). Ohmic heating assisted synthesis of coumarinyl porphyrin derivatives. *RSC Advances, 5*(81), 66192–66199.

El Darra, N., Nabil, G., Vorobiev, E., Louka, N., & Maroun, R. (2013). Extraction of polyphenols from red grape pomace assisted by pulsed ohmic heating. *Food and Bioprocess Technology, 6*(5), 1281–1289.

Engchuan, W., Jittanit, W., & Garnjanagoonchorn, W. (2014). The ohmic heating of meat ball: Modeling and quality determination. *Innovative Food Science and Emerging Technologies, 23*, 121–130.

Fetterman, J. C. (1928). The electrical conductivity method of processing milk. *Agricultural Engineering, 9*, 107–108.

Gally, T., Rouaud, O., Jury, V., & Le-Bail, A. (2016). Bread baking using ohmic heating technology; a comprehensive study based on experiments and modelling. *Journal of Food Engineering, 190*, 176–184.

Gally, T., Le-Bail, A., Ogé, A., Havet, M., Rouaud, O., & Jury, V. (2017). Proofing of bread dough assisted by ohmic heating. *Innovative Food Science & Emerging Technologies, 39*, 55–62.

Gavahian, M., Chu, Y. H., & Sastry, S. K. (2018). Extraction from food and natural products by moderate electric field: Mechanisms, benefits, and potential industrial applications. *Comprehensive Reviews in Food Science and Food Safety, 17*(4), 1040–1052.

Gavahian, M., & Farahnaky, A. (2018). Ohmic-assisted hydrodistillation technology: A review. *Trends in Food Science & Technology, 72*, 153–161.

Gavahian, M., & Tiwari, B. K. (2020). Moderate electric fields and ohmic heating as promising fermentation tools. *Innovative Food Science & Emerging Technologies, 64*(November 2019), 102422.

Geada, P., Rodrigues, R., Loureiro, L., Pereira, R., Fernandes, B., Teixeira, J. A., … Vicente, A. A. (2018). Electrotechnologies applied to microalgal biotechnology—Applications, techniques and future trends. *Renewable and Sustainable Energy Reviews, 94*, 656–668.

Getchell, B. E. (1935). Electric pasteurization of milk. *Agricultural Engineering, 16*(10), 408–410.

Golding, M. (2019). Exploring and exploiting the role of food structure in digestion. In O. Gouseti, G. M. Bornhorst, S. Bakalis, & A. Mackie (Eds.), *Interdisciplinary approaches to food digestion* (pp. 81–128). Cham: Springer International Publishing.

Goullieux, A., & Pain, J.-P. (2014). Chapter 22—Ohmic heating. In *Emerging technologies for food processing* (2nd ed.). Elsevier Ltd.

Guida, V., Ferrari, G., Pataro, G., Di Maro, A., & Parente, A. (2013). The effects of ohmic and conventional blanching on the nutritional, bioactive compounds and quality parameters of artichoke heads. *LWT—Food Science and Technology, 53*(2), 569–579.

Hradecky, J., Kludska, E., Belkova, B., Wagner, M., & Hajslova, J. (2017). Ohmic heating: A promising technology to reduce furan formation in sterilized vegetable and vegetable/meat baby foods. *Innovative Food Science and Emerging Technologies, 43*(February), 1–6.

Icier, F., & Bozkurt, H. (2011). Ohmic heating of liquid whole egg: Rheological behaviour and fluid dynamics. *Food and Bioprocess Technology, 4*(7), 1253–1263.

Jaeger, H., Roth, A., Toepfl, S., Holzhauser, T., Engel, K-H., Knorr, D., … Steinberg, P. (2016). Opinion on the use of ohmic heating for the treatment of foods. *Trends in Food Science & Technology, 55*, 84–97. http://www.sciencedirect.com/science/article/pii/S0924224416301066.

Jaeschke, D. P., Menengol, T., Rech, R., Mercali, G. D., & Marczak, L. D. F. (2016). Carotenoid and lipid extraction from *Heterochlorella luteoviridis* using moderate electric field and ethanol. *Process Biochemistry, 51*(10), 1636–1643.

Jones, F. (1897). *Apparatus for electrically treating liquids.* US Patent 592735.

Kanjanapongkul, K. (2017). Rice cooking using ohmic heating: Determination of electrical conductivity, water diffusion and cooking energy. *Journal of Food Engineering, 192*, 1–10.

Kanjanapongkul, K., Tia, S., Wongsa-Ngasri, P., & Yoovidhya, T. (2009). Coagulation of protein in surimi wastewater using a continuous ohmic heater. *Journal of Food Engineering, 91*(2), 341–346.

Knirsch, M. C., dos Santos, C. A., Vicente, A. A., & Penna, T. C. V. (2010). Ohmic heating—A review. *Trends in Food Science & Technology, 21*(9), 436–441.

Lakkakula, N. R., Lima, M., & Walker, T. (2004). Rice bran stabilization and rice bran oil extraction using ohmic heating. *Bioresource Technology, 92*(2), 157–161.

Lascorz, D., Torella, E., Lyng, J. G., & Arroyo, C. (2016). The potential of ohmic heating as an alternative to steam for heat processing shrimps. *Innovative Food Science and Emerging Technologies, 37*, 329–335.

Lebovka, N. I., Praporscic, I., Ghnimi, S., & Vorobiev, E. (2005). Does electroporation occur during the ohmic heating of food? *Journal of Food Science, 70*(5), E308–E311.

Lima, M., Henkitt, B. F., & Sastry, S. K. (2001). Diffusion of beet dye during electrical and conventional heating at steady-state temperature. *Journal of Food Process Engineering, 24*(5), 331–340.

Loghavi, L., Sastry, S. K., & Yousef, A. E. (2007). Effect of moderate electric field on the metabolic activity and growth kinetics of *Lactobacillus acidophilus*. *Biotechnology and Bioengineering, 98*(4), 872–881.

Loghavi, L., Sastry, S. K., & Yousef, A. E. (2008). Effect of moderate electric field frequency on growth kinetics and metabolic activity of *Lactobacillus acidophilus*. *Biotechnology Progress, 24*(1), 148–153.

Loghavi, L., Sastry, S. K., & Yousef, A. E. (2009). Effect of moderate electric field frequency and growth stage on the cell membrane permeability of *Lactobacillus acidophilus*. *Biotechnology Progress, 25*(1), 85–94.

Loypimai, P., Moongngarm, A., Chottanom, P., & Moontree, T. (2015). Ohmic heating-assisted extraction of anthocyanins from black rice bran to prepare a natural food colourant. *Innovative Food Science & Emerging Technologies, 27*, 102–110.

Lyng, J. G., McKenna, B. M., & Arroyo, C. (2018). Chapter 3. Ohmic heating of foods. In A. Proctor (Ed.), *Alternatives to Conventional Food Processing* (2nd ed., pp. 95–137). Royal Society Chemistry.

Machado, L. F., Pereira, R. N., Martins, R. C., Teixeira, J. A., & Vicente, A. A. (2010). Moderate electric fields can inactivate *Escherichia coli* at room temperature. *Journal of Food Engineering, 96*(4), 520–527.

Mesías, M., Wagner, M., George, S., & Morales, F. J. (2016). Impact of conventional sterilization and ohmic heating on the amino acid profile in vegetable baby foods. *Innovative Food Science & Emerging Technologies, 34*, 24–28.

Mittelmann, E. (1949). *US1468871A—Method and apparatus for pasteurizing milk—Google Patents*. https://patents.google.com/patent/US1468871A/en. November 11, 2020.

Mizrahi, S., Kopelman, I. J., & Perlman, J. (2007). Blanching by electro-conductive heating. *International Journal of Food Science & Technology, 10*(3), 281–288. http://doi.wiley.com/10.1111/j.1365-2621.1975.tb00031.x. November 11, 2020.

Moreno, J., Simpson, R., Pizarro, N., Pavez, C., Dorvil, F., Petzold, G., & Bugueño, G. (2013). Influence of ohmic heating/osmotic dehydration treatments on polyphenoloxidase inactivation, physical properties and microbial stability of apples (cv. Granny Smith). *Innovative Food Science & Emerging Technologies, 20*, 198–207.

Palaniappan, S., & Sastry, S. K. (1991). Electrical conducitivty of selected juices: Influences of temperature, solids content, applied voltage, and particle size. *Journal of Food Process Engineering, 14*(4), 247–260. https://onlinelibrary.wiley.com/doi/full/10.1111/j.1745-4530.1991.tb00135.x. November 11, 2020.

Pataro, G., Barca, G. M. J., Pereira, R. N., Vicente, A. A., Teixeira, J. A., & Ferrari, G. (2014). Quantification of metal release from stainless steel electrodes during conventional and pulsed ohmic heating. *Innovative Food Science and Emerging Technologies, 21*, 66–73.

Pereira, R. N., Martins, R. C., & Vicente, A. A. (2008). Goat milk free fatty acid characterization during conventional and ohmic heating pasteurization. *Journal of Dairy Science, 91*(8), 2925–2937.

Pereira, R. N., Rodrigues, R. M., Genisheva, Z., Oliveira, H., Freitas, V., Teixeira, J. A., & Vicente, A. A. (2016). Effects of ohmic heating on extraction of food-grade phytochemicals from colored potato. *LWT—Food Science and Technology, 74*, 493–503.

Pereira, R. N., Rodrigues, R. M., Ramos, Ó. L., Xavier-Malcata, F., Teixeira, J. A., & Vicente, A. A. (2016b). Production of whey protein-based aggregates under ohmic heating. *Food and Bioprocess Technology, 9*(4), 576–587.

Pereira, R. N., Rodrigues, R. M., Ramos, O. L., Pinheiro, A. C., Martins, J. T., Teixeira, J. T., & Vicente, A. A. (2018). Electric field processing: Novel perspectives on allergenicity of milk proteins. *Journal of Agricultural and Food Chemistry, 66*(43), 11227–11233.

Pereira, R. N., Teixeira, J. A., Vicente, A. A., Cappato, L. P., da Silva Ferreira, M. V., da Silva Rocha, R., & da Cruz, A. G. (2018). Ohmic heating for the dairy industry: A potential technology to develop probiotic dairy foods in association with modifications of whey protein structure. *Current Opinion in Food Science, 22*, 95–101.

Pereira, R. N., & Vicente, A. A. (2010). Environmental impact of novel thermal and non-thermal technologies in food processing. *Food Research International, 43*(7), 1936–1943.

Pereira, R. N., Rodrigues, R. M., Altinok, E., Ramos, O. L., Xavier-Malcata, F., Maresca, P., … Vicente, A. A. (2017). Development of iron-rich whey protein hydrogels following application of ohmic heating—Effects of moderate electric fields. *Food Research International, 99*, 435–443.

Pereira, R. N., Costa, J., Rodrigues, R. M., Villa, C., Machado, L., Mafra, I., & Vicente, A. A. (2020a). Effects of ohmic heating on the immunoreactivity of β-lactoglobulin—A relationship towards structural aspects. *Food & Function, 11*, 4002–4013.

Pereira, R. N., Coelho, M., Genisheva, Z., Fernandes, J. M., Vicente, A. A., Pintado, M. E., & Teixeira, J. A. (2020b). Using ohmic heating effect on grape skins as a pretreatment for anthocyanins extraction. *Food and Bioproducts Processing, 124*, 320–328.

Pinto, J., Silva, V. L. M., Silva, A. M., Silva, A. M. S., Costa q, J. C. S, Santos, L. M. N. B. F., ... Teixeira, J. A. (2013). Ohmic heating as a new efficient process for organic synthesis in water. *Green Chemistry, 15*(4), 970.

Praporscic, I., Ghnimi, S., & Vorobiev, E. (2005). Enhancement of pressing of sugar beet cuts by combined ohmic heating and pulsed electric field treatment. *Journal of Food Processing and Preservation, 29*(5–6), 378–389.

Richardson, G. H., & Cornwell, E. H. (1968). *US3394011A—Continuous production of cheese curd.*

Rocha, C. M. R., Genisheva, Z., Ferreira-Santos, P., Rodrigues, R., Vicente, A., A., Teixeira, J., A., & Pereira, R., N. (2018). Electric field-based technologies for valorization of bioresources. *Bioresource Technology, 254*, 325–339.

Rodrigues, R. M., Avelar, Z., Machado, L., Pereira, R. N., & Vicente, A. A. (2020). Electric field effects on proteins—Novel perspectives on food and potential health implications. *Food Research International, 137*(109709), 1–14.

Rodrigues, R. M., Avelar, Z., Vicente, A. A., Petersen, S. B., & Pereira, R. N. (2020). Influence of moderate electric fields in β-lactoglobulin thermal unfolding and interactions. *Food Chemistry, 304*(125442), 1–8.

Rodrigues, R. M., Fasolin, L. H., Avelar, Z., Petersen, S. B., Vicente, A. A., & Pereira, R. N. (2020). Effects of moderate electric fields on cold-set gelation of whey proteins—From molecular interactions to functional properties. *Food Hydrocolloids, 101*(105505), 1–10.

Rodrigues, R. M., Martins, A. J., Ramos, O. L., Xavier-Malcata, F., Teixeira, J. A., & Vicente, A. A. (2015). Influence of moderate electric fields on gelation of whey protein isolate. *Food Hydrocolloids, 43*, 329–339.

Rodrigues, R. M., Genisheva, Z., Rocha, C. M. R., Teixeira, J. A., Vicente, A. A., & Pereira, R. N. (2019). 6—Ohmic heating for preservation, transformation, and extraction. In F. Chemat, & E. Vorobiev (Eds.), *Green food processing techniques- Preservation, transformation and extraction* (pp. 159–191). Academic Press. B T—Green Food Processing Techniques.

Rodrigues, R. M., Vicente, A. A., Petersen, S. B., & Pereira, R. N. (2019). Electric field effects on β-lactoglobulin thermal unfolding as a function of PH—Impact on protein functionality. *Innovative Food Science and Emerging Technologies, 52*, 1–7.

Roos, Y. H., Fryer, P. J., Knorr, D., Schuchmann, H., Schroën, K., Schutyser, M. A. I., ... Windhab, E. J. (2016). Food engineering at multiple scales: Case studies, challenges and the future—A European perspective. *Food Engineering Reviews, 8*(2), 91–115.

Roux, S., Courel, M., Birlouez-Aragon, I., Municino, F., Massa, M., & Pain, J-P. (2016). Comparative thermal impact of two UHT technologies, continuous ohmic heating and direct steam injection, on the nutritional properties of liquid infant formula. *Journal of Food Engineering, 179*, 36–43.

Saberian, H., Hamidi-Esfahani, Z., Gavlighi, H. A., Banakar, A., & Barzegar, M. (2018). The potential of ohmic heating for pectin Extraction from orange waste. *Journal of Food Processing and Preservation, 42*(2), e13458.

Sakr, M., & Liu, S. (2014). A comprehensive review on applications of ohmic heating (OH). *Renewable and Sustainable Energy Reviews, 39*, 262–269.

Salengke, S., Waris, S. A., & Mochtar, A. H. (2016). Design and optimization of pilot scale ohmic-based carrageenan extraction technology. In *14th food engineering conference.* Melbourn.

Samaranayake, C. P., & Sastry, S. K. (2016). Effects of controlled-frequency moderate electric fields on pectin methylesterase and polygalacturonase activities in tomato homogenate. *Food Chemistry, 199*(Supplement C), 265–272.

Samaranayake, C. P., & Sastry, S. K. (2018). In-situ activity of α-amylase in the presence of controlled-frequency moderate electric fields. *LWT—Food Science and Technology, 90*(October 2017), 448–454.

Samaranayake, C. P., Sastry, S. K., & Zhang, H. (2005). Pulsed ohmic heating-a novel technique for minimization of electrochemical reactions during processing. *Journal of Food Science, 70*(8), e460–e465. https://doi.org/10.1111/j.1365-2621.2005.tb11515.x. November 11, 2020.

Sastry, S. K. (2004). Advances in ohmic heating and moderate electric field (MEF) processing. In G. V. Barbosa-Canovas, M. S. Tapia, & P. Cano (Eds.), *Novel food processing technologies* (pp. 491–499). CRC Press.

Seidi Damyeh, M., & Niakousari, M. (2016). Impact of Ohmic-assisted hydrodistillation on kinetics data, physicochemical and biological properties of *Prangos ferulacea* Lindle. Essential oil: Comparison with conventional hydrodistillation. *Innovative Food Science & Emerging Technologies, 33*, 387–396.

Sensoy, I. (2014). A review on the relationship between food structure, processing, and bioavailability. *Critical Reviews in Food Science and Nutrition, 54*(7), 902–909.

Sensoy, I., & Sastry, S. K. (2004). Extraction using moderate electric fields. *Journal of Food Science, 69*(1), 7–13.

Silva, V. L. M., Santos, L. M. N. B. F., & Silva, A. M. S. (2017). Ohmic heating: An emerging concept in organic synthesis. *Chemistry—A European Journal, 23*(33), 7853–7865.

Silva, E. S., et al. (2017). In D. Miklavčič (Ed.), *3 Handbook of Electroporation.* Cham: Springer International Publishing.

Simpson, R., Ramírez, C, Almonacid, A., Moreno, J., Nuñez, H., & Jaques, A. (2015). Diffusion mechanisms during the osmotic dehydration of granny smith apples subjected to a moderate electric field. *Journal of Food Engineering, 166*, 204–211.

Souza, B. W. S., Cerqueira, M. A., Casariego, A., Lima, A. M. P., Teixeira, J. A., & Vicente, A. A. (2009). Effect of moderate electric fields in the permeation properties of chitosan coatings. *Food Hydrocolloids, 23*(8), 2110–2115.

Souza, B. W. S., Cerqueira, M. A., Martins, J. T., Casariego, A., Teixeira, J. A., & Vicente, A, A. (2010). Influence of electric fields on the structure of chitosan edible coatings. *Food Hydrocolloids, 24*(4), 330–335.

Stone, G. E. (1909). Influence of electricity on microorganisms. *Botanical Gazette, 48*(5), 359–379.

Sung-Won, Y., & Ki-Myung, K. (2002). Leakage of cellular materials from *Saccharomyces cerevisiae* by ohmic heating. *Journal of Microbiology and Biotechnology, 12*(2), 183–188.

Tadpitchayangkoon, P., Park, J. W., & Yongsawatdigul, J. (2012). Gelation characteristics of tropical surimi under water bath and ohmic heating. *LWT—Food Science and Technology, 46*(1), 97–103.

Torkian Boldaji, M., Borghei, A. M., Beheshti, B., & Hosseini, S. E. (2015). The process of producing tomato paste by ohmic heating method. *Journal of Food Science and Technology, 52*(6), 3598–3606.

Varghese, K. S., Pandey, M. C., Radhakrishna, K., & Bawa, A. S. (2014). Technology, applications and modelling of ohmic heating: A review. *Journal of Food Science and Technology, 51*(10), 2304–2317.

Wongsa-Ngasri, P., & Sastry, S. K. (2015). Effect of ohmic heating on tomato peeling. *LWT—Food Science and Technology, 61*(2), 269–274.

Yin, Z., Hoffmann, M., & Jiang, S. (2018). Sludge disinfection using electrical thermal treatment: The role of ohmic heating. *Science of the Total Environment, 615*, 262–271.

Zell, M., Lyng, J. G., Cronin, D. A., & Morgan, D. J. (2010). Ohmic cooking of whole beef muscle—Evaluation of the impact of a novel rapid ohmic cooking method on product quality. *Meat Science, 86*(2), 258–263.

CHAPTER 8

Pulsed Electric Fields in Sustainable Food

MIRIAN PATEIRO[A] • RUBÉN DOMÍNGUEZ[A] • IGOR TOMASEVIC[B] • PAULO EDUARDO SICHETTI MUNEKATA[A] • MOHAMMED GAGAOUA[C] • JOSÉ MANUEL LORENZO[A]

[a]Galician Meat Technology Center, Galicia Technology Park, Ourense, Spain, [b]Department of Animal Source Food Technology, Faculty of Agriculture, University of Belgrade, Belgrade, Serbia, [c]Food Quality and Sensory Science Department, Teagasc Ashtown Food Research Centre, Dublin, Ireland

8.1 INTRODUCTION

Just over 100 years ago when the concept of PEF emerged as a promising preservation technique (Alirezalu et al., 2020), but it was not until the 1940s and 1950s when the first applications were made in the food industry (Toepfl, Kinsella, & Parniakov, 2020). Its implementation in processing industries (alcoholic beverages, dairy products, fruit juices, and liquid eggs) arises from the loss of quality (i.e., antioxidants, bioactive compounds, health-promoting compounds, vitamins) that occurs in heat-treated products (Alirezalu, Munekata, et al., 2020).

This technology is based on the electroporation of membranes. This biophysical phenomenon supposes the increase of cell membrane permeability through the externally application of short and intense electric pulses during a short time (from nanoseconds to milliseconds), which involves the application of 0.1 to 50 kV/cm to the food (Fig. 8.1). Depending on the electric field strength, promote transient or permanent pore formation, resulting in a structural modification that allows the release of intracellular substances, and the cell disruption, respectively (Barba et al., 2015). Moreover, depending on the electric field strength and the number of pulses, PEF can gives rise to different effects in the applied product (Dziadek et al., 2019). So far, there are many food products in which this technology has been applied such as alcoholic beverages, dairy, eggs, fruits and vegetables, meat and fish products (Alirezalu, Munekata, et al., 2020; He, Yin, Yan, & Wang, 2017; Puértolas, Saldaña, & Raso, 2017; Yang, Huang, Lyu, & Wang, 2016; Yogesh, 2016).

Although there are several parameters involved in the effectiveness of this technique (sample characteristics, type of equipment, especial energy input, process temperature, pulse intensity, and strength of field), PEF conditions must be adjusted to the application desired (Alirezalu, Munekata, et al., 2020). The systems developed so far have evolved remarkably from hand-built laboratory equipment in the 1990s to the present day. Today, several PEF systems have been specifically designed for industrial use, responding to commercial demand to facilitate the development of new applications (Kempkes, 2017).

In general, shelf-life extension, quality preservation and nutrient retention are among the main benefits obtained of its application on foods. At microbiological level, PEF causes the modification of the phospholipid bilayer of the cytoplasmic membrane of the microorganism present in the product, which results in the loss of structure, impermeability, cellular constituents, and functions of the membrane. Consequently, there is a significant loss of viability and cell death. Regarding the quality of the product, minerals, proteins, vitamins, and flavors are not modified by PEF due to the little increase of temperature that occurs during treatment, which implies the minimization in the degradation of heat sensitive compounds (Alirezalu et al., 2020).

Although there are many advantages associated with this technology (Fig. 8.2), its application is restricted to products with low electrical conductivity and no air bubbles, since a dielectric breakdown could occur (Lammerskitten, Wiktor, Parniakov, & Lebovka, 2020). This effect promotes undesired electrochemical reactions with the corresponding formation of high reactive species (e.g., free radicals), and the appearance of operational limitations such as the evolution of gas bubbles and the increase of pressure. As a result, the under processing of food products, the reduction of the treatment efficiency and the lifetime of the PEF equipment could occur (Chauhan & Unni, 2015). Although in many food products the conductivity is determined

Sustainable Production Technology in Food. https://doi.org/10.1016/B978-0-12-821233-2.00002-2
Copyright © 2021 Elsevier Inc. All rights reserved.

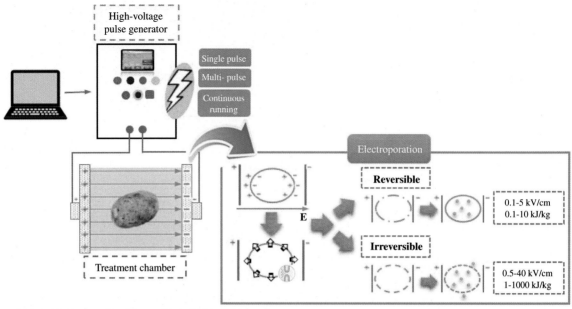

FIG. 8.1 Electroporation: How pulsed electric fields work.

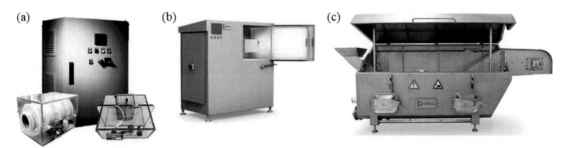

FIG. 8.2 PEF systems: Equipment at laboratory (A), pilot (B), and industrial (C) scale.

by the composition of the food itself, it is recommended, whenever possible, to add salt once the product has been treated (Amiali & Ngadi, 2012) (Fig. 8.3).

This chapter reflects the applicability of this emerging technology in different foods, providing detailed information about the effects and considerations for a successful industrial implementation, as well as the potential benefits for both the consumers and the industry (Fig. 8.3).

8.2   IMPLICATION OF PEF IN THE PRODUCTION OF SAFE FOOD

8.2.1   Mitigation of Toxic Compounds

Food provides the energy and nutrients necessary for life. However, some foods may contain toxins that

are naturally present or generated by contamination, spoilage, or poor storage. The new regulations, which control in a more restrictive way the presence of contaminants in foods, have led to the use of alternative or complementary non-thermal treatments in order to prevent the appearance of changes during processing and storage (Dourado et al., 2019). PEF has been proposed by several authors as a promising technology to reduce the concentrations of some toxins (e.g., acrylamide, mycotoxins) from food matrix without altering their sensory and nutritional properties (Pallarés, Berrada, Fernández-Franzón, & Ferrer, 2020). The effectiveness of this decontamination depends on the operational conditions (process duration, pulse width, electrical conductivity and pH of food, and voltage intensity).

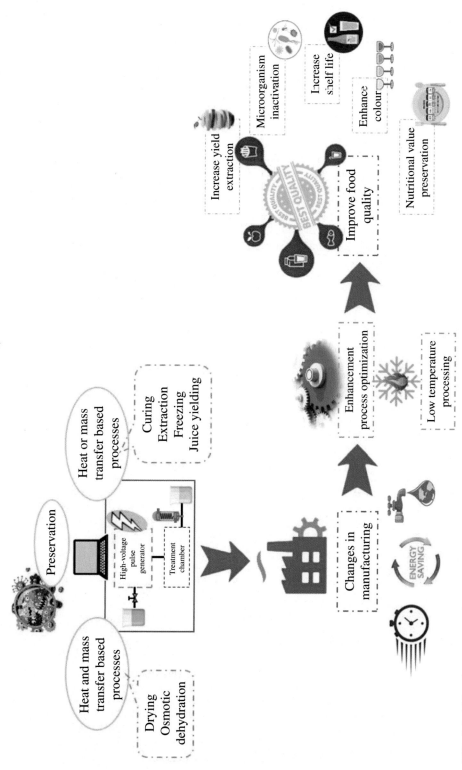

FIG. 8.3 Main advantages and applications of PEF.

Potato (*Solanum tuberosum* L.) is one of the most commonly consumed vegetables throughout the world, ranking third in Europe and fifth in global agricultural production behind crops such as maize and wheat (FAO, 2020a, 2020b). Acrylamide is one of these contaminants that concern the potato industry, since its consumption could endanger the health of consumers (EFSA, 2015). In fact, mitigation measures and benchmark levels for the reduction of the presence of acrylamide in food are regulated in the European Union (OJEU, 2017). This natural by-product is formed during the cooking process when starchy foods are cooked at high temperatures (above 120°C), being the Maillard reaction the main responsible for its formation.

Food industry has made great efforts to identify and implement measures to reduce the levels of acrylamide in food. Despite the effectiveness of conventional treatments, such as hot water blanching, the drastic conditions used can modify the quality of the final product (Genovese et al., 2019). This has led to the search for new technologies that allow to extract Maillard reaction substrates without altering the sensory attributes of the product. In this regard, the application of PEF reduces the formation of acrylamide as it allows to reduce the contents of sugars and amino acids used as substrate in the Maillard reaction. Genovese et al. (2019) observed that the application of an electric field strength of 1.5 kV/cm, 1000 pulses at fixed pulse width (10 µs), frequency (100 Hz), and repetition time (10 ms) in potato chips allowed to improve, through an irreversible electroporation, the diffusion of Maillard reaction substrates. Moreover, PEF pre-treatment led to the achievement of the reduction of the initial contents of fructose and asparagine (5.4% and 48%, respectively), which resulted in a reduction of 31% of acrylamide compared to 17% found with conventional blanched treatments. Despite this depletion, it was not possible to be below the permitted levels for this compound (0.75 mg/kg) (OJEU, 2017), which suggest the possibility of combining PEF with mild blanching treatments. Additional studies have revealed a reduction in the contents of acrylamide when there are low levels of reducing sugars and the frying time is short (52%, 57% and 19% for sweet potato, beetroots and carrot chips, respectively) (Siemer, Töpfl, Witt, & Ostermeier, 2018).

Another compound formed during the heating or conservation processes of food (neoformed contaminants), which could also be harmful to human, is 5-hydroxymethylfurfural (HMF). This furanic compound, which has also attracted the attention of the scientific community in recent years, results from the Maillard reaction and from direct dehydration of sugars under acidic conditions (caramelization) (Capuano & Fogliano, 2011). Although it can be found in foods like bakery products, fruit juices, malt and vinegar, bread, and coffee are the most important contributors to dietary HMF intake. There are studies that show the efficacy of PEF in reducing the formation of HMF (Khaneghah et al., 2020), which would prevent its conversion in vivo into 5-sulfoxymethylfurfural (SMF) with genotoxic properties (Capuano & Fogliano, 2011). Akdemir Evrendilek, Celik, Agcam, and Akyildiz (2017) observed that, unlike heat treatments, the application of PEF in apple juice did not result in significant formation of furfural and HMF. In addition, the contents of both compounds depended on the treatment conditions. In this regard, their formation was observed at moderate level of electric field strength (10–20 kV/cm) and high level of energy (80–140 J/s), while lower PEF intensities resulted in lower contents. This behavior was also observed by other authors in strawberry juice (Aguiló-Aguayo, Oms-Oliu, Soliva-Fortuny, & Martín-Belloso, 2009).

Significant differences were also observed in *Emblica officinalis* juice between PEF (26 kV/cm/500 µs) and heat treatments (90°C/60 s), which resulted in HMF contents of ≅ 1.5 mg/L and ≅ 8.0 mg/L, respectively (Bansal, Sharma, Ghanshyam, Singla, & Kim, 2015). The application of high-intensity pulsed electric field (HIPEF) in date juice (35 kV/cm/1000 µs) also displayed lower HMF contents compared to those found in thermally pasteurized sample (5.90 vs 6.90 mg/L, respectively) (Mtaoua, Sánchez-Vega, Ferchichi, & Martín-Belloso, 2017). This could be due to thermal treatments would favor Maillard's reaction in the presence of reducer sugars (glucose and fructose) and acids; while HIPEF would retain ascorbic acid in the juice avoiding its oxidation and the supply of reactive carbonyl groups, which would act as precursors of non-enzymatic browning reactions that lead to the formation of HMF (Aguiló-Aguayo et al., 2009). The positive effects of PEF treatment were also observed by Mtaoua et al. (2017) during the shelf-life of date juice, since there was a slight increase in HMF content during 5 weeks of storage compared to the trend showed in thermally treated samples (≅ 7.5 vs ≅ 9.0 mg/L, respectively).

Other compounds that pose a serious risk to human health are fungal toxins (mycotoxins). These substances appear throughout the entire food chain, from field crops to processed foods, through feed and raw or unprocessed foods. They are produced by several hundred species of molds (*Alternaria, Aspergillus, Claviceps, Fusarium, Penicillium,* and *Stachybotrys*), which can grow on food under specific conditions of humidity and temperature (Adebo et al., 2021). In the case of processed foods, cereal-based products (bread, pasta, breakfast cereals, etc.), beverages (wine, coffee, cocoa, beer, and juices), foods of animal origin (milk, cheese) and baby foods are important sources of exposure to mycotoxins. Even though, these compounds are relatively sensible to temperature, in most cases they are not eliminated with the application of high temperatures.

The application of PEF has an important role in mycotoxin reduction. In this regard, Pallarés, Barba, Berrada, Tolosa, and Ferrer (2020) evaluated the effect of PEF on enniatins and beauvericin contents in juice and smoothies. The operational conditions applied (voltage 30 kV, field strength 3 kV/cm, specific energy 500 kJ/kg, 238 pulses, and processing time 5 min) allowed to obtain reduction percentages from 43% to 53% in juices, and from 56% to 70% in smoothies, respectively. These percentages could be related with the sensitivity of these compounds to temperature, and so it was demonstrated by the authors since 26% of degradation was related with the temperature reached during the treatment (around 70°C). For these reasons, the combination of this technology with thermal ones can be considered for the mycotoxin reduction (Subramanian, Shanmugam, Ranganathan, Kumar, & Reddy, 2017). In addition, pH of the media also has an important effect on mycotoxins mitigation, being the alkaline conditions ($10 \leq pH \leq 12.03$) the most effective (Vijayalakshmi, Nadanasabhapathi, Kumar, & Sunny Kumar, 2018). Good results were also found with acidic conditions ($pH \leq 4.0$), reflected in the results obtained by Pallarés, Barba, et al. (2020) related with the pH of juices and smoothies (3.88 and 3.94, respectively).

8.2.2 Control of Food Allergy

Food allergies are abnormal responses of the immune system to certain food components (Taylor, 2021). It is one of the major public health concerns, since it affects a considerable part of the population; more than 10% of the population suffers from at least one food allergy (Warren, Jiang, & Gupta, 2020). Moreover, it is estimated that more than 170 foods could cause allergic reactions, proteins being those, that are associated with a higher percentage of incidences (Shriver & Yang, 2011).

Food processing allows to enhance nutritional quality, extend shelf-life, and ensure safety. However, sometimes the temperatures involved in these processes can lead to the degradation of some nutrients and native structures, which could influence allergenicity (Ekezie, Sun, & Cheng, 2017). Therefore, the possibility of using non-thermal techniques to develop hypoallergenic foods has been evaluated. In this regard, it has been demonstrated that PEF could induce changes in the structural characteristics of food allergens, contributing to allergen mitigation (Sun, Yu, Zeng, Yang, & Jia, 2011).

Eggs are considered as the second most allergenic food (Yang, Tu, Wang, Li, & Tian, 2017). Although sterilization is the main application of PEF in egg products (Yogesh, 2016), recent studies have evaluated the effect of PEF on the structure and stability of egg white proteins (Wu, Zhao, Yang, & Yan, 2015). In this regard, Yang, Tu, Wang, Zhang, et al. (2017) evaluated the effect of PEF on structural properties of ovalbumin, a glycoprotein considered as the dominant allergen in egg white. The authors applied low-intensity pulsed electric fields (LPEF, below 25 kV/cm, for 180 μs) or short time (< 60 μs, at 35 kV/cm), and HIPEF (exceeding 25 kV/cm, for 180 μs) or long time (> 60 μs, at 35 kV/cm) in the ovalbumin isolated from fresh hen eggs. The immunogenic properties were determined, since the decrease of IgG and IgE binding abilities would reduce significantly the potential allergenicity of ovalbumin. In the first case, IgG and IgE binding capacities gradually increased, while the application of a higher treatment for a long time (35 kV/cm, 180 μs) resulted in a significant reduction due to the aggregation of ovalbumin and a loss of the α-helix structure. Therefore, PEF could change the secondary and tertiary structures of egg white proteins, influencing the immunogenic properties and therefore, it could be used to decrease egg allergenicity.

Peanut is also associated with the apparition of food allergy, being 2S albumins, like Ara h 2 and Ara h 6, the predominant allergens in children from USA and Europe compared to PR-10 (Ara h 8) and Lipid Transfer Protein (LTP) (Ara h 9) sensitizations (Ballmer-Weber et al., 2015; Tontini et al., 2017). The studies carried out to date present no clear results about the effect of PEF in the reduction of these allergens. Vanga et al. (2016) observed a modification of α-helix structure in peanut flour samples treated with 10, 15, and 20 kV for 60–180 min. The effects found were higher in the samples subjected to longer treatments. In contrast, no significant changes were observed by Johnson et al. (2010) on native peanut Ara h 2, 6 and in other allergens present in apple (Mal d 3 and Mal d 1b).

PEF could also be used in wine industry to avoid the use of sulphites, since these additives cause health damage associated with allergic reactions, headaches, asthma, abdominal pain, and bronchoconstriction (van Wyk, Silva, & Farid, 2019). The application of 31, 40, and 50 kV/cm treatments allowed to reduce the presence of microorganisms as *Brettanomyces* linked to negative impacts in wine quality and shelf-life. *D*-values (time required to achieve 1 log reduction) reflected that the effect was depended on the intensity applied, observing higher effects in samples treated with 50 kV/cm (181.8, 36.1 and 13.0 μs for 31, 40 and 50 kV/cm treatments, respectively). Moreover, this electric field intensity allowed to duplicate the inactivation from 1.5 to 3.0 log reductions (cfu/mL), probably due to the increase in temperature ($\cong 10\,^{\circ}$C) that occurs during the treatment.

With that being stated, it seems that the modification achieved by PEF on food allergens is minimal compared to the effects observed with other non-thermal treatments. Therefore, it is necessary to continue optimizing the conditions that allow to mitigate food allergens.

8.2.3 Preservation of Foods

The use of PEF also offers new opportunities for obtaining safe food in a sustainable way (Dziadek et al., 2019). It is known that thermal processes (pasteurization and sterilization) reduce the microbial load by inactivating microorganisms during processing of foods, which ensures their viability and extend their shelf-life. However, this entails a loss of nutritional (bioactive compounds, vitamins), sensorial quality (aroma losses, non-enzymatic browning, texture) and even the possible beneficial effects on health (especially those related with the presence of phenolic compounds), which leads to a loss of attractiveness of the product for consumers (Leong, Burritt, & Oey, 2016; Mtaoua et al., 2017).

The use of PEF is becoming more common, considered appropriate for the pre-treatment of liquid and semi-solid edible products (Ricci, Parpinello, & Versari, 2018). In this regard, the juice processing industries are one of the sectors where it is being used. Depending on the purpose of PEF implementation, it could be applied either after the crushing or after the pressing to improve the extraction or the preservation of the juice, respectively. This makes the equipment (PEF chamber) used different. Mtaoua et al. (2017) reported that the use of HIPEF allows to preserve or even increase the quality of juices (Table 8.1). In the particular case of Tunisian date juice, the application of HIPEF (35 kV/cm for 1000 μs using pulses of 4 μs pulses at 100 Hz in bipolar mode) resulted in higher contents of polyphenol contents compared to those observed in samples subjected to heat treatment (90 °C/60 s) and in untreated juices (569.55 mg/L GAE vs 494.35 and 483.32 mg/L GAE, respectively). This fact is related with the electroporation effect produced during the treatment with PEF, which

leads to the permeabilization of cells enhancing the extraction of intracellular bioactive compounds. In addition, these contents are preserved during the refrigerated storage. Other positive effect of PEF treatment was the reduction of HMF contents (5.25 and 5.90 mg/L for HIPEF and pasteurization, respectively), probably due to high retention of acid in the juice. Moreover, no significant effects were observed in color parameters.

Similar effects were observed in apple juice (Dziadek et al., 2019). In this case, PEF treatment allowed to maintain vitamin C contents (21.61 mg/100 mL), probably due to the low processing temperature associated to this treatment and the acidic conditions of apple juice that would protect this thermolabile compound. Moreover, an increase of total polyphenol contents and a reduction of microbial load (mesophilic bacteria, microscopic fungi, and yeasts) were reported by the aforementioned authors. After 72 h of refrigerated storage, the samples subjected to a greater treatment (400 pulses at 30 kV/cm) were those that showed the best results for total polyphenols (295.66 vs 234.83 mg/100 mL) and antioxidant activity (16.69 vs 12.50 μmol Trolox/mL) together with a total inhibition of the spoilage microorganisms compared to those found control in samples.

In dairy products, PEF is applied to inactivate the spoilage and pathogenic bacteria (*Bacillus cereus, Cronobacter sakazakii, Escherichia coli, Mycobacterium paratuberculosis, Pseudomonas* spp., *Salmonella* spp., *Staphylococcus aureus,* and *Listeria* spp.), usually present in this type of products (Buckow, Chandry, Ng, McAuley, & Swanson, 2014; Sharma, Bremer, et al., 2014). The application of HIPEF (15–60 kV/cm for few microseconds) in the food matrix contributes to the milk quality preservation due to the formation and stabilization of membrane pores (transmembrane polarization). In addition to membrane permeabilization, the promising effects of PEF could also be associated with the loss of cell wall integrity in vegetative bacteria and the changes in coat, cortex, and core in spores (Pillet, Formosa-Dague, Baaziz, Dague, & Rols, 2016; Soltanzadeh et al., 2020). However, the effectiveness of the treatment depends on the microorganism and the vegetative and spore forms (Pillet et al., 2016; Walter et al., 2016). In general, the results found in the literature showed that Gram-negative bacteria were more sensitive to PEF-induced inactivation compared to Gram-positive bacteria (Sharma, Bremer, et al., 2014).

According to the information reported in previous research on dairy products, the impacts of PEF technology resulted from the balance between electrical and thermal effects, which allow to improve the results of

TABLE 8.1
Effect of PEF Treatment on the Inactivation of Microorganisms in Foods.

Product	Microorganism	PEF Conditions	Outcomes	Reference
Apple juice	Mesophilic, psychrophilic, fungi, yeast, coliform, enterococci, *Salmonella*, and *Staphylococcus aureus*	Electric field (30 kV/cm) Number of pulses (400 pulses)	Reduction of microbial load. Maintain levels of vitamin C (21.61 mg/100 mL) Increase of polyphenol content	Dziadek et al. (2019)
Liquid whey protein	*Listeria innocua*	Electric field (32 kV/cm) Flow rate (7.1 kg/h) Pulse width (3 μs) Pre-heating (20°C) 2% whey protein at pH 4	Inverse relationship between temperature and inactive rate. Decrease of 6.5 $\log_{10}$.	Schottroff et al. (2019)
Tunisian date juice	–	Electric field (35 kV/cm) Pulse width (4 μs) Frequency (100 Hz) Pulse duration (1000 μs)	Increase of polyphenol content (569.55 mg/L GAE). Decrease of HMF (5.25 mg/L)	Mtaoua et al. (2017)
Whole milk	*Escherichia coli* *Staphylococcus aureus* *Listeria innocua* *Pseudomonas aeruginosa*	Electric field (18–28 kV/cm) Flow rate (4.2 mL/s) Pulse width (20 μs) Frequency (10–140 Hz) Pulse duration (17–235 μs) Pre-heating (55°C for 24 s)	Results similar than those found in pasteurization at optimal conditions ($\geq$ 23 kV/cm, 55°C for 24 s)	Sharma, Bremer, Oey, and Everett (2014)
Whole milk	*E. coli, S. aureus* and *Pseudomonas fluorescens*	Electric field (40 kV/cm) Flow rate (142 mL/min) Pulse width (89 μs) Pre-heating (32.5°C)	Decrease of 5, 5.2 and 5.3 $\log_{10}$ cycle for *E. Coli, S. aureus* and *Pseudomonas fluorescens*	Cregenzán-Alberti, Halpin, Whyte, Lyng, and Noci (2015)
UHT whole milk	*E. coli* and *Pseudomonas fluorescens*	Electric field (35 kV/cm) Energy input (240 kJ/kg) Flow rate (120 mL/min) Pulse width (40 μs) Pre-heating (55°C)	Decrease of 6 $\log_{10}$ cycle for *E. coli*, and *Pseudomonas fluorescens*	Walter, Knight, Ng, and Buckow (2016)

HIPEF, high-PEF; HMF, hydroxymethylfurfural; *SMUF*, simulated milk ultra-filtrated.

microbial inactivation as the temperature increases (Alirezalu, Munekata, et al., 2020). Therefore, this suggest that from an application standpoint, PEF must be combined with other technologies (i.e., mild heating below pasteurization conditions) to enhance its impact on milk microorganisms. Moreover, this would allow to ensure the safety and extend the shelf-life of dairy products with a minimal impact on nutrients and technological properties.

In this regard, Sharma, Bremer, et al. (2014) evaluated the possibility of combining PEF (18–28 kV/cm, 10–140 Hz, 17–235 μs) with a pre-heating (4–55 °C for 24 s) and an intermediate cooling step to obtain similar effects than those obtained with pasteurization in whole milk. The authors reported that *Pseudomonas*

aeruginosa counts were reduced 5–6 log10 applying 22–28 kV/cm for 17–101 μs at 50°C. In the case of *Escherichia coli, Staphylococcus aureus,* and *Listeria innocua,* the counts were reduced at 55°C below the detection limit. Therefore, the bacterial inactivation obtained with milk pre-heating at 55°C for 24 s and PEF treatment ($\geq$ 23 kV/cm) could be compared to the levels achieved with pasteurization inactivation. This temperature would be also responsible for 6 $\log_{10}$ cycle reduction of *Escherichia coli* and *Pseudomonas fluorescens* observed by Walter et al. (2016) in UHT whole milk. Moreover, the treatment conditions applied (35 kV/cm, 40 μs, 240 kJ/kg, 55°C) would also preserve important aspects such as fat globule integrity, protein bioactivity/structure, and vitamins (Buckow et al., 2014).

The temperature was also the one that influenced the results obtained by Cregenzán-Alberti et al. (2015). These authors evaluated the application of PEF to inactivate *Escherichia coli*, *Staphylococcus aureus*, and *Pseudomonas fluorescens* in milk. A treatment of 40 V/cm with a flow rate of 142 mL/min applied at 32.5°C for 89 μs resulted in 5 log_{10} cycle reduction. The increase of the fluidity of membranes at mild heating could be the reason of these higher inactivation levels. However, sometimes this combination may not produce the expected result, since the presence of bubbles, conductivity, or even pH could condition its effectiveness. In this regard, Schottroff et al. (2019) observed an inverse relationship between temperature and inactivation rate of *Listeria innocua* in liquid whey protein, being protein concentration, pH, and temperature the variables that highly influenced the decontamination. Among them, pH has a great effect on inactivation, with higher rates under acidic (pH 4) than basic (pH 7) conditions. The optimal conditions (32 kV/cm during 3 μs, 2% whey protein, pH 4, and 20°C) allowed to obtain a reduction of 6.5 log_{10}.

On the other hand, it is important to take into account dormant bacterial endospores, since they are associated with foodborne outbreaks. PEF combined with elevated temperatures have been reported as a useful technology to inactivate several bacterial endospores. This would be the case of *Bacillus subtilis* endospores considered as one of the most resistant in the dairy industry. Cregenzán-Alberti, Arroyo, Dorozko, Whyte, and Lyng (2017) evaluated the possibility of applying HIPEF to inactivate them in reconstituted skimmed milk powder. The results were very promising since the application of 38 kV/cm for 10 μs and 466 Hz at maximum processing temperatures (up to 123°C) would allow to get inactivation values greater than 4.5 log_{10} cycles. These results could be due to the internal damage caused by PEF on the endospore core at frequencies higher than 1 kHz and the fracture of the coat at frequencies lower than 300 Hz (Choi et al., 2008). Another endospore of *Bacillus* genus that could appear in the dairy plant and that needs to be controlled is *Bacillus cereus*. Bermúdez-Aguirre, Dunne, and Barbosa-Cánovas (2012) found that PEF combined with a mild thermal treatment allowed to extend the shelf-life of whole and skim milk. The authors observed that treatments were more efficient with increasing temperature. Thus, temperatures below 40°C induced the germination of dormant bacterial spores, the use of 50°C improved the spore death and 65°C increased the spore reduction. Therefore, a treatment at 40 kV/cm after 276 pulses and 65°C resulted in 2.5 and 2.0 log_{10}

reductions in skim and whole milk, respectively. Moreover, the application of nisin (50 IU/mL) would improve the results in skim milk, reducing spore counts by 3.6 log_{10} when it was combined with 40 kV/cm, 144 pulses and 65°C.

Something similar phenomenon happens with enzymes, either endogenous or exogenous, produced by microorganisms, since they could also cause the deterioration of the product (Sharma, Oey, Bremer, & Everett, 2018). Electric fields strengths above 10 kV/cm, long processing times (> 250 μs) and high temperatures (> 40°C) give place to electrostatic interactions in the structure of the protein, resulting in conformational changes in the structure of the enzyme and in its inactivation (Terefe, Buckow, & Versteeg, 2015). In this regard, it was reported that treatments with PEF at temperatures below 40°C would induce the enzyme activity, whereas treatment above 60–65°C favors their degradation. This behavior was observed by McAuley, Singh, Haro-Maza, Williams, and Buckow (2016) in the activity of plasmin. Although no significant reductions were observed at 52°C after the application of 30 kV/cm for 22 μs with a flow rate of 2.4 L/min, the activity was increased at 42°C (Table 8.2).

In the case of milk, alkaline phosphatase, catalase, lactoperoxidase, lysozyme, ribonuclease, and xanthine oxidase could condition its quality. Sharma et al. (2018) observed a decrease of 96%–97% in the activity of alkaline phosphatase when PEF (25.7 kV/cm for 34 μs) was applied after a preheating (55°C for 24 s) in bovine whole milk. This result was similar to those found after thermal treatments (63°C for 30 min or 73°C for 15 s). Lower reductions were observed in the activities of xanthine oxidase (30%) and plasmin (7%). Similar values were also found by Sharma, Oey, et al. (2014). Lipase is other enzyme that is hardly inactivated after PEF processing, achieving percentages of inactivation higher than 60% (Sharma et al., 2018; Vega-Mercado et al., 2019).

Eggs are also a favorable growth medium for several microorganisms; therefore, pasteurization is mandatory in liquid egg products to ensure their storability and their safety. It has been reported that PEF effectively reduces the activity of various microorganisms in several egg products. Therefore, PEF could be used as a safe alternative of conventional egg processing technologies (Yogesh, 2016). However, PEF treatment can also alter the structural and functional properties when a threshold level of 250 kJ/kg is reached, since this would correspond to a temperature ~65°C (Heinz, Álvarez, Angersbach, & Knorr, 2001). In contrast, a lower level of energy does not affect these functional properties, but the complete microbial inactivation is not achieved

TABLE 8.2
Effect of PEF Treatment on Enzyme Inactivation in Foods.

Product	Enzyme	PEF Conditions	Outcomes	Reference
SMUF	Lipase and plasmin	Lipase: 27.4–37.3 kV/cm, 314.5 μs, 34–35°C Plasmin: 30 kV/cm, 100 μs	Lipase: 62.1% and 13% inactivation in batch and continuous treatment Plasmin: 90% inactivation at 15°C in continuous treatment	Vega-Mercado et al. (2019)
Whole milk	Alkaline phosphatase, lipase, plasmin, xanthine oxidase	Electric field (15.9–26.1 kV/cm) Flow rate (4.2 mL/min) Pulse width (20 μs) Frequency (10–60 Hz) Pulse duration (34–101 μs) Pre-heating (55°C)	EFI dependent activity. Optimal EFI at 26.1 kV/cm: 97%, 82%, 12% and 32% inactivation for alkaline phosphatase, lipase, plasmin, xanthine oxidase, respectively	Sharma, Oey, et al. (2014)
	Lipase, plasmin, and xanthine oxidase	Electric field (25.7 kV/cm) Pulse duration (34 μs) Pre-heating (55°C for 24 s)	Lipase: 38% inactivation. Plasmin: 7% inactivation. Xanthine oxidase: 30% inactivation	Sharma et al. (2018)
	Plasmin	Electric field (30 kV/cm) Flow rate (2.4 L/min) Pulse duration (22 μs) Pre-heating (42°C or 52°C)	No significant decline at 52°C and slight increase at 42°C	McAuley et al. (2016)
	Xanthine oxidase	Electric field (20–26 kV/cm) Pulse width (20 μs) Pulse duration (34 μs) Frequency (20 Hz) Pre-heating (55°C for 24 s)	PEF preserve the functionality of native proteins, especially enzymes. Denaturation ≅ 13% less compared to thermal treatment (63°C/30 min or 73°C/55 s)	Sharma, Oey, and Everett (2016)

EFI, electrical field intensity; *SMUF*, simulated milk ultra-filtrated.

(Monfort, Gayán, Raso, Condón, & Álvarez, 2010). This makes necessary the combination of PEF with other technologies to increase the efficiency of PEF at low electric field level.

In contrast, the bactericidal effects of PEF are more difficult to achieve in solid foods, probably due to heterogeneity of many foods and their richness in ions (Gudmundsson & Hafsteinsson, 2005). In these foods, it would be necessary to apply treatments of greater intensity (above 10 kV/cm), which would alter their texture (Guerrero-Beltrán & Welti-Chanes, 2016). Therefore, there are hardly any studies on the effect of PEF on the pathogens present in fish and meat.

8.3 PEF FOR THE PRODUCTION OF HEALTHIER FOOD

Between the positive effects that electroporation has in the foods treated with PEF is the reduction in the use of additives, since after treatment the partial disruption

of cellular tissues and an effect in the mass transfer processes are produced (Gómez et al., 2019). This is especially important for the meat industry, since there are many meat products that require a curing process for the production of safe products with characteristic sensory properties (flavor and odor), and to extend their shelf-life (Pérez-Santaescolástica et al., 2019). Among these preservatives are nitrites, phosphates, and salt whose use is questioned due to the health problems associated with their consumption. Although there are different strategies that allow to reduce their content, recent trends show the possibility of using noninvasive technologies such as PEF, ultrasound and high-pressure processing (Pinton et al., 2020).

Although there are still few studies in this field, the studies conducted to date show that PEF treatment accelerates salting due to the better diffusion of NaCl (Pinton et al., 2020), which would allow to reduce the curing time without negative effects on important physicochemical parameters as color, water-binding capacity or texture (Bhat, Morton, Mason, & Bekhit, 2019a; McDonnell, Allen, Chardonnereau, Arimi, & Lyng, 2014). The fragmentation of myofibrils after treatment and the breakdown of the muscle structure would explain these effects. In this regard, Bhat, Morton, Mason, and Bekhit (2020) demonstrated the applicability of PEF to reduce salt in beef jerky. The results showed that the resulted low-sodium meat product had a sodium content 34% lower than the control sample (9886 vs 15,044 mg/kg for samples with 1.2% NaCl along with PEF treatment and control samples with 2% NaCl, respectively), while sensory attributes displayed similar scores (6.66 vs 6.17 for saltiness, and 6.31 vs 6.24 for overall acceptability in control and treated samples, respectively). This demonstrates that PEF leads to better saltiness perception during chewing without significant changes in the quality of the product (color, oxidative, and microbial stability).

Phosphates are also additives commonly used in the manufacture of processed products such as meat products, cheese, canned fish, beverages, and baked products Although their use is safe, the increase in the consumption of these products has caused the acceptable daily intake of phosphorus to be exceeded. This is joined by the growing demand for "clean label" products. Even though several authors have suggested the potential use of PEF to reduce the use of phosphates in food (Pinton et al., 2020; Thangavelu, Kerry, Tiwari, & McDonnell, 2019), there are hardly any studies to prove it. Among them, the research carried out by Toepfl, Mathys, Heinz, and Knorr (2006) stands out. The authors demonstrated that PEF could improve the WHC, and therefore it has positive effects on the texture of injected hams containing phosphate.

The improvement of the diffusion characteristics of the product would also allow to develop low-fat products in response to the demands of consumers who are increasingly aware of the relationship between diet and health (Ignat, Manzocco, Brunton, Nicoli, & Lyng, 2015). This would have a great impact on the snack industry since it is recognized that the fat from deep-fried products can contribute to non-communicable diseases (Ostermeier, Hill, Töpfl, & Jäger, 2020). In addition to reducing production costs, this technology would make it possible to obtain a healthier product by reducing the amount of oil needed to fry potato crisps (chips) and maintaining the desired mouth feel. Reductions between 25% and 30% of oil intake are achieved, since this technology favors cutting resulting in thinner and smoother slices. Additionally, frying times can be shortened due to faster water leakage. As a result, the quality of the product could even be improved, increasing

crispiness and flexibility with a smoother surface, which gives a uniform and brighter color.

Moreover, it is important to highlight the green concept of this technology, since its use as pre-treatment would avoid the use of large amounts of water and energy, usually necessary during the blanching step (65°C/45 min) to which potatoes are subjected before cutting, which also results in lower microbial load (Botero-Uribe, Fitzgerald, Gilbert, & Midgley, 2017; Siemer et al., 2018). This technology is not only applied to potatoes but also recent applications show the possibility of using PEF in vegetable chips from beetroot, carrot, cassava, and pumpkin (Xu et al., 2020).

8.4 EFFECT OF PEF ON THE TECHNOLOGICAL PROPERTIES OF FOOD

8.4.1 Improvement of Drying Food Process

Drying is one of the most widely used techniques for preserving food (Dziki, 2020). In the past, foods such as fruits, vegetables, meat, and fish were dried in the sun, through trial and error, to have food in times of scarcity. Commercially, this technique, which involves the removal of water by heat treatment, converts fresh food into dehydrated food, adding value to the raw material since it allows to maintain the properties and the stability of the products during both storage and processing, protecting them from exposure to oxygen and light, pH, temperature, and storage time.

Freeze-drying is the most efficient and widely used drying method as it is capable of complete removal of water from food (Oyinloye & Yoon, 2020). However, this method of preservation has associated high processing costs. Therefore, recent research suggest the possibility of using a pre-treatment to accelerate the drying step and improve the quality of the resulted product (Dziki, 2020). PEF is one the technologies that can be used for the pre-treatment of food that favor these improvements during drying process (Table 8.3), resulting in a more attractive product (Ghosh, Gillis, Levkov, Vitkin, & Golberg, 2020; Siemer et al., 2018; Wu, Wang, & Guo, 2020).

Lammerskitten et al. (2020) demonstrated that the application of PEF pre-treatment (1.07 kV/cm, 1 kJ/kg, 40 μs) in strawberry dices before freeze-drying (45°C and 1 mbar) allows a more homogenous drying, improving the microstructure (uniform shape and inhibition of shrinkage) and the texture (more porous and crispier) of the fruit. Therefore, the resulted product is characterized by a better visual quality (preservation of color). Electroporation phenomenon is the main responsible of these effects, probably because it allows

TABLE 8.3
Application of PEF as pre-treatment in freeze-drying process.

Source	PEF Conditions	Freeze-drying	Gradient	Reference
Apple	Electric field (1.07 kV/cm) Energy input (0.5, 1 kJ/kg) Pulse width (10 ms) Pulse duration (40 ms)	40°C, 100 Pa	Reduced drying time (57%), increased water diffusion coefficient (44%), and higher crystallinity of treated samples (35.5%) Better reconstitution properties, and unchanged SSL	Lammerskitten et al. (2019)
Carrot	Electric field (0, 1.85, 5 kV/cm) Energy input (5.63, 8, 80 kJ/kg) Number of pulses (10, 50, 100)	Convective drying (70°C, air velocity 2 m/s)	Optimal conditions (5 kV/cm, 8 kJ/kg, 10 pulses): reduced drying time (6.9%–8.2%), increased water diffusion coefficient (16.7%). Decreased L^* (25.3%), retention of carotenoids	Wiktor, Dadan, Nowacka, Rybak, and Witrowa-Rajchert (2019)
	Electric field (0.6 kV/cm) Pulse duration (0.1 s)	Vacuum drying (0.3 bar; 25, 50, 75, and 90°C)	Reduced drying time (33–55%). β-carotene retention. Smaller changes in ΔE	Liu, Pirozzi, Ferrari, Vorobiev, and Grimi (2020)
Onion	Electric field (0.36–1.07 kV/cm) Energy input (0.2–20.0 kJ/kg)	Convective drying (45°C, 60°C, and 75°C, air velocity 0.2 m/s, 5 h)	Optimal conditions (1.07 kV/cm, 4 kJ/kg): reduced drying time (6.4%–30.3%)	Ostermeier, Giersemehl, Siemer, Töpfl, and Jäger (2018)
Potato	Electric field (1000, 1250, 1500 V/cm) Pulse width (60, 90, 120 μs) Pulses number (15, 30, 45) Pulse duration (500 ms)	75°C, 40–45 Pa	Optimal conditions (1500 V/cm, 120 μs, 45 pulses): reduced drying time (31.5%), improved drying rate (14.3%)	Wu and Zhang (2014)
Sea cucumber	Electric field (22.5 kV) Frequency (70 Hz) Pulse duration (5 min)	Vacuum freeze (until 13% moisture content)	Reduced drying time (16.43%), improved rehydration ratio (11.7%) No significant differences in shrinkage, texture, protein, and acid mucopolysaccharide content	Bai and Luan (2018)
Red bell pepper	Electric field (1.0 kV/cm) Number of pulses (20–200) Pulse duration (2.0–28.6 ms)	Freeze-drying (−40°C, 0.5 mbar, 72 h)	Higher rehydration capacity (>50%) Improvement of mechanical properties (firmness reduction >60%)	Fauster, Giancaterino, Pittia, and Jaeger (2020)
Strawberries	Electric field (1.07 kV/cm) Energy input (1 kJ/kg) Pulse width (0.5 s) Frequency (2 Hz) Pulse duration (40 μs)	Freeze-drying (−45°C, 1 mbar)	Better visual quality: more preserved color (lower L^* and higher a^* values), more uniform shape, crispier	Lammerskitten, Wiktor, Parniakov, and Lebovka (2020)

a^*, redness; ΔE, total color difference; L^*, lightness; *SSL*, soluble solid loss.

a better distribution of sugar and water inside the sample. Moreover, Wu and Zhang (2019) reported the same effects on apple.

In addition to the dehydration of the product, its subsequent rehydration is also very important in the quality of the product. Ideally, the resulting product should have characteristics as similar as possible to the fresh product. In this regard, Taiwo, Angersbach, and Knorr (2002) demonstrated that the use of HIPEF as a pre-drying treatment (20 pulses, 48 J/kg) before an osmotic dehydration allowed a higher rehydration capacity using low temperatures (24 °C and 45 °C).

8.4.2 Modification of Texture Properties of Foods

PEF treatment can be a good strategy to change the textural properties during the production and processing of numerous products, since the modifications in the tissue structure can largely affect the product characteristics. This would be the case of meat products, where tenderness is one of the parameters that determine the sensory attributes and therefore, the acceptance of the product by the consumer (Gómez et al., 2019). This has led to the evaluation of the possibility of using PEF to meat tenderization and meat ageing. This practice, commonly used to improve the consistency of beef tenderness, would allow to meet the demands of a highly demanding consumer who look for products with an optimal degree of tenderness, especially in products from beef (Verbeke et al., 2010).

Traditionally, the maturation of meat is done at refrigeration temperatures and in vacuum conditions, but this entails high costs due to the time necessary to act the endogenous proteolytic enzymes responsible for the natural maturation process (Bhat, Morton, Mason, & Bekhit, 2018a). Moreover, the age of the animals, the type of piece, the amount of connective tissues, the activity of calpains, and the extent of proteolytic degradation condition the duration of this process (Alahakoon, Oey, Bremer, & Silcock, 2019). In addition, it would be necessary to calculate the time that elapses from packaging to the moment the meat is placed at point of sale, since maturation does not stop during this period.

The application of PEF would accelerate this process, since the permeabilization of cells would improve the diffusion of the enzymes and/or the incorporation of substances for the curing/marinating of meat products (Table 8.4). Moreover, PEF leads to a myofibril fragmentation, which results in a break of the structure of the muscle. Consequently, the water holding capacity and the texture of the products could be modified (O'Dowd, Arimi, Noci, Cronin, & Lyng, 2013). In addition,

a pre-treatment with PEF would facilitate the activation of calpains, which together with a subsequent post-mortem ageing period would improve PEF effect due to the increase of proteolysis of troponin-T and desmin (Bekhit, Suwandy, Carne, van de Ven, & Hopkins, 2016). In fact, the degradation of these structural proteins of the muscle is used as a marker of myofibrillar protein degradation and meat tenderization (Sun et al., 2014). In contrast, the release of cathepsins from lysosomes in pre-rigor muscles would induces the glycolysis (Bhat, Morton, Mason, & Bekhit, 2019a).

PEF allows a uniform treatment of *pre-rigor* and *post-rigor* muscles (Bekhit et al., 2016). However, it is necessary to take into account that the temperature of *pre-rigor* storage, the intensity and the repetition of PEF treatment, and the treated piece influence the tenderization (Bhat, Morton, Mason, & Bekhit, 2018a; Suwandy et al., 2015c). In this regard, Suwandy et al. (2015c) observed that cold-boned and hot-boned meat displayed opposite effects when the number of treatments is repeated. In the first case, a decrease in the shear force values were observed, probably due to the negative effect of high temperature on *post-mortem* proteolysis (Kim, Stuart, Nygaard, & Rosenvold, 2012). Regarding the muscle to be treated, Bekhit et al. (2016) observed that *longissimus lumborum* (LL) and *semimembranosus* (SM) beef muscles displayed a different behavior after the application of a PEF pre-treatment (10 kV, 90 Hz) and an ageing period of 21 days. The results showed a reduction of tenderness in LL, while SM was not affected by PEF treatments during ageing.

The effect of species was observed when the conditions optimized for beef were applied in deer. In this regard, no significant effect was observed on venison tenderization (Bhat et al., 2019). Although the application of T_1 (2.5 kV, 50 Hz) and T_2 (10 kV, 90 Hz) in cold-boned *longissimus dorsi* induced the tenderization due to the slight increase in the calpain activity and proteolysis of troponin-T, no significant effects were found in shear force values after 21 days of ageing. However, a significant increase was observed in the protein digestibility of samples treated with PEF, highlighting the positive influence on the digestion process. Higher concentrations of most free amino acids were observed for PEF-treated samples (Bhat, Morton, Mason, Bekhit, & Mungure, 2019). A similar trend was observed by Arroyo et al. (2015) in fresh (1.2 kV/cm, 10 Hz) and frozen (2.1 kV/cm, 10 Hz) turkey breast when a PEF treatment was applied. PEF-treated samples were different from controls (22.5 vs 18.9 N and 21.1 vs 21.9 N in fresh and frozen samples, respectively). In contrast, venison loins treated with HIPEF (10 kV, 50 Hz) improved meat tenderness by

TABLE 8.4
Application of PEF as Pretreatment for Meat Tenderization.

Source	Muscle Status	PEF Conditions	Ageing Period	Outcomes	Reference
Red deer (*Cervus elaphus*) *Longissimus dorsi*	*Post-rigor*	Electric field (0.2 and 0.5 kV/cm) Energy input (1.93 and 70.2 kJ/kg) Pulse width (20 μs) Pulse duration (30 s)	21 days at 4°C	Improvement of calpain activity and proteolysis of troponin-T. No significant tenderization response after an extended ageing	Bhat et al. (2019)
		LPEF: 2.5 kV, 1.93 kJ/kg HIPEF: 10.0 kV, 70.2 kJ/kg Frequency (50 Hz) Pulse width (20 μs)	21 days at 4°C Chiller conditions: air velocity of 1.5 m/s and with RH set at 80%	HIPEF improved meat tenderness by 9% DAWL was affected by PEF intensity (24.66, 27.09 and 29.62 for control, LPEF and HPEF, respectively) CL was not affected by PEF treatments	Mungure et al. (2020)
Beef *Longissimus et lumborum*	*Post-rigor*	10 kV, 90 Hz, 20 μs, 6 h	–	PEF improved calpain activity and increased troponin-T and desmin proteolysis, especially in low-pH meats No significant effect on water loss and shear force	Suwandy et al. (2015a)
		Electric field (0.23 to 0.68 kV/cm) Voltage (2.5 and 10 kV) Frequency (200 Hz) Pulse width (20 μs)	Vacuum packaged and stored for 14 days at 4°C	Ultrastructural changes were observed at LPEF and HIPEF. HIPEF samples had higher shear force values than LPEF (100 N vs 55 N after 14 days of post-treatment)	Khan et al. (2017)
Beef *Longissimus lumborum* and *semimembranosus*	*Post-rigor*	Electric field (0.27–0.56 kV/cm) Voltages (5 kV, 10 kV) Frequency (20, 50, 90 Hz)	–	Texture benefits in LL (19.5% reduction) regarding electric input. SM tenderness increases with frequency (4.1%, 10.4% and 19.1% reduction in the shear force at 20, 50 and 90 Hz, respectively)	Bekhit, van de Ven, Suwandy, Fahri, and Hopkins (2014)
		Electric field (5 kV, 10 kV) Frequency (20, 50, 90 Hz)		Inverse relationship between tenderness and treatment intensity in LL. Opposite behavior was observed in SM (21.6% decline)	Suwandy et al. (2015b)
		Repeated (1 ×, 2 ×, 3 ×) PEF treatment (10 kV, 90 Hz, 20 μs)		Increased tenderness by 2.5 N with every extra treatment in LL. In contrast, less proteolysis was observed with increasing number of PEF applications	Suwandy et al. (2015c)
	Pre-rigor	Repeated (1 ×, 2 ×, 3 ×) PEF treatment (10 kV, 90 Hz, 20 μs)	–	3 × PEF treatment reduced tenderness in LL. Opposite behavior was observed in SM	Bekhit et al. (2016)

Continued

TABLE 8.4
Application of PEF as Pretreatment for Meat Tenderization—cont'd

Source	Muscle Status	PEF Conditions	Ageing Period	Outcomes	Reference
Beef *semimembranosus*	*Post-rigor*	Electric field (1.4 kV/cm) Energy input (250 kJ/kg) Frequency (50 Hz) Pulse width (20 μs)	Vacuum packaged and stored for 7 days at −20°C	The combination of PEF with freezing-thawing improved tenderness (4.50 vs 6.01 and 5.77 kgF for frozen-thawed PEF treated samples, untreated frozen-thawed and fresh controls, respectively)	Faridnia et al. (2015)
	Post-rigor	T_1 (5 kV–0.36 kV/cm–90 Hz) T_2 (10 kV–0.60 kV/cm–20 Hz)	Vacuum packaged and stored for 14 days at 4°C	PEF improved calpain activity and increased troponin-T and desmin proteolysis. No significant effect on shear force and myofibrillar fragmentation index Limited application in muscles from older animals	Bhat et al. (2019b)
Turkey breast meat	*Post-rigor*	Electric field (7.5, 10, and 12.5 kV for fresh meat; 14, 20, and 25 kV for frozen meat) Energy input (11 to 94 kJ/kg) Frequency (10, 55 and 110 Hz) Pulse number (100, 200 and 300) Pulse width (20 μs)	–	No significant differences in weight loss, CL, and texture either in fresh or frozen samples Slightly differences in sensory evaluation for texture	Arroyo et al. (2015)
Whole briskets (beef deep *pectoralis* muscle)	*Post-rigor*	Electric field (1 and 1.5 kV/cm) Energy input (40–50 and 90–100 kJ/kg) Voltages (15–35 kV) Frequency (50 Hz) Pulse width (20 μs)	Vacuum packed and stored at −18°C	PEF improve tenderness and decrease cooking time of collagen-rich meat cuts	Alahakoon, Oey, Silcock, and Bremer (2017)

CL, cooking loss; *DAWL*, dry ageing weight loss; *EFS*, electric field strength; *HIPEF*, high-intensity pulsed electric field; *LL, Longissimus lumborum*; *LPEF*, low-intensity pulsed electric field; *SM, Semimembranosus* muscle.

9%, probably due to the myofibril disruption in myofibrillar protein structure (Mungure et al., 2020). Even though most of the differences are due to the ageing method (wet and dry aged), dry ageing weight loss was also affected by PEF intensity (24.66, 27.09 and 29.62 for control, LPEF and HIPEF, respectively). Therefore, it seems that a minimum electric field intensity would be necessary to induce an irreversible electroporation in animal tissues during an ageing period (Bhat, Morton, Mason, & Bekhit, 2018b).

In general, the results obtained with PEF showed an average reduction of 20% in the shear force values (Suwandy et al., 2015b). These values would comply with the reductions needed to induce a significant impact on tenderness of muscles from younger animals (Bekhit et al., 2014). Lower values were observed in venison loins despite the combination of ageing process and HIPEF. Moreover, the obtained values are not as good as those observed with hydrodynamic pressure or hydrodyne processing treatment (Mungure et al., 2020).

Contradictory outcomes were also noticed by some authors because in some cases PEF is not be able to tenderize meat or leads to negative effects on tenderness. In the first case, Bhat, Morton, Mason, and Bekhit (2019b) observed that PEF would not be able to tenderize pieces from culled dairy animals, since these muscles are characterized by being excessively tough. On the other hand, Suwandy et al. (2015a) found that the application of a high-intensity PEF (10 kV, 90 Hz) on hot-boned LL resulted in tougher meat, probably linked to the protein and enzyme denaturation caused by the ohmic heating that occurs during treatment. The combination of PEF with freezing-thawing could solve this negative effect, since it accelerates proteolysis and reduces shear force values (Faridnia et al., 2015).

8.5 LIMITATIONS AND CHALLENGES

Summarizing the aforementioned, the application of PEF has several advantages associated with obtaining healthier products, reducing toxic compounds and additives, and preserving nutrients and food safety in a sustainable way. However, there is still some drawbacks that have not yet been resolved such as the high capital cost, the controversy regarding the inactivation of microorganisms and enzymes in solid and semi-solid samples, the dielectric breakdown caused by bubble generation, challenging to use with conductive materials, and the limited protocols for its scaled-up (Priyadarshini, Rajauria, O'Donnell, & Tiwari, 2019).

Moreover, despite PEF is considered a non-thermal food processing technology, sometimes its application can increase the temperature, which could affect heat-sensitive compounds such as lipids rich in polyunsaturated fatty acids (PUFAs) and proteins (Barba et al., 2015). In addition, the modifications caused in the cell permeability make lipids more susceptible to oxidation, favoring the reactions between unsaturated fatty acids and prooxidant substances (Arroyo et al., 2015). As a result of these reactions, volatile compounds are formed conditioning the shelf-life of the product and the consumer acceptability (Faridnia et al., 2015). It is important to note that there are hardly any studies in which the effect of electroporation on lipid and protein oxidation reactions are studied (Cropotova et al., 2021). But also, a discoloration of the product caused by the damage of pigments, denaturation of proteins (myoglobin), degradation of bioactive peptides and vitamins, with the consequent decrease of the nutritional value of the final product (Gómez et al., 2019). All these effects will depend on PEF processing parameters. Indeed, PEF treatments with high energy input (> 50 kJ/kg) lead to a decrease in luminosity. In contrast, low intensity treatments (< 5 kJ/kg) resulted in small changes of lightness and yellowness, while redness was not affected (Baldi et al., 2021).

Another important aspect to take into account is related to the safety of the treated product, since sometimes the electrode corrosion can occur. This is due to some transition metal ions have catalytic effects for certain food reactions, such as lipid oxidation. This would lead to corrosion of the electrode which could impact the sensory attributes of processed food products. Therefore, it would be necessary to modify the pulse generator systems selecting more stable electrode materials to reduce the amount of electrochemical reactions that take place, improving the technical feasibility of this technology and the chemical safety of food in view of a future application of PEF technology at industrial level (Arshad et al., 2020; Pataro, Falcone, Donsì, & Ferrari, 2014).

Finally, all these limitations would make it difficult the scaling-up of PEF to industrial scale. In fact, although this technology has been investigated extensively and there already are commercial PEF systems working in different countries, most of the results obtained still refer to laboratory results.

ACKNOWLEDGMENTS

Thanks to GAIN (Axencia Galega de Innovación) for supporting this study (grant number IN607A2019/01). Mirian Pateiro, Rubén Domínguez, Paulo E.S. Munekata, and José M. Lorenzo are members of the HealthyMeat network, funded by CYTED (ref. 119RT0568).

REFERENCES

Adebo, O. A., Molelekoa, T., Makhuvele, R., Adebiyi, J. A., Oyedeji, A. B., Gbashi, S., et al. (2021). A review on novel non-thermal food processing techniques for mycotoxin reduction. *International Journal of Food Science & Technology*, 56(1), 13–27. https://doi.org/10.1111/ijfs.14734.

Aguiló-Aguayo, I., Oms-Oliu, G., Soliva-Fortuny, R., & Martín-Belloso, O. (2009). Changes in quality attributes throughout storage of strawberry juice processed by high-intensity pulsed electric fields or heat treatments. *LWT—Food Science and Technology*, 42(4), 813–818. https://doi.org/10.1016/j.lwt.2008.11.008.

Akdemir Evrendilek, G., Celik, P., Agcam, E., & Akyildiz, A. (2017). Assessing impacts of pulsed electric fields on quality attributes and furfural and hydroxymethylfurfural formations in apple juice. *Journal of Food Process Engineering*, 40(5). https://doi.org/10.1111/jfpe.12524.

Alahakoon, A. U., Oey, I., Bremer, P., & Silcock, P. (2019). Quality and safety considerations of incorporating post-PEF ageing into the pulsed electric fields and sous vide processing chain. *Food and Bioprocess Technology*, 12(5), 852–864. https://doi.org/10.1007/s11947-019-02254-6.

Alahakoon, A. U., Oey, I., Silcock, P., & Bremer, P. (2017). Understanding the effect of pulsed electric fields on thermostability of connective tissue isolated from beef pectoralis muscle using a model system. *Food Research International*, 100, 261–267. https://doi.org/10.1016/j.foodres.2017.08.025.

Alirezalu, K., Munekata, P. E. S., Parniakov, O., Barba, F. J., Witt, J., Toepfl, S., et al. (2020). Pulsed electric field and mild heating for milk processing: A review on recent advances. *Journal of the Science of Food and Agriculture*, 100(1), 16–24. https://doi.org/10.1002/jsfa.9942.

Alirezalu, K., Pateiro, M., Yaghoubi, M., Alirezalu, A., Peighambardoust, S. H., & Lorenzo, J. M. (2020). Phytochemical constituents, advanced extraction technologies and techno-functional properties of selected Mediterranean plants for use in meat products. A comprehensive review. *Trends in Food Science and Technology*, 100, 292–306. https://doi.org/10.1016/j.tifs.2020.04.010.

Amiali, M., & Ngadi, M. O. (2012). Microbial decontamination of food by pulsed electric fields (PEFs). In A. Demirci, & M. O. Ngadi (Eds.), *Microbial decontamination in the food industry. Novel methods and applications* (pp. 407–449). Cambridge, UK: Woodhead Publishing Limited.

Arroyo, C., Eslami, S., Brunton, N. P., Arimi, J. M., Noci, F., & Lyng, J. G. (2015). An assessment of the impact of pulsed electric fields processing factors on oxidation, color, texture, and sensory attributes of Turkey breast meat. *Poultry Science*, 94(5), 1088–1095. https://doi.org/10.3382/ps/pev097.

Arshad, R. N., Abdul-Malek, Z., Munir, A., Buntat, Z., Ahmad, M. H., Jusoh, Y. M. M., et al. (2020, October 1). Electrical systems for pulsed electric field applications in the food industry: An engineering perspective. *Trends in Food Science and Technology*, 104, 1–13. https://doi.org/10.1016/j.tifs.2020.07.008.

Bai, Y., & Luan, Z. (2018). The effect of high-pulsed electric field pretreatment on vacuum freeze drying of sea cucumber. *International Journal of Applied Electromagnetics and Mechanics*, 57(2), 247–256. https://doi.org/10.3233/JAE-180009.

Baldi, G., D'Elia, F., Soglia, F., Tappi, S., Petracci, M., & Rocculi, P. (2021). Exploring the effect of pulsed electric fields on the technological properties of chicken meat. *Food*, 10(2), 241. https://doi.org/10.3390/foods10020241.

Ballmer-Weber, B. K., Lidholm, J., Fernández-Rivas, M., Seneviratne, S., Hanschmann, K.-M., Vogel, L., et al. (2015). IgE recognition patterns in peanut allergy are age dependent: Perspectives of the EuroPrevall study. *Allergy*, 70(4), 391–407. https://doi.org/10.1111/all.12574.

Bansal, V., Sharma, A., Ghanshyam, C., Singla, M. L., & Kim, K. H. (2015). Influence of pulsed electric field and heat treatment on Emblica officinalis juice inoculated with Zygosaccharomyces bailii. *Food and Bioproducts Processing*, 95, 146–154. https://doi.org/10.1016/j.fbp.2015.05.005.

Barba, F. J., Parniakov, O., Pereira, S. A., Wiktor, A., Grimi, N., Boussetta, N., et al. (2015). Current applications and new opportunities for the use of pulsed electric fields in food science and industry. *Food Research International*, 77, 773–798. https://doi.org/10.1016/j.foodres.2015.09.015.

Bekhit, A. E. D. A., Suwandy, V., Carne, A., van de Ven, R., & Hopkins, D. L. (2016). Effect of repeated pulsed electric field treatment on the quality of hot-boned beef loins and topsides. *Meat Science*, 111, 139–146. https://doi.org/10.1016/j.meatsci.2015.09.001.

Bekhit, A. E. D. A., van de Ven, R., Suwandy, V., Fahri, F., & Hopkins, D. L. (2014). Effect of pulsed electric field treatment on cold-boned muscles of different potential tenderness. *Food and Bioprocess Technology*, 7(11), 3136–3146. https://doi.org/10.1007/s11947-014-1324-8.

Bermúdez-Aguirre, D., Dunne, C. P., & Barbosa-Cánovas, G. V. (2012). Effect of processing parameters on inactivation of Bacillus cereus spores in milk using pulsed electric fields. *International Dairy Journal*, 24(1), 13–21. https://doi.org/10.1016/j.idairyj.2011.11.003.

Bhat, Z. F., Morton, J. D., Mason, S. L., & Bekhit, A. E.-D. A. (2018a). Applied and emerging methods for meat tenderization: A comparative perspective. *Comprehensive Reviews in Food Science and Food Safety*, 17(4), 841–859. https://doi.org/10.1111/1541-4337.12356.

Bhat, Z. F., Morton, J. D., Mason, S. L., & Bekhit, A. E. D. A. (2018b). Role of calpain system in meat tenderness: A review. *Food Science and Human Wellness*, 7, 196–204. https://doi.org/10.1016/j.fshw.2018.08.002.

Bhat, Z. F., Morton, J. D., Mason, S. L., & Bekhit, A. E. D. A. (2019a). Current and future prospects for the use of pulsed electric field in the meat industry. *Critical Reviews in Food Science and Nutrition*, 59(10), 1660–1674. https://doi.org/10.1080/10408398.2018.1425825.

Bhat, Z. F., Morton, J. D., Mason, S. L., & Bekhit, A. E. D. A. (2019b). Does pulsed electric field have a potential to improve the quality of beef from older animals and how? *Innovative Food Science and Emerging Technologies*, 56, 102194. https://doi.org/10.1016/j.ifset.2019.102194.

Bhat, Z. F., Morton, J. D., Mason, S. L., & Bekhit, A. E. D. A. (2020). The application of pulsed electric field as a sodium reducing strategy for meat products. *Food Chemistry*, *306*(October 2019), 125622. https://doi.org/10.1016/j.foodchem.2019.125622.

Bhat, Z. F., Morton, J. D., Mason, S. L., Bekhit, A. E. D. A., & Mungure, T. E. (2019). Pulsed electric field: Effect on in-vitro simulated gastrointestinal protein digestion of deer longissimus dorsi. *Food Research International*, *120*, 793–799. https://doi.org/10.1016/j.foodres.2018.11.040.

Bhat, Z. F., Morton, J. D., Mason, S. L., Mungure, T. E., Jayawardena, S. R., & Bekhit, A. E. D. A. (2019). Effect of pulsed electric field on calpain activity and proteolysis of venison. *Innovative Food Science and Emerging Technologies*, *52*, 131–135. https://doi.org/10.1016/j.ifset.2018.11.006.

Botero-Uribe, M., Fitzgerald, M., Gilbert, R. G., & Midgley, J. (2017, September 1). Effect of pulsed electrical fields on the structural properties that affect french fry texture during processing. *Trends in Food Science and Technology*, *67*, 1–11. https://doi.org/10.1016/j.tifs.2017.05.016.

Buckow, R., Chandry, P. S., Ng, S. Y., McAuley, C. M., & Swanson, B. G. (2014). Opportunities and challenges in pulsed electric field processing of dairy products. *International Dairy Journal*, *34*(2), 199–212. https://doi.org/10.1016/j.idairyj.2013.09.002.

Capuano, E., & Fogliano, V. (2011, May 1). Acrylamide and 5-hydroxymethylfurfural (HMF): A review on metabolism, toxicity, occurrence in food and mitigation strategies. *LWT—Food Science and Technology*, *44*, 793–810. https://doi.org/10.1016/j.lwt.2010.11.002.

Chauhan, O. P., & Unni, L. E. (2015). Pulsed electric field (PEF) processing of foods and its combination with electron beam processing. In S. D. Pillai, & S. Shayanfar (Eds.), *Electron beam pasteurization and complementary food processing technologies* (pp. 157–184). https://doi.org/10.1533/9781782421085.2.157.

Choi, J., Wang, D., Namihira, T., Katsuki, S., Akiyama, H., Lin, X., et al. (2008). Inactivation of spores using pulsed electric field in a pressurized flow system. *Journal of Applied Physics*, *104*(9). https://doi.org/10.1063/1.3006440, 094701.

Cregenzán-Alberti, O., Arroyo, C., Dorozko, A., Whyte, P., & Lyng, J. G. (2017). Thermal characterization of Bacillus subtilis endospores and a comparative study of their resistance to high temperature pulsed electric fields (HTPEF) and thermal-only treatments. *Food Control*, *73*, 1490–1498. https://doi.org/10.1016/j.foodcont.2016.11.012.

Cregenzán-Alberti, O., Halpin, R. M., Whyte, P., Lyng, J. G., & Noci, F. (2015). Study of the suitability of the central composite design to predict the inactivation kinetics by pulsed electric fields (PEF) in Escherichia coli, Staphylococcus aureus and Pseudomonas fluorescens in milk. *Food and Bioproducts Processing*, *95*, 313–322. https://doi.org/10.1016/j.fbp.2014.10.012.

Cropotova, J., Tappi, S., Genovese, J., Rocculi, P., Dalla Rosa, M., & Rustad, T. (2021). The combined effect of pulsed electric field treatment and brine salting on changes in the oxidative stability of lipids and proteins and color characteristics of sea bass (Dicentrarchus labrax). *Heliyon*, *7*(1). https://doi.org/10.1016/j.heliyon.2021.e05947, e05947.

Dourado, C., Pinto, C., Barba, F. J., Lorenzo, J. M., Delgadillo, I., & Saraiva, J. A. (2019, June 1). Innovative non-thermal technologies affecting potato tuber and fried potato quality. *Trends in Food Science and Technology*, *88*, 274–289. https://doi.org/10.1016/j.tifs.2019.03.015.

Dziadek, K., Kopeć, A., Dróżdż, T., Kiełbasa, P., Ostafin, M., Bulski, K., et al. (2019). Effect of pulsed electric field treatment on shelf life and nutritional value of apple juice. *Journal of Food Science and Technology*, *56*(3), 1184–1191. https://doi.org/10.1007/s13197-019-03581-4.

Dziki, D. (2020). Recent trends in pretreatment of food before freeze-drying. *PRO*, *8*(12), 1661. https://doi.org/10.3390/pr8121661.

EFSA. (2015). *Acrylamide*. Retrieved January 19, 2021, from Chemical contaminants website. https://www.efsa.europa.eu/en/topics/topic/acrylamide.

Ekezie, F. G. C., Sun, D. W., & Cheng, J. H. (2017). A review on recent advances in cold plasma technology for the food industry: Current applications and future trends. *Trends in Food Science and Technology*, *69*, 46–58. https://doi.org/10.1016/j.tifs.2017.08.007.

FAO. (2020a). *Crops. FAOSTAT*. Retrieved January 19, 2021, from http://www.fao.org/faostat/en/#data/QC.

FAO. (2020b). *Crops. Statistical pocketbook. World food and agriculture 2020*. Rome, Italy: Food and Agriculture Organization of the United Nations (FAO).

Faridnia, F., Ma, Q. L., Bremer, P. J., Burritt, D. J., Hamid, N., & Oey, I. (2015). Effect of freezing as pre-treatment prior to pulsed electric field processing on quality traits of beef muscles. *Innovative Food Science and Emerging Technologies*, *29*, 31–40. https://doi.org/10.1016/j.ifset.2014.09.007.

Fauster, T., Giancaterino, M., Pittia, P., & Jaeger, H. (2020). Effect of pulsed electric field pretreatment on shrinkage, rehydration capacity and texture of freeze-dried plant materials. *LWT—Food Science and Technology*, *121*, 108937. https://doi.org/10.1016/j.lwt.2019.108937.

Genovese, J., Tappi, S., Luo, W., Tylewicz, U., Marzocchi, S., Marziali, S., et al. (2019). Important factors to consider for acrylamide mitigation in potato crisps using pulsed electric fields. *Innovative Food Science and Emerging Technologies*, *55*, 18–26. https://doi.org/10.1016/j.ifset.2019.05.008.

Ghosh, S., Gillis, A., Levkov, K., Vitkin, E., & Golberg, A. (2020). Saving energy on meat air convection drying with pulsed electric field coupled to mechanical press water removal. *Innovative Food Science and Emerging Technologies*, *66*, 102509. https://doi.org/10.1016/j.ifset.2020.102509.

Gómez, B., Munekata, P. E. S., Gavahian, M., Barba, F. J., Martí-Quijal, F. J., Bolumar, T., et al. (2019). Application of pulsed electric fields in meat and fish processing industries: An overview. *Food Research International*, *123*(April), 95–105. https://doi.org/10.1016/j.foodres.2019.04.047.

Gudmundsson, M., & Hafsteinsson, H. (2005). Effect of high intensity electric field pulses on solid foods. In D.-W. Sun (Ed.), *Emerging technologies for food processing* (pp. 141–153). https://doi.org/10.1016/B978-012676757-5/50008-6.

Guerrero-Beltrán, J. A., & Welti-Chanes, J. (2016). Pulse electric fields. In B. Caballero, P. M. Finglas, & F. Toldrá (Eds.), *Encyclopedia of food and health* (pp. 561–565). Academic Press.

He, G., Yin, Y., Yan, X., & Wang, Y. (2017). Application of pulsed electric field for treatment of fish and seafood. In D. Miklavčič (Ed.), *Handbook of electroporation* (pp. 2637–2655). https://doi.org/10.1007/978-3-319-26779-1.

Heinz, V., Álvarez, I., Angersbach, A., & Knorr, D. (2001). Preservation of liquid foods by high intensity pulsed electric fields – basic concepts for process design. *Trends in Food Science & Technology, 12*(3–4), 103–111. https://doi.org/10.1016/S0924-2244(01)00064-4.

Ignat, A., Manzocco, L., Brunton, N. P., Nicoli, M. C., & Lyng, J. G. (2015). The effect of pulsed electric field pre-treatments prior to deep-fat frying on quality aspects of potato fries. *Innovative Food Science and Emerging Technologies, 29*, 65–69. https://doi.org/10.1016/j.ifset.2014.07.003.

Johnson, P. E., Van der Plancken, I., Balasa, A., Husband, F. A., Grauwet, T., Hendrickx, M., et al. (2010). High pressure, thermal and pulsed electric-field-induced structural changes in selected food allergens. *Molecular Nutrition & Food Research, 54*(12), 1701–1710. https://doi.org/10.1002/mnfr.201000006.

Kempkes, M. A. (2017). Industrial pulsed electric field systems. In D. Miklavcic (Ed.), *Handbook of electroporation* (pp. 1–21). Springer International Publishing.

Khan, A. A., Randhawa, M. A., Carne, A., Mohamed Ahmed, I. A., Barr, D., Reid, M., et al. (2017). Effect of low and high pulsed electric field on the quality and nutritional minerals in cold boned beef M. longissimus et lumborum. *Innovative Food Science and Emerging Technologies, 41*, 135–143. https://doi.org/10.1016/j.ifset.2017.03.002.

Khaneghah, A. M., Gavahian, M., Xia, Q., Denoya, G. I., Roselló-Soto, E., & Barba, F. J. (2020). Effect of pulsed electric field on Maillard reaction and hydroxymethylfurfural production. In *Pulsed electric fields to obtain healthier and sustainable food for tomorrow* (pp. 129–140). https://doi.org/10.1016/b978-0-12-816402-0.00006-9.

Kim, Y. H. B., Stuart, A., Nygaard, G., & Rosenvold, K. (2012). High pre rigor temperature limits the ageing potential of beef that is not completely overcome by electrical stimulation and muscle restraining. *Meat Science, 91*(1), 62–68. https://doi.org/10.1016/j.meatsci.2011.12.007.

Lammerskitten, A., Mykhailyk, V., Wiktor, A., Toepfl, S., Nowacka, M., Bialik, M., et al. (2019). Impact of pulsed electric fields on physical properties of freeze-dried apple tissue. *Innovative Food Science and Emerging Technologies, 57*, 102211. https://doi.org/10.1016/j.ifset.2019.102211.

Lammerskitten, A., Wiktor, A., Mykhailyk, V., Samborska, K., Gondek, E., Witrowa-Rajchert, D., et al. (2020). Pulsed electric field pre-treatment improves microstructure and crunchiness of freeze-dried plant materials: Case of strawberry. *LWT—Food Science and Technology, 134*, 110266. https://doi.org/10.1016/j.lwt.2020.110266.

Lammerskitten, A., Wiktor, A., Parniakov, O., & Lebovka, N. (2020). An overview of the potential applications to produce healthy food products based on pulsed electric field treatment. In *Pulsed electric fields to obtain healthier and sustainable food for tomorrow*. https://doi.org/10.1016/b978-0-12-816402-0.00002-1.

Leong, S. Y., Burritt, D. J., & Oey, I. (2016). Evaluation of the anthocyanin release and health-promoting properties of pinot noir grape juices after pulsed electric fields. *Food Chemistry, 196*, 833–841. https://doi.org/10.1016/j.foodchem.2015.10.025.

Liu, C., Pirozzi, A., Ferrari, G., Vorobiev, E., & Grimi, N. (2020). Effects of pulsed electric fields on vacuum drying and quality characteristics of dried carrot. *Food and Bioprocess Technology, 13*(1), 45–52. https://doi.org/10.1007/s11947-019-02364-1.

McAuley, C. M., Singh, T. K., Haro-Maza, J. F., Williams, R., & Buckow, R. (2016). Microbiological and physicochemical stability of raw, pasteurised or pulsed electric field-treated milk. *Innovative Food Science and Emerging Technologies, 38*, 365–373. https://doi.org/10.1016/j.ifset.2016.09.030.

McDonnell, C. K., Allen, P., Chardonnereau, F. S., Arimi, J. M., & Lyng, J. G. (2014). The use of pulsed electric fields for accelerating the salting of pork. *LWT—Food Science and Technology, 59*(2P1), 1054–1060. https://doi.org/10.1016/j.lwt.2014.05.053.

Monfort, S., Gayán, E., Raso, J., Condón, S., & Álvarez, I. (2010). Evaluation of pulsed electric fields technology for liquid whole egg pasteurization. *Food Microbiology, 27*(7), 845–852. https://doi.org/10.1016/j.fm.2010.05.011.

Mtaoua, H., Sánchez-Vega, R., Ferchichi, A., & Martín-Belloso, O. (2017). Impact of high-intensity pulsed electric fields or thermal treatment on the quality attributes of date juice through storage. *Journal of Food Processing and Preservation, 41*(4). https://doi.org/10.1111/jfpp.13052, e13052.

Mungure, T. E., Farouk, M. M., Birch, E. J., Carne, A., Staincliffe, M., Stewart, I., et al. (2020). Effect of PEF treatment on meat quality attributes, ultrastructure and metabolite profiles of wet and dry aged venison longissimus dorsi muscle. *Innovative Food Science and Emerging Technologies, 65*, 102457. https://doi.org/10.1016/j.ifset.2020.102457.

O'Dowd, L. P., Arimi, J. M., Noci, F., Cronin, D. A., & Lyng, J. G. (2013). An assessment of the effect of pulsed electrical fields on tenderness and selected quality attributes of post rigour beef muscle. *Meat Science, 93*(2), 303–309. https://doi.org/10.1016/j.meatsci.2012.09.010.

OJEU. (2017). *Commission Regulation (EU) 2017/2158 of 20 November 2017 establishing mitigation measures and benchmark levels for the reduction of the presence of acrylamide in food.* (p. L304/24-44). p. L304/24-44 Official Journal of the European Communities.

Ostermeier, R., Giersemehl, P., Siemer, C., Töpfl, S., & Jäger, H. (2018, November 1). Influence of pulsed electric field (PEF) pre-treatment on the convective drying kinetics of onions. *Journal of Food Engineering, 237*, 110–117. https://doi.org/10.1016/j.jfoodeng.2018.05.010.

Ostermeier, R., Hill, K., Töpfl, S., & Jäger, H. (2020). Pulsed electric field as a sustainable tool for the production of healthy snacks. In *Pulsed electric fields to obtain healthier and sustainable food for tomorrow* (pp. 103–128). https://doi.org/10.1016/b978-0-12-816402-0.00005-7 (2016).

Oyinloye, T. M., & Yoon, W. B. (2020). Effect of freeze-drying on quality and grinding process of food produce: A review. *PRO*, *8*, 354. https://doi.org/10.3390/PR8030354.

Pallarés, N., Barba, F. J., Berrada, H., Tolosa, J., & Ferrer, E. (2020). Pulsed electric fields (PEF) to mitigate emerging mycotoxins in juices and smoothies. *Applied Sciences*, *10*(19), 6989. https://doi.org/10.3390/app10196989.

Pallarés, N., Berrada, H., Fernández-Franzón, M., & Ferrer, E. (2020). Risk assessment and mitigation of the mycotoxin content in medicinal plants by the infusion process. *Plant Foods for Human Nutrition*, *75*(3), 362–368. https://doi.org/10.1007/s11130-020-00820-4.

Pataro, G., Falcone, M., Donsì, G., & Ferrari, G. (2014). Metal release from stainless steel electrodes of a PEF treatment chamber: Effects of electrical parameters and food composition. *Innovative Food Science and Emerging Technologies*, *21*, 58–65. https://doi.org/10.1016/j.ifset.2013.10.005.

Pérez-Santaescolástica, C., Fraeye, I., Barba, F. J., Gómez, B., Tomasevic, I., Romero, A., et al. (2019, April 1). Application of non-invasive technologies in dry-cured ham: An overview. *Trends in Food Science and Technology*, *86*, 360–374. https://doi.org/10.1016/j.tifs.2019.02.011.

Pillet, F., Formosa-Dague, C., Baaziz, H., Dague, E., & Rols, M. P. (2016). Cell wall as a target for bacteria inactivation by pulsed electric fields. *Scientific Reports*, *6*(1), 1–8. https://doi.org/10.1038/srep19778.

Pinton, M. B., dos Santos, B. A., Lorenzo, J. M., Cichoski, A. J., Boeira, C. P., & Campagnol, P. C. B. (2020). Green technologies as a strategy to reduce NaCl and phosphate in meat products: An overview. *Current Opinion in Food Science*, *40*, 1–5. https://doi.org/10.1016/j.cofs.2020.03.011.

Priyadarshini, A., Rajauria, G., O'Donnell, C. P., & Tiwari, B. K. (2019). Emerging food processing technologies and factors impacting their industrial adoption. *Critical Reviews in Food Science and Nutrition*, *59*(19), 3082–3101. https://doi.org/10.1080/10408398.2018.1483890.

Puértolas, E., Saldaña, G., & Raso, J. (2017). Pulsed electric field treatment for fruit and vegetable processing. In D. Miklavčič (Ed.), *Handbook of electroporation* (pp. 2495–2515). https://doi.org/10.1007/978-3-319-32886-7.

Ricci, A., Parpinello, G. P., & Versari, A. (2018). Recent advances and applications of pulsed electric fields (PEF) to improve polyphenol extraction and color release during red winemaking. *Beverages*, *4*(1), 18. https://doi.org/10.3390/beverages4010018.

Schottroff, F., Gratz, M., Krottenthaler, A., Johnson, N. B., Bédard, M. F., & Jaeger, H. (2019). Pulsed electric field preservation of liquid whey protein formulations—Influence of process parameters, pH, and protein content on the inactivation of Listeria innocua and the retention of bioactive ingredients. *Journal of Food Engineering*, *243*, 142–152. https://doi.org/10.1016/j.jfoodeng.2018.09.003.

Sharma, P., Bremer, P., Oey, I., & Everett, D. W. (2014). Bacterial inactivation in whole milk using pulsed electric field processing. *International Dairy Journal*, *35*(1), 49–56. https://doi.org/10.1016/j.idairyj.2013.10.005.

Sharma, P., Oey, I., Bremer, P., & Everett, D. W. (2014). Reduction of bacterial counts and inactivation of enzymes in bovine whole milk using pulsed electric fields. *International Dairy Journal*, *39*(1), 146–156. https://doi.org/10.1016/j.idairyj.2014.06.003.

Sharma, P., Oey, I., Bremer, P., & Everett, D. W. (2018). Microbiological and enzymatic activity of bovine whole milk treated by pulsed electric fields. *International Journal of Dairy Technology*, *71*(1), 10–19. https://doi.org/10.1111/1471-0307.12379.

Sharma, P., Oey, I., & Everett, D. W. (2016). Thermal properties of milk fat, xanthine oxidase, caseins and whey proteins in pulsed electric field-treated bovine whole milk. *Food Chemistry*, *207*, 34–42. https://doi.org/10.1016/j.foodchem.2016.03.076.

Shriver, S. K., & Yang, W. W. (2011). Thermal and nonthermal methods for food allergen control. *Food Engineering Reviews*, *3*, 26–43. https://doi.org/10.1007/s12393-011-9033-9.

Siemer, C., Töpfl, S., Witt, J., & Ostermeier, R. (2018). *Use of pulsed electric fields (PEF) in the food industry*. DLG Expert report 5/2018.

Soltanzadeh, M., Peighambardoust, S. H., Gullon, P., Hesari, J., Gullón, B., Alirezalu, K., et al. (2020). Quality aspects and safety of pulsed electric field (PEF) processing on dairy products: A comprehensive review. *Food Reviews International*. https://doi.org/10.1080/87559129.2020.1849273.

Subramanian, V., Shanmugam, N., Ranganathan, K., Kumar, S., & Reddy, R. (2017). Effect of combination processing on aflatoxin reduction: Process optimization by response surface methodology. *Journal of Food Processing and Preservation*, *41*(6). https://doi.org/10.1111/jfpp.13230, e13230.

Sun, X., Chen, K. J., Berg, E. P., Newman, D. J., Schwartz, C. A., Keller, W. L., et al. (2014). Prediction of troponin-T degradation using color image texture features in 10d aged beef longissimus steaks. *Meat Science*, *96*(2), 837–842. https://doi.org/10.1016/j.meatsci.2013.09.012.

Sun, W.-W., Yu, S.-J., Zeng, X.-A., Yang, X.-Q., & Jia, X. (2011). Properties of whey protein isolate-dextran conjugate prepared using pulsed electric field. *Food Research International*, *44*(4), 1052–1058. https://doi.org/10.1016/j.foodres.2011.03.020.

Suwandy, V., Carne, A., van de Ven, R., Bekhit, A. E. D. A., & Hopkins, D. L. (2015a). Effect of pulsed eectric field treatment on the eating and keeping qualities of cold-boned beef loins: Impact of initial pH and fibre orientation. *Food and Bioprocess Technology*, *8*(6), 1355–1365. https://doi.org/10.1007/s11947-015-1498-8.

Suwandy, V., Carne, A., van de Ven, R., Bekhit, A. E. D. A., & Hopkins, D. L. (2015b). Effect of pulsed electric field treatment on hot-boned muscles of different potential tenderness. *Meat Science*, *105*, 25–31. https://doi.org/10.1016/j.meatsci.2015.02.009.

Suwandy, V., Carne, A., van de Ven, R., Bekhit, A. E. D. A., & Hopkins, D. L. (2015c). Effect of repeated pulsed electric field treatment on the quality of cold-boned beef loins and topsides. *Food and Bioprocess Technology*, *8*(6), 1218–1228. https://doi.org/10.1007/s11947-015-1485-0.

Taiwo, K. A., Angersbach, A., & Knorr, D. (2002). Influence of high intensity electric field pulses and osmotic dehydration on the rehydration characteristics of apple slices at different temperatures. *Journal of Food Engineering, 52*(2), 185–192. https://doi.org/10.1016/S0260-8774(01)00102-9.

Taylor, S. L. (2021). *Emerging problems with food allergens.* Retrieved March 24, 2021, from Food, Nutrition and Agriculture Food and Agriculture Organization website. http://www.fao.org/3/x7133m/x7133m03.htm.

Terefe, N. S., Buckow, R., & Versteeg, C. (2015). Quality-related enzymes in plant-based products: Effects of novel food processing technologies part 2: Pulsed electric field processing. *Critical Reviews in Food Science and Nutrition, 55,* 1–15. https://doi.org/10.1080/10408398.2012.701253.

Thangavelu, K. P., Kerry, J. P., Tiwari, B. K., & McDonnell, C. K. (2019, December 1). Novel processing technologies and ingredient strategies for the reduction of phosphate additives in processed meat. *Trends in Food Science and Technology, 94,* 43–53. https://doi.org/10.1016/j.tifs.2019.10.001.

Toepfl, S., Kinsella, J., & Parniakov, O. (2020). Industrial scale equipment, patents, and commercial applications. In F. Barba, O. Parniakov, & A. Wiktor (Eds.), *Pulsed electric fields to obtain healthier and sustainable food for tomorrow* (pp. 269–281). https://doi.org/10.1016/b978-0-12-816402-0.00012-4.

Toepfl, S., Mathys, A., Heinz, V., & Knorr, D. (2006). Review: Potential of high hydrostatic pressure and pulsed electric fields for energy efficient and environmentally friendly food processing. *Food Reviews International, 22*(4), 405–423. https://doi.org/10.1080/87559120600865164.

Tontini, C., Marinangeli, L., Maiello, N., Abbadessa, S., Villalta, D., & Antonicelli, L. (2017). Ara h 6 sensitization in peanut allergy: Friend, foe or innocent bystander? *European Annals of Allergy and Clinical Immunology, 49*(1), 18–21.

van Wyk, S., Silva, F. V. M., & Farid, M. M. (2019). Pulsed electric field treatment of red wine: Inactivation of Brettanomyces and potential hazard caused by metal ion dissolution. *Innovative Food Science and Emerging Technologies, 52,* 57–65. https://doi.org/10.1016/j.ifset.2018.11.001.

Vanga, S. K., Singh, A., Kalkan, F., Gariepy, Y., Orsat, V., & Raghavan, V. (2016). Effect of thermal and high electric fields on secondary structure of Peanut protein. *International Journal of Food Properties, 19*(6), 1259–1271. https://doi.org/10.1080/10942912.2015.1071841.

Vega-Mercado, H., Powers, J. R., Martín-Belloso, O., Luedecke, L., Barbosa-Cánovas, G. V., & Swanson, B. G. (2019). Change in susceptibility of proteins to proteolysis and the inactivation of an extracellular protease from Pseudomonas fluorescens M3/6 when exposed to pulsed electric fields. In G. V. Barbosa-Cánovas, Q. H. Zhang, & G. Tabilo-Munizaga (Eds.), *Pulsed electric fields in food processing* (pp. 105–120). https://doi.org/10.1201/9780429133459-7.

Verbeke, W., Van Wezemael, L., de Barcellos, M. D., Kügler, J. O., Hocquette, J. F., Ueland, Ø., et al. (2010). European beef consumers' interest in a beef eating-quality guarantee. Insights from a qualitative study in four EU countries. *Appetite, 54*(2), 289–296. https://doi.org/10.1016/j.appet.2009.11.013.

Vijayalakshmi, S., Nadanasabhapathi, S., Kumar, R., & Sunny Kumar, S. (2018). Effect of pH and pulsed electric field

process parameters on the aflatoxin reduction in model system using response surface methodology: Effect of pH and PEF on aflatoxin reduction. *Journal of Food Science and Technology, 55*(3), 868–878. https://doi.org/10.1007/s13197-017-2939-3.

Walter, L., Knight, G., Ng, S. Y., & Buckow, R. (2016). Kinetic models for pulsed electric field and thermal inactivation of Escherichia coli and Pseudomonas fluorescens in whole milk. *International Dairy Journal, 57,* 7–14. https://doi.org/10.1016/j.idairyj.2016.01.027.

Warren, C. M., Jiang, J., & Gupta, R. S. (2020). Epidemiology and burden of food allergy. *Current Allergy and Asthma Reports, 20,* 1–9. https://doi.org/10.1007/s11882-020-0898-7.

Wiktor, A., Dadan, M., Nowacka, M., Rybak, K., & Witrowa-Rajchert, D. (2019). The impact of combination of pulsed electric field and ultrasound treatment on air drying kinetics and quality of carrot tissue. *LWT—Food Science and Technology, 110,* 71–79. https://doi.org/10.1016/j.lwt.2019.04.060.

Wu, X., Wang, C., & Guo, Y. (2020). Effects of the high-pulsed electric field pretreatment on the mechanical properties of fruits and vegetables. *Journal of Food Engineering, 274,* 109837. https://doi.org/10.1016/j.jfoodeng.2019.109837.

Wu, Y., & Zhang, D. (2014). Effect of pulsed electric field on freeze-drying of potato tissue. *International Journal of Food Engineering, 10*(4), 857–862. https://doi.org/10.1515/ijfe-2014-0149.

Wu, Y., & Zhang, D. (2019). Pulsed electric field enhanced freeze-drying of apple tissue. *Czech Journal of Food Sciences, 37*(6), 432–438. https://doi.org/10.17221/230/2018-CJFS.

Wu, L., Zhao, W., Yang, R., & Yan, W. (2015). Pulsed electric field (PEF)-induced aggregation between lysozyme, ovalbumin and ovotransferrin in multi-protein system. *Food Chemistry, 175,* 115–120. https://doi.org/10.1016/j.foodchem.2014.11.136.

Xu, Z., Leong, S. Y., Farid, M., Silcock, P., Bremer, P., & Oey, I. (2020). Understanding the frying process of plant-based foods pretreated with pulsed electric fields using frying models. *Food, 9*(7), 949. https://doi.org/10.3390/foods9070949.

Yang, N., Huang, K., Lyu, C., & Wang, J. (2016). Pulsed electric field technology in the manufacturing processes of wine, beer, and rice wine: A review. *Food Control, 61,* 28–38. https://doi.org/10.1016/j.foodcont.2015.09.022.

Yang, W.-H., Tu, Z.-C., Wang, H., Li, X., & Tian, M. (2017). High-intensity ultrasound enhances the immunoglobulin (Ig) G and IgE binding of ovalbumin. *Journal of the Science of Food and Agriculture, 97*(9), 2714–2720. https://doi.org/10.1002/jsfa.8095.

Yang, W., Tu, Z., Wang, H., Zhang, L., Gao, Y., Li, X., et al. (2017). Immunogenic and structural properties of ovalbumin treated by pulsed electric fields. *International Journal of Food Properties, 20*(Suppl. 3), S3164–S3176. https://doi.org/10.1080/10942912.2017.1396479.

Yogesh, K. (2016, February 1). Pulsed electric field processing of egg products: A review. *Journal of Food Science and Technology, 53,* 934–945. https://doi.org/10.1007/s13197-015-2061-3.

Innovative Technologies in Sustainable Food Production: High Pressure Processing

SVEN KARLOVIĆ • TOMISLAV BOSILJKOV • DAMIR JEŽEK •
MARINELA NUTRIZIO • ANET REŽEK JAMBRAK
Faculty of Food Technology and Biotechnology, University of Zagreb, Zagreb, Croatia

9.1 INTRODUCTION

The novel mostly non-thermal technologies such as high hydrostatic pressure, high-intensity ultrasonics, cold plasma, pulsed electric fields and many others are currently trending in scientific research and have significant propagation through the food industry. High hydrostatic pressure (HHP) is currently most proved technology for food processing and has the most significant share in the food industry of all novel technologies. Almost all branches of food industry such as fruit and vegetable, meat, dairy, as well as other product processing use HHP today. As with all non-thermal processes, a crucial advantage of HHP is processing at ambient temperatures, which can ensure sensory and nutritive quality of processed foods. Using high pressure instead of high temperatures can also eliminate pathogen bacteria, ensure safety, and even prolong shelf life. Transmitting high pressures through pressure medium (usually water) requires that processed foods are placed in the flexible packaging beforehand, which has some advantages and disadvantages. After treatment packaging is simply dried and it is ready for labeling and storage or transport. Such processing procedure eliminates any possibility of subsequent contamination during handling, filling, or other operations.

Nowadays, both industry and consumers demand sustainable production with minimally-processed food products with extended shelf-life and characteristics of a fresh product (Atuonwu & Tassou, 2018). The high pressure treatment of food dates since 1899 when Hite investigated it as a new method for milk pasteurization (Hite, 1899). Until today, the high pressure processing (HPP) is so developed that it has become one of the most successful cases of the implementation of innovative nonthermal technologies in the industry. The reason is that optimal combination of applied pressure (up to 600 MPa), temperature (usually at room temperature) and treatment time is capable to inactivate microorganisms and enzymes without causing significant changes to sensorial, nutritive, and structural characteristics of the treated food, unlike heating (Barba, Koubaa, et al., 2017; Tsevdou, Gogou, & Taoukis, 2019).

Successful application of HPP has been reported for many food products with improved color, texture, flavor, yield and nutritional properties (Atuonwu & Tassou, 2018; Chauhan et al., 2015; Gong et al., 2015). Furthermore, the HPP has also been explored in extractions and improved bioavailability of high-added value compounds (such as polyphenols, essential oils, carotenoids, fatty acids, etc.) from various food sources (Barba, Mariutti, et al., 2017; Tsikrika, O'Brien, & Rai, 2019). Therefore, the main advantages of HPP include improved extraction yield, reduced processing time, use of green solvents, product shelf-life extension with minimum impact to the product quality, which presents HPP with a great potential to reduce waste during processing and for more sustainable application (Atuonwu & Tassou, 2018; Chemat et al., 2020).

On the other side, HPP with its advantages uses a significant amount of energy. Some comparative studies have compared energy consumption of HPP with conventional thermal processing of food. The results showed that HPP consumes a significantly higher amount of energy compared to thermal processing which is, finally, remarkably influencing overall costs of HPP per unit (Aganovic et al., 2017; Rodriguez-Gonzalez et al., 2015; Sampedro et al., 2014). For that reason, current challenge for researchers and technologists is to reduce the energy consumption during HPP. Most studies on energy efficiency were based on experimental analysis under specific conditions or modelling with lack of necessary level of details or process optimization. Therefore, mathematical models

Sustainable Production Technology in Food. https://doi.org/10.1016/B978-0-12-821233-2.00011-3
Copyright © 2021 Elsevier Inc. All rights reserved.

that stimulate the behavior of HPP systems and predicting the influence of processing parameters on energy consumption have been developed. The model developed by Atuonwu and Tassou (2018) could help HPP operators to determine optimal processing conditions considering energy consumption, but also satisfying product quality and safety requirements; providing a basis for improved process automation (Atuonwu & Tassou, 2018).

The environmental and economic evaluation should also take in account life cycle assessment (LCA) and life cycle costing (LCC) methodologies. Cacace et al. (2020) have evaluated economic and environmental sustainability of HPP of foods using mentioned approaches compared to thermal pasteurization of orange juice and modified atmosphere packaging of sliced Parma ham. As a result, HPP was presented as more expansive than thermal processing, but showed a lower environmental impact in almost all impact categories. Compared to modified atmosphere, HPP was both less expensive and had a lower environmental impact in most categories (Cacace et al., 2020).

The market of high pressure equipment has expanded and it is available for professionals from both agriculture and food sectors, with Hiperbaric company as a leader in the production of equipment for HPP (Chemat et al., 2020). Although the market has spread, the correct evaluation of cost and environmental impact, digitalization, and optimization of the process have a potential to increase the application of HPP in food industry all over the world. This paragraph presents the use of HPP as innovative technology for food processing and presents sustainable examples of application of HPP.

9.2 INNOVATIVE TECHNOLOGIES IN FOOD PRODUCTION—HIGH PRESSURE PROCESSING (HPP)

9.2.1 High Hydrostatic Pressure (HHP) Processing

Primary demands for the finished food products are safety and quality. The safety chain includes producing, processing, packaging, labeling, and finally selling finished food products. Its goal is to prevent contamination of food and food-borne illnesses. Today, the main principle of elimination of pathogenic bacteria is using high temperatures in conventional pasteurization. As this has negative impact on the nutritional and sensory qualities of food products, HHP is introduced in the processing part of the chain with the goal of non-thermal inactivation of bacteria and retaining of fresh food quality. Other disadvantage of thermal processing is its cost. Drying, freezing, pasteurization, and

refrigeration are crucial to ensure safe food, but these processes consume most energy in the food industry (Picart-Palmade, Cunault, Chevalier-Lucia, Belleville, & Marchesseau, 2018). Development of new non-thermal technologies as a part of green technologies also decrease costs related to those unit operations.

9.2.1.1 Typical HHP process

HHP is relatively simple process used for treatment of liquid or solid food materials, and consists of a few steps. Before initiation of the processing, food product is packaged into adequate flexible packaging (such as bottles, pouches, and vacuum packaging). Prerequisite for packaging is to be at least 20% elastic to allow for volume change during processing which will result from increase in pressure. Packaging, therefore, must withstand the mechanical stress caused by elevated pressure and maintain physical integrity during and after processing. Most plastic packaging used today, such as PVC, PP, PE, as well as others, are usually suitable for HHP processing. While it is theoretically possible to use rigid materials such as glass in combination with elastic materials, they are deemed not suitable for processing. There is possibility, and there are already some products on the market which use modified atmosphere packaging (MAP) in HHP process. This is a viable process; however, it is not very efficient. Due to high compressibility of gasses, there should always be minimal amount of gas (headspace) inside of packaging. More gas leads to longer pressurization time, which increases costs. MAP packaging is also much more abundant in volume than vacuum packaging for the same amount of product. Such packaging consequently occupies larger volume in basket and pressure vessel, which also leads to increase in cost, as larger volume of individual items leads to fewer processed items per batch (Fig. 9.1).

In the first step, packaged products are placed in the basket which is subsequently submerged in the pressure-transmitting fluid (which is almost always water in the industry setting, but it could be propylene-glycol, as well as some other liquids more suitable for some processes such as high-pressure freezing) inside of pressure vessel. Alternatively, in horizontal systems basket is rolled into empty pressure vessel, and the water fills in after closing of vessel. Based on the preset parameters (pressure, time, temperature, speed of pressurization/depressurization, number of steps/ramps, pulses) HPP equipment process inserted products. In the last step products are taken out of the vessel and dried. As HHP operation is last phase in the chain of unit operations performed on food products, there only remains to label packaging and storage.

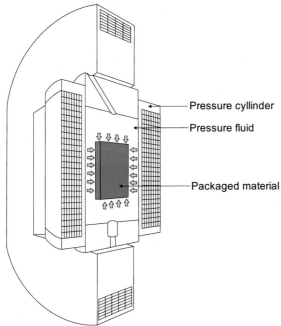

Pressure cyllinder

Pressure fluid

Packaged material

FIG. 9.1 Cross–section of the vertical loading pressure vessel.

Processing parameters. Pressures used in the laboratory setting are going up to 1200 MPa, which is currently useful only for scientific research, as there is practically no industrial equipment which can go over 600 MPa. The range of pressures from 300 to 600 MPa is adequate for almost all food processing needs. Further increase in pressure will just put additional strain on pumps, seals and other components and increase production costs, with no visible benefits. According to thermodynamic principles such as Le chateliers law of equilibrium and isostatic rule which govern HHP, pressure is quasi-instantaneously transferred through the whole volume of processed product, which ensures simultaneous and equal processing in every point of the volume. This is opposed to conventional thermal processing where outside layers are heated first, while heating of deeper layers depends on the conduction. That property of pressure processing allows for simultaneous processing of food of various sizes and shapes and ensures a more straightforward scale-up of process independently of volume. High pressure also promotes or discourages physical and chemical phenomena based on the decrease in volume. The mentioned law of equilibrium for HHP equates that when the system at equilibrium is subjected to rise in pressure, system will change to a new equilibrium. A consequence of

this is that the system will counteract to new change by decreasing volume to decrease pressure. Decrease leads to inhibition of all chemical and physical reactions which could increase volume, and also promotes favorable phase changes such as condensation or freezing. It also promotes some changes in molecular conformation, especially on the proteins. Due to the isostatic compression, adiabatic heating phenomena occurs where the pressure of liquid and processed material is increased from work done on it by compressing piston or other compression methods. Adiabatic heating leads to increase in temperature based on the material properties, which could be in the range of 3 °C to 8 °C for every 100 MPa exerted.

Temperature inside of HHP chamber can usually be set from −50 °C up to 130 °C, depending on the type of process. For the most extraction and pasteurization processes, ambient temperature is adequate. HHP freezing process naturally requires temperatures at the lower end of the range, which usually makes this process too expensive and not used in the industry, despite many advantages. Higher temperatures can speed up extraction process, at the expense of longer processing time (and increase in costs). It can also contribute to inactivation of some more pressure-resistant microorganism spores.

There are various times linked to the HHP process, and all of them could have an impact on the quality of final product. Besides rarely changed pressurization time (adjustable, but approximately 10 MPa s^{-1}) and depressurization time (from instant to few minutes), preset pressure hold time is crucial factor. It can range from seconds to hours (even days). For most uses, it is usually from 1 min to 10–15 min. Depending on the process parameters used, heating time should also be included in the calculations. It doesn't have any impact on the quality of final product, but in can significantly increase total processing time, and with that, costs of processing Fig. 9.2.

There is still yet not investigated enough possibility of incorporation of some other parameters in the process. While one full cycle is usually enough for the treatment of food, there is possibility to use multiple pressurization—depressurization cycles (which also multiply costs), multiple stage pressurization for inactivation of spores, multiple pulsed treatment, as well as some other possibilities.

9.2.1.2 Main goal of HHP
Microbial inactivation is necessary to ensure safe food, which makes it most important goal of HHP processing. However, this can be achieved with already established

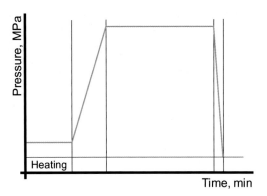

FIG. 9.2 Pressure—time flow of usual HHP process (*red line*—temperature; *blue line*—pressure).

(and usually cheaper) technologies such as conventional thermal pasteurization. How can then producers and consumers benefit from HHP technology? The keywords are non-thermal, sustainable, and green technology. Usage of pressure instead of the heat allows inactivation of unwanted microorganisms while retaining freshness and all sensory properties such as smell, taste, color, and texture of fresh products. It also enables retaining of almost all nutritional properties, so there is minimal or no loss of vitamins, antioxidants, and other beneficial compounds that usually degrade during thermal processing.

For the inactivation of vegetative microorganisms with HHP and prolonging of shelf life, high to medium acid food are required. Pressures for inactivation of microorganisms are in the range from 200 to 600 MPa, based on the target microorganisms, pH, and product type and food laws of particular country (Toepfl, Mathys, Heinz, & Knorr, 2006). Most bacteria must undergo at least log 5 reductions for food to be deemed safe for consumption. Usual target bacteria are *Salmonella, Listeria monocytogenes, Escherichia coli, Clostridium botulinum, Staphylococcus aureus*, as well as some others.

As consumers are increasingly conscious of environmental impact of their choices, they also expect it and dictate from other industries such as food industry. They expect novel, better products, increase of shelf life, decrease of food waste, decrease of additives and other chemicals, clean labeling, less pollution. All of those factors are achievable goals by using HHP technology as replacement for today's conventional methods.

9.3   SUSTAINABLE FOOD PRODUCTION—TOOLS AND GUIDES

Sustainability is nowadays great buzzword. It is a result of distance from the nature for many, many years. As

we know about the 1st, 2nd, and 3rd industrial revolutions, industrial production, machine production, development of robots for food processing, packaging etc. There what huge increase in production and transport of goods, there were massive production of waste, energy usage, large amount of carbon emission, and huge impact to the environment. Humans realized that we need to come back to nature and live accordingly, not to be self-centered. Sustainable food production come along with agriculture and cannot be separated. There are strategies in European Union and action plans, in different national development, that requires setting up a process and a functioning institutional structure. As stated by Agenda 2030. Sustainable Development, adopted by all United Nations Member States in 2015, there are 17 Sustainable Development Goals (SDGs), which are an urgent call for action by all countries It call for economic, environment, and social impact and peace and prosperity for people and the planet, now and into the future (Fig. 9.3). It is of crucial importance to engage sustainable food and agriculture with the broader SDG strategy and process in each country. The main idea is to raise awareness of the SDGs and their implications on food and agriculture (Gimenez-Escalante & Rahimifard, 2019). There should be focus on agriculture production, implementation of elements of industry 4.0, use of Internet of Things (IoT) in agricultural production, usage of sensors to monitor important factors in agricultural cropping etc. The idea is to use less energy, productive, and effective production with strategy and smart production, than to use less water, lower time consuming process, lower carbon emission production, lower impact to the environment etc. (Cacace et al., 2020; Nikmaram & Rosentrater, 2019; Pardo & Zufía, 2012). There are 17 sustainable development goals and therefore, guides and tools for sustainable food production should be focused on digitalization, from farms, crops, through processing, preservation, and packaging. The guides should be oriented to energy savings, economic impact with equality for all, with respect to social goals. The idea is also to engage stakeholders in cross-sectoral and multidisciplinary dialogue on SDGs towards SDG implementation. There is process outlined, by Food and Agricultural Organization—FAO (2018), in the UNDG Reference Guide on Mainstreaming the 2030 Agenda for Sustainable Development, which offers a common platform for SDG work at country level.

The approach in food processing and production is moved towards experimental research and implementation of nonthermal processing in processing of fruit and vegetables juices, seafood products etc. High pressure

Economic Pillar

Environmental Pillar

Social Pillar

FIG. 9.3 Three pillars of sustainable development goals (SDG) (Kostoska & Kocarev, 2019).

processing (HPP) is one of the nonthermal processing techniques that is mostly researched and mostly implemented in food industry. The main impact of high-pressure processing food preservation, often called "cold pressurization" or "cold pascalization." There are many efficient applications of HPP in food preservation and the process was approved by Food and drug administration (FDA). HPP was also used in homogenization and many pre-treatment processes like before freezing, drying etc. The main goals and strategies of usage of high-pressure processing is to have fast processing, low energy consumption, low usage of water, low by-products, waste etc. The main issue are high initial investments, but this also is returned (return of investment—ROI) in less than 5 years. The application of high-pressure processing can be considered as green processing, clean processing, clean label as asked by consumers.

The importance of consumer need is of high importance. Food scientists and technologists should address the consumer needs in optimizing the production and quality of food products. The production of food goes from agriculture crops, from processing, preservation, transportation to consumers. The whole process should be overviewed and calculated through different tools considering life cycle assessment (LCA), total quality deployment (TQD), quality function analysis (QFA) etc. (Djekic et al., 2017; Jambrak et al., 2019; Notarnicola et al., 2017; Stasiulaitiene et al., 2016). There are also some research papers that are reporting energy

consumption of non-thermal processing (Magureanu et al., 2018; Pardo & Zufía, 2012; Stasiulaitiene et al., 2016). The idea is to have real research approach in finding the exact multivariable results in application of nonthermal processing, like cold plasma, high pressure processing, pulsed light, pulsed electric fields, ultrasound, hydrodynamic cavitation, etc. The positive effect should be valuable as low energy consumption in reaching the preservation effect, or higher yield of extraction or other best output result. The best result should report low water usage, lower impact to the environment, reuse of by-products, reduce of waste, recycle of energy, water, material in processing line.

The reasonable approach in implementation of nonthermal processing and among them, high pressure processing (HPP), is to move forward in developing digitalization, sustainable production, smart sensors, Industry of things (IoT) with the focus on founding smart factories (Barba et al., 2016; Meng et al., 2018; Savastano et al., 2019; Srai et al., 2016). Elements of Industry 4.0 should go in line with high pressure processing and other nonthermal processing. To understand Industry 4.0, it is essential to see the full value chain which includes suppliers and the origins of the materials and components needed for various forms of smart manufacturing. Nonthermal processing should include end-to-end digital supply chain and the final destination of production, regardless of the number of intermediary steps and the end customer. Industry 4.0

FIG. 9.4 The visual perception of connection between high pressure processing, Industry 4.0 and sustainable development goals (SDG).

is the information intensive transformation that should go to nonthermal processing, and should be connected to the environment of data, people, processes, services, systems, and IoT-enabled industrial assets. The visual perception of connection between high pressure processing, Industry 4.0 and sustainable development goals (SDG) are presented in Fig. 9.4.

It is important to utilize actionable data and information as a way and means to realize smart industry and ecosystems of industrial innovation and collaboration (Srai et al., 2016; Zhong et al., 2017). A key role is ruled by the IoT, in the scope of Industry 4.0. Industry 4.0 includes cloud computing (and cloud platforms), big data (advanced data analytics, data lakes, edge intelligence) with (related) artificial intelligence, data analysis, storage and compute power at the edge of networks (edge computing), mobile, data communication/network technologies, manufacturing execution systems, enterprise resource planning (ERP), programmable logic controllers (PLC), sensors, and actuators etc. This also should be implemented and put in the strategy of nonthermal processing, validation, and implementation in future smart factories.

9.4 EXAMPLES IN THE SUSTAINABLE FOOD PRODUCTION BY MEANS OF HIGH PRESSURE PROCESSING (HPP)

By one definition, sustainable food production uses economically efficient non-polluting processes and systems and conserve non-renewable energy and natural resources. High hydrostatic pressure processing fits well under all aspects of sustainable food production. It is zero-waste technology which uses less energy for the same process compared to conventional technologies. Using HHP in wide array of processing areas ensures its adaptability on current and future demands from food industry. As HHP is used in microbial inactivation, extraction, freezing, as well as in the production of novel foods, it is versatile enough to encompass processing of raw material, finished products, and food waste. If we use PP/EI ratio, HHP is way over any other currently used technology. HHP product performance (PP) is high as its products are unparalleled, no matter if we compare nutritional quality, sensory properties such as taste, smell, texture and color, or environmental impact index (EI) based on the amount of produced food and water waste and other factors.

As Dong et al. (2016), Spira, Bisconsin-Junior, Rosenthal, and Monteiro (2018) and various other researchers show, shelf life of HHP products is prolonged compared to conventional treatments. This prevents additional costs of discarding of expired products and food losses, so HHP can significantly contribute to the sustainability of whole food production process. Prolonged shelf life can be also obtained without using of various additives. Removing the preservatives and other additives from the picture can ensure clean label which is highly sought after. This also eliminate environmental and health impact. Speaking of health, another not so direct, but still significant aspect is avoidance of some health problems associated with lower nutritional quality, high chemical content food. As HHP ensure uncompromised quality of final food products, consumers can reap same full benefits as they would from the fresh unprocessed food.

Processing of fresh fruit and vegetables as raw material generates large amount of food waste and food by-products. This can range from peels, seeds, leaves, and less desirable fruit which fail to comply to industry standards. Such fruit and vegetable byproducts present environmental problem, as it can only partially be reused as compost, while the rest simply ends as a waste. It is estimated that 25% of raw material become waste in the postharvest chain, and it can rise up to 50% if we include produced and expired foods. This mostly consists of fruit, vegetables, and cereal products with short shelf life (Boye & Arcand, 2013). This is also a problem from economic perspective, as producer must pay for transport, disposal cost, increased resource cost of food processing and production, as well as increased greenhouse gas (GHG) emissions. There is also lack of profit based on the smaller volume of processed raw material.

9.4.1 Extraction From Food Waste and Byproducts

It is thoroughly investigated and proved that the HHP is most efficient technology for the extraction of bioactive compounds from various plant material. Extraction produces higher yield during shorter extraction time, and all that on lower (usually room) temperatures. Process can also use more environmentally friendly solvents, and in much lower quantities compared to conventional thermal, microwave or high-intensity ultrasonic extraction. Recent research show that byproducts and waste could be further processed as it is rich in valuable nutritive and other compounds. One of the compounds which can be extracted from sugar beet, tomato, banana, papaya, carrot, and other waste is pectin. Conventional extraction of pectin from citrus and other types of peels takes a long time and leads to degradation of target compounds. Pressures used in HHP processing disrupts structure of the matrix and increase cell permeability, which in turn increase mass transfer of compounds during extraction. Enhancing extraction process using HHP ensures shorter processing times, lower temperatures, and higher yields. According to Ninčević, Ježek, Karlović, and Bosiljkov (2019), yields from tomato peel waste are up to 20% larger compared to conventional extraction. This is also confirmed by Guo et al. (2012). Jun (2006) proposed feasible HHP process to reuse tomato paste waste rich in lycopene, with best extraction results obtained during extremely short processing time of 1 min, while extraction yield is at 92%. Another viable application of HHP technology is extraction of natural dyes. Natural dyes are nontoxic and present in the food and beverage industry byproducts, and as such ends as cattle feed. Due to the reusing of food waste as feed, it must not be contaminated with organic solvents, and only possible solvent for extraction of dyes is a water. HHP significantly increase extraction yields of tannins and potentially other bioactive compounds from waste produced after processing of raspberries (Suthanthangjai & Kajda, 2004), black currant, grapes (Morata et al., 2015; Teixeira et al., 2014), black tea (Briones-Labarca, Giovagnoli-Vicuña, & Chacana-Ojeda, 2019), onions (Roldán-Marín et al., 2009), and other raw materials. This HHP enhanced processes have significantly shorter (minutes instead of hours) processing times at ambient or slightly elevated temperatures instead of usual 60–95°C. Those extraction processes are shown to be viable at lower range of pressures (approximately 300 MPa), with further increase in the yield as the pressure rises to 500 MPa. This significantly decrease costs of final product and obtain valuable resource which would instead finish as feed. Those same berries, grape, and citrus waste are excellent commercial source of bioactive compounds such and polyphenols. Extraction is today regularly carried out using long time/high temperature conventional extraction where temperature have negative impact on the quality of the extracts. Using ethanol and HPP extraction (or another frequently used term HPAE—high pressure assisted extraction) at 600 MPa yielded significantly more extracts (Corrales et al., 2008).

9.5 CONCLUSIONS AND FUTURE REMARKS

Just those few examples show why businesses are planning to significantly increase spend on sustainability and novel technologies. Costly investment in technologies such as HHP (based on the processing volume, in the range of few hundred thousand to millions of euros) can lead to increase of final costs of products. On the other hand, such technologies can ensure products which have an impact beyond just the cost of processing. As Nielsen company states in 2019, 73% of global consumers would change their consumption habits, and pay more to reduce their impact on the environment. Consumers with all current eco, green, organic, and other trends are also becoming more educated and aware of what they put into their bodies, and if those products also contribute to the environment, even better. HHP and some other novel technologies have a bright future and they are guaranteed to last.

ACKNOWLEDGMENTS

Authors would like to thank Croatian Science Foundation for funding the project "High voltage discharges for green solvent extraction of bioactive compounds from Mediterranean herbs (IP-2016-06-1913)." Marinela Nutrizio and Anet Režek Jambrak would like to thank to "Young researchers' career development project—training of doctoral students" of the Croatian Science Foundation funded by the European Union from the European Social Fund.

REFERENCES

Aganovic, K., et al. (2017). Pilot scale thermal and alternative pasteurization of tomato and watermelon juice: An energy comparison and life cycle assessment. *Journal of Cleaner Production, 141*, 514–525. https://doi.org/10.1016/j.jclepro.2016.09.015. Elsevier Ltd.

Atuonwu, J. C., & Tassou, S. A. (2018). Model-based energy performance analysis of high pressure processing systems. *Innovative Food Science and Emerging Technologies, 47*, 214–224. https://doi.org/10.1016/j.ifset.2018.02.017. Elsevier Ltd.

Barba, F. J., Koubaa, M., et al. (2017). Mild processing applied to the inactivation of the main foodborne bacterial pathogens: A review. *Trends in Food Science and Technology*, 20–35. https://doi.org/10.1016/j.tifs.2017.05.011. Elsevier Ltd.

Barba, F. J., Mariutti, L. R. B., et al. (2017). Bioaccessibility of bioactive compounds from fruits and vegetables after thermal and nonthermal processing. *Trends in Food Science and Technology*, 195–206. https://doi.org/10.1016/j.tifs.2017.07.006. Elsevier Ltd.

Barba, F. J., et al. (2016). Implementation of emerging technologies. In *Innovation strategies in the food industry: Tools for implementation*. https://doi.org/10.1016/B978-0-12-803751-5.00007-6.

Boye, J. I., & Arcand, Y. (2013). Current trends in green technologies in food production and processing. *Food Engineering Reviews, 5*, 1–17.

Briones-Labarca, V., Giovagnoli-Vicuña, C., & Chacana-Ojeda, M. (2019). High pressure extraction increases the antioxidant potential and in vitro bio-accessibility of bioactive compounds from discarded blueberries. *CyTA—Journal of Food, 17*(1), 622–631. https://doi.org/10.1080/19476337.2019.1624622.

Cacace, F., et al. (2020). Evaluation of the economic and environmental sustainability of high pressure processing of foods. *Innovative Food Science and Emerging Technologies, 60*, 102281. https://doi.org/10.1016/j.ifset.2019.102281. Elsevier Ltd.

Chauhan, O. P., et al. (2015). Effect of high pressure processing on yield, quality and storage stability of peanut paneer. *International Journal of Food Science & Technology, 50*(6), 1515–1521. https://doi.org/10.1111/ijfs.12782. Blackwell Publishing Ltd.

Chemat, F., et al. (2020). A review of sustainable and intensified techniques for extraction of food and natural products. *Green Chemistry*. https://doi.org/10.1039/C9GC03878G. The Royal Society of Chemistry.

Corrales, M., et al. (2008). Extraction of anthocyanins from grape by-products assisted by ultrasonics, high hydrostatic pressure or pulsed electric fields: A comparison. *Innovative Food Science and Emerging Technologies, 9*, 85–91. 2008.

Djekic, I., et al. (2017). Review on environmental models in the food chain—Current status and future perspectives. *Journal of Cleaner Production*. https://doi.org/10.1016/j.jclepro.2017.11.241.

Dong, C., Shaoxiang, P., Jun, C., Xueli, P., Xingfeng, G., Lin, G., et al. (2016). Comparing the effects of high hydrostatic pressure and ultrahigh temperature on quality and shelf life of cloudy ginger juice. *Food and Bioprocess Technology, 9*, 1779–1793.FAO, 2018FAO. (2018). *Food and agriculture: Key to achieving the 2030 agenda for sustainable development* (un.org).

Gimenez-Escalante, P., & Rahimifard, S. (2019). A methodology to assess the suitability of food processing technologies for distributed localised manufacturing. *Sustainability, 11*(12), 3383. https://doi.org/10.3390/su11123383. MDPI AG.

Gong, Y., et al. (2015). Comparative study of the microbial stability and quality of carrot juice treated by high-pressure processing combined with mild temperature and conventional heat treatment. *Journal of Food Process Engineering, 38*(4), 395–404. https://doi.org/10.1111/jfpe.12170. Blackwell Publishing Inc.

Guo, X., Han, H. D., Xi, H., Rao, L., Liao, X., Hu, X., et al. (2012). Extraction of pectin from navel orange peel assisted by ultra-high pressure, microwave or traditional heating: A comparison. *Carbohydrate Polymers, 88*, 441–448.

Hite, B. H. (1899). *The effect of pressure in the preservation of milk: A preliminary report*. Morgantown, WV: West Virginia Agricultural and Forestry Experiment Station Bulletins. https://doi.org/10.33915/agnic.58.

Jambrak, A. R., et al. (2019). Impact of novel nonthermal processing on food quality: Sustainability, modelling, and negative aspects. *Journal of Food Quality, 2019*, 2171375.

https://doi.org/10.1155/2019/2171375. Hindawi Publishing Corporation.

Jun, X. (2006). Application of high hydrostatic pressure processing of food to extracting lycopene from tomato paste waste. *High Pressure Research: An International Journal*, *26*(1), 33–41.

Kostoska, O., & Kocarev, L. (2019). A novel ICT framework for sustainable development goals. *Sustainability*, *11*(7), 1961. https://doi.org/10.3390/su11071961. MDPI AG.

Magureanu, M., et al. (2018). High efficiency plasma treatment of water contaminated with organic compounds. Study of the degradation of ibuprofen. *Plasma Processes and Polymers*, *15*(6). https://doi.org/10.1002/ppap.201700201. Wiley-VCH Verlag.

Meng, Y., et al. (2018). Enhancing sustainability and energy efficiency in smart factories: A review. *Sustainability (Switzerland)*, *10*(12). https://doi.org/10.3390/su10124779. MDPI AG.

Morata, A., et al. (2015). Grape processing by high hydrostatic pressure: Effect on microbial populations, phenol extraction and wine quality. *Food and Bioprocess Technology*, 8. https://doi.org/10.1007/s11947-014-1405-8.

Nikmaram, N., & Rosentrater, K. A. (2019). Overview of some recent advances in improving water and energy efficiencies in food processing factories. *Frontiers in Nutrition*. https://doi.org/10.3389/fnut.2019.00020. Frontiers Media S.A.

Ninčević, G. A., Ježek, D., Karlović, S., & Bosiljkov, T. (2019). *Application of high hydrostatic pressure for pectin extraction from agro-food waste and by-products*. Las Vegas: Open Access Books.

Notarnicola, B., et al. (2017). The role of life cycle assessment in supporting sustainable agri-food systems: A review of the challenges. *Journal of Cleaner Production*, *140*, 399–409. https://doi.org/10.1016/j.jclepro.2016.06.071.

Pardo, G., & Zufía, J. (2012). Life cycle assessment of food-preservation technologies. *Journal of Cleaner Production*, *28*, 198–207. https://doi.org/10.1016/j.jclepro.2011.10.016.

Picart-Palmade, L., Cunault, C., Chevalier-Lucia, D., Belleville, M.-P., & Marchesseau, S. (2018). Potentialities and limits of some non-thermal technologies to improve sustainability of food processing. *Frontiers in Nutrition*, *5*, 130.

Rodriguez-Gonzalez, O., et al. (2015). Energy requirements for alternative food processing technologies-principles, assumptions, and evaluation of efficiency. *Comprehensive Reviews in Food Science and Food Safety*, *14*(5), 536–554. https://doi.org/10.1111/1541-4337.12142. Blackwell Publishing Inc.

Roldán-Marín, E., et al. (2009). Onion high-pressure processing: Flavonol content and antioxidant activity. *LWT—Food Science and Technology*, *42*, 835–841. https://doi.org/10.1016/j.lwt.2008.11.013.

Sampedro, F., et al. (2014). Cost analysis and environmental impact of pulsed electric fields and high pressure processing in comparison with thermal pasteurization. *Food and Bioprocess Technology*, *7*(7), 1928–1937. https://doi.org/10.1007/s11947-014-1298-6. Springer New York LLC.

Savastano, M., et al. (2019). Contextual impacts on industrial processes brought by the digital transformation of manufacturing: A systematic review. *Sustainability (Switzerland)*, *11*(3). https://doi.org/10.3390/su11030891. MDPI AG.

Spira, P., Bisconsin-Junior, A., Rosenthal, A., & Monteiro, M. (2018). Effects of high hydrostatic pressure on the overall quality of Pêra-Rio orange juice during shelf life. *Food Science and Technology International*. https://doi.org/10.1177/1082013218768997.

Srai, J. S., et al. (2016). Distributed manufacturing: Scope, challenges and opportunities. *International Journal of Production Research*, *54*(23), 6917–6935. https://doi.org/10.1080/00207543.2016.1192302. Taylor and Francis Ltd.

Stasiulaitiene, I., et al. (2016). Comparative life cycle assessment of plasma-based and traditional exhaust gas treatment technologies. *Journal of Cleaner Production*, *112*, 1804–1812. https://doi.org/10.1016/j.jclepro.2015.01.062. Elsevier Ltd.

Suthanthangjai, W., & Kajda, P. (2004). The effect of high hydrostatic pressure on the anthocyanins of raspberry (Rubus idaeus). *Food Chemistry*, *90*, 193–197. https://doi.org/10.1016/j.foodchem.2004.03.050.

Teixeira, A., et al. (2014). Natural bioactive compounds from winery by-products as health promoters: A review. *International Journal of Molecular Sciences*, *15*(9), 15638–15678. https://doi.org/10.3390/ijms150915638.

Toepfl, S., Mathys, A., Heinz, V., & Knorr, D. (2006). Review: Potential of high hydrostatic pressure and pulsed electric fields for energy efficient and environmentally friendly food processing. *Food Reviews International*, *22*, 405–423.

Tsevdou, M., Gogou, E., & Taoukis, P. (2019). High hydrostatic pressure processing of foods. In *Green Food Processing Techniques* (pp. 87–137). Elsevier. https://doi.org/10.1016/b978-0-12-815353-6.00004-5.

Tsikrika, K., O'Brien, N., & Rai, D. K. (2019). The effect of high pressure processing on polyphenol oxidase activity, phytochemicals and proximate composition of Irish potato cultivars. *Foods*, *8*(10). https://doi.org/10.3390/foods8100517. MDPI Multidisciplinary Digital Publishing Institute.

Zhong, R. Y., et al. (2017). Intelligent manufacturing in the context of industry 4.0: A review. *Engineering*, *3*(5), 616–630. https://doi.org/10.1016/J.ENG.2017.05.015. Elsevier Ltd.

CHAPTER 10

Ultrasound Processing: A Sustainable Alternative

NOELIA PALLARÉS[A] • HOUDA BERRADA[A] • EMILIA FERRER[A] • JIANJUN ZHOU[A] • MIN WANG[A] • FRANCISCO J. BARBA[A] • MLADEN BRNČIĆ[B]
[a]Department of Preventive Medicine and Public Health, Food Science, Toxicology and Forensic Medicine, Faculty of Pharmacy, Universitat de València, València, Spain, [b]Faculty of Food Technology and Biotechnology, University of Zagreb, Zagreb, Croatia

10.1 INTRODUCTION

Sustainable development is related to meet the needs of the present without compromising the ability of future generations to meet their needs. Sustainability can be also defined, as consumption that can continue indefinitely without natural, physical, human, and intellectual capital degradation. It is based in a three-dimension approach: (i) environmental sustainability, (ii) social solidarity, and (iii) economic efficiency profit. The principle of fairness between present and future generations should be taken into account in the use of environmental, economic, and social resources (García-Serna, Pérez-Barrigón, & Cocero, 2007).

Sustainable development goals (SDGs; 2015–2030), constitute an extension of millennium development goals (MDGs; 2000–2015), and were adopted by all United Nations Member States as universal call to action for poverty eradication, global prosperity, and environment protection (Sachs, 2012; World Health Organization., 2015).

The MDGs played an important part in securing progress against poverty, hunger, and diseases, but in a world under climate change and other serious environmental problems, it was understood that worldwide environmental objectives needed more attention (Sachs, 2012). In this sense, the world's governments adopted a new round of global initiatives to follow the post 2015 MDG goals. In June 2021, at the United Nations Conference on Sustainable Development (Rio + 20), the Member States adopted a document in which they decided to launch a process to develop a set of SDGs to build upon the MDGs. In January 2015, the General Assembly began the negotiation process on the post-2015 development agenda and the process culminated with the establishment of the "2030 Agenda for Sustainable Development" at the UN Sustainable Development Summit in September 2015. The 2030 Agenda is based in 17 sustainable development goals (SDGs) (Desa, 2016; United Nations, 2021).

The UN's 2030 program for sustainable development, and the society growing demand of greener alternatives, encourage global food management to implement sustainable food production systems. The current food systems must fight against different concerns such as climate change, waste management, water scarcity, soil preservation, as well as security of food supply, safety, quality, and accessibility. In this context, it has promoted a growing interest of industry to develop affordable, safe, effective, and ecologically innovative techniques (Arshad et al., 2021; Desa, 2016). These techniques such as ultrasound, sub-critical and supercritical fluid extraction, microwave, accelerated solvent extraction, high-pressure processing, and pulsed electric field, have been proposed as alternative to conventional ones. These alternative techniques offer the potential to reduce the use of toxic chemical solvents, while improving process efficiency. Furthermore, these techniques operate under low temperature so does not affect the stability of food beneficial compounds (Picart-Palmade, Cunault, Chevalier-Lucia, Belleville, & Marchesseau, 2019).

Among them, ultrasound technique (USN) links with green chemistry and eco-friendly characteristics. This technology does not involve chemical solvents and their environmental and health risks associated, being a sustainable alternative to industry in multiple ways: enhancing extraction yield, enhancing aqueous extraction processes, providing the opportunity to use alternative clean and/or green solvents and, enhancing extraction of heat-sensitive components under conditions that would otherwise have low or unacceptable yields, improving energy efficiency,

Sustainable Production Technology in Food. https://doi.org/10.1016/B978-0-12-821233-2.00006-X
Copyright © 2021 Elsevier Inc. All rights reserved.

TABLE 10.1
Ultrasound (USN) Processing Towards Sustainability.

Sustainable Goals	US Application	Principal Contributions Towards Sustainability	References
Food safety and security	Microorganism's decontamination	USN allows to reduce or eliminate important pathogenic microorganisms from food	Barba, Koubaa, do Prado-Silva, Orlien, and Sant'Ana (2017); de São José et al. (2014)
	Enzyme activities	USN can alter enzyme characteristics, promoting or inhibiting its activities	Terefe et al. (2009)
	Reducing contaminants	USN has potential to be applied in pesticides and mycotoxins decontamination	Gavahian, Pallares, Al Khawli, Ferrer, and Barba (2020)
Environmental sustainability	Wastewater treatment	USN technology has promising applications for the degradation of persistent organic contaminant compounds in wastewater	Adewuyi (2005)
	Recovery of bioactive compounds from food wastes	USN can effectively recover valuable compounds such as polysaccharides, polyphenols, essential oils, pigments, proteins, flavor compounds, enzymes, and dietary fibers from food wastes or by-products	Roselló-Soto et al. (2015)
	Biomass valorization	USN can improve the biomass conversion into added-value products in the production of fine chemicals and energy	Flores, Cravotto, Bizzi, Santos, and Iop (2021)
Economic sustainability	USN and clean extraction	USN technology is a simple, efficient, and cost-effective alternative to be applied in the industry	Clodoveo (2019)
	USN uses in drying	USN application in drying constitutes an interesting alternative reducing the time, which may involve an energy saving process	Kowalski and Pawłowski (2015)

improving water recovery, and reducing wastes and residues (Cintas, 2016; Misra et al., 2017; Roselló-Soto et al., 2015). Table 10.1 summarizes USN contributions in sustainable processing. Ultrasounds is being applied in different fields such as chemistry, chemical engineering, medicine and pharmacology, material sciences, physics, biochemistry and molecular biology, environmental science, and agricultural science (Chatel, 2018).

10.2 ULTRASOUND (USN) CHARACTERISTICS AND SUSTAINABILITY

USN is considered as an innovative food processing alternative. This technology is based on the application of ultrasound waves comprised between 20 kHz and 100 MHz. The waves travel through a medium, resulting in a series of compression and rarefaction. At sufficiently power, the rarefaction exceeds the attractive forces between molecules in a liquid phase, which produces the formation of cavitation bubbles. The constant growth of gas bubbles in the medium, results in bubbles collapse and cavitation. This phenomenon produces the break-down of liquid-solid interfaces (Arvanitoyannis, Kotsanopoulos, & Savva, 2017; Roohi et al., 2019). USN can be classified into higher-power ultrasound and lower power ultrasound. Higher-power ultrasounds at lower frequencies (20–100 kHz), can cause cavitation (implosion of gas bubbles) with sound intensities of 10–1000 W/cm^2. It is able to modify material properties, producing the disruption of the physical integrity or accelerating chemical reactions by the generation of immense pressure, shear, and temperature gradient in the medium. In contrast, lower power ultrasound involves the use of high frequencies (2–10 MHz) usually at low power (up to 10 W). It does not cause alterations in the properties of the material (Rastogi, 2011). The frequency range will determine USN applications in food processing, analysis, and quality control. USN has various applications, such as the assistance of thermal treatments, the improvement of mass transfer processes, the food preservation, and the texture manipulation. Moreover, shows some advantages such as efficiency, low solvent consumption and high level of automation (Hashemi et al., 2018; Knorr et al., 2011; Koubaa et al., 2016). USN is regarded as an environmental-friendly technology. Comparing ultrasound with conventional processes, it reduces the processing time and cost, simplifies the manipulation, gives a higher purity of the final product obtained, eliminates post-treatment of wastewater and consumes only a fraction of the energy and time normally needed in conventional processes. Fig. 10.1 shows the potential contributions of USN to reach the global goals for sustainable development. In the food industry USN is applied efficiently in several processes such as cutting, freezing, drying, tempering, bleaching, sterilization, and extraction (Barba et al., 2017; Chemat & Khan, 2011; Gavahian, Chen, Mousavi Khaneghah, Barba, & Yang, 2018).

10.2.1 USN Applications in Food Safety and Security

10.2.1.1 Effect of USN on microorganism decontamination

Some innovative food preservation technologies, such as USN, alone or combined with other preservation techniques (i.e., mild temperature) allows to reduce or eliminate important spoilage and/or pathogenic microorganisms from food (Barba et al., 2017; de São José et al., 2014).

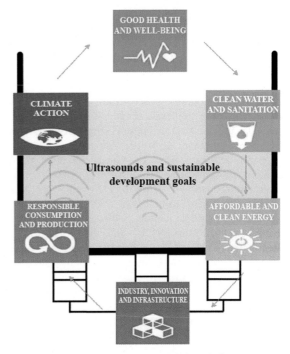

FIG. 10.1 Ultrasounds and the global goals for sustainable development.

Regarding USN mechanism explaining its impact on microorganism, the primary targets of USN in cells are cell walls and membranes. Cavitation and micronization are some of its action mechanisms. Temperature and pressure also have an important role in its antimicrobial potential (Misra et al., 2018; Zinoviadou et al., 2015).

Microorganism´ sensitivity to USN differs according to different factors, such as the size and shape of the cell and its external layers morphology. In general resistance decreases as follows: spores > fungi (molds) > yeasts > Gram-positive bacterias > Gram-negative bacterias, being bacterial spores the most resistant (Barba et al., 2017; Zinoviadou et al., 2015). In this line, several studies have reported the USN decontamination effect on *L. monocytogenes*, *Salmonella spp.*, *E. coli*, *S. aureus*, *B. subtilis* and some other microorganisms, especially when combined to other techniques (Barba et al., 2017). For instance, Sarkinas et al. (2018) investigated the effect of high intensity ultrasound treatment, under 300 W and 600 W, 28 kHz and 10–30 min, on pathogenic bacteria (*Listeria monocytogenes*, *Bacillus cereus*, *Escherichia coli*, *Salmonella typhimurium*) and some phytoviruses. These authors observed that the treatment was effective to eliminate vegetative cells of Gram-positive and Gram negative bacteria from 1.59 to 3.4 log in bacterial suspensions and some phytoviruses in fruits. Moreover, Gram positive bacteria resulted more susceptible to USN technology than the Gram negative ones. In other study, Haughton et al. (2012) evaluated the potential of sonication and thermosonication to reduce the levels of *Campylobacter, enterobacteriaceae* and total viable counts on raw poultry. After 16 min of treatment of broiler skin pieces in a high-intensity USN unit, no viable *Campylobacter* or *enterobacteriaceae* were observed, while total viable counts were reduced up to 2.49 log10 CFU/g. USN has been also widely studied to inactivate milk microorganism. For example, *Listeria innocua* cells were studied under batch pasteurization (63°C) and thermo-sonication treatments (63°C, 24 kHz, 129 mW/mL) in milk products. After, the thermo-sonicated treatment *Listeria innocua* cells showed the formation of pores. After 30 min USN treatment allowed cells rupture (Bermúdez-Aguirre, Mawson, & Barbosa-Cánovas, 2011). In other study, D'amico, Silk, Wu, and Guo (2006) studied the effect of USN treatment (20 kHz, 100% power level, 150 W acoustic power and 118 W/cm^2 acoustic intensity) on reducing *Listeria monocytogenes* levels in milk. Continuous flow USN treatment combined with mild heat at 57°C during 18 min resulted in a reduction of 5-log for *L. monocytogenes* in ultrahigh-temperature milk, and of 5-log reduction in total aerobic bacteria in raw milk.

Finally, in fruits and vegetables, Sagong et al. (2011) obtained maximum reductions of 2.75, 3.18, and 2.87 log CFU/g for *E. coli, S. typhimurium,* and *L. monocytogenes,* respectively in lettuce after combining USN treatment (40 kHz) with 2% of organic acid in 5 min. The combination of USN treatment with organic acids was more effective in pathogen reduction comparing to individual treatments.

10.2.1.2 Effect of USN on enzyme activities

USN can alter enzymes characteristics, promoting changes on enzymes, substrates, and reactions between enzymes and substrates. Moreover, it provides an optimal environment for the reactions (Zinoviadou et al., 2015). Regarding USN action on enzymes inactivation, Terefe et al. (2009) investigated the USN inactivation kinetics of enzymes polygalacturonase (PG) and pectin methylesterase (PME) in tomato juice under frequency of 20 kHz, and temperatures between 50°C and 75°C. They observed that the degradation of pectin by the synergistic action of PME and PG during vegetable processing, may result in a decrease of viscosity in the final

product, decreasing tomato products quality. These authors also observed that thermosonication promotes a complete inactivation of PME and 72% inactivation of PG after 4 min of treatment improving tomato juice rheological properties. The inactivation of PME and PG was also previously studied by Wu, Gamage, Vilkhu, Simons, and Mawson (2008), for quality improvement of tomato juices. Tomato juice was treated under thermosonication (24 kHz), at amplitudes of 25, 50 and 75 μm. The treatment at 60°C and 65°C was suitable to obtain tomato juice with a low residual PME activity and high viscosity.

10.2.1.3 Effect of USN on food contaminants

Food contaminants include natural toxins, food processing, and environmental contaminants, such as pesticides and toxins. These contaminants may affect food quality and produce adverse human and animal health effects. Food industries are interested in the development of processes and practices that prevent and reduce the presence of those contaminants. In the last years, the potential of USN in the reduction of pesticides and mycotoxins contaminants has been reported by several authors (Gavahian et al., 2020).

Mycotoxin removal by USN. USN has been applied by various researchers to reduce mycotoxins producer fungi and mycotoxins levels. For instance, Rudik, Morgunova, and Krasnikova (2020) proposed USN at low frequencies 24–26 kHz and intensity of no more than 1 W/cm^2 as an effective treatment to reduce mold fungi content in grains, and subsequently mycotoxins production. In other study, Liu, Li, Bai, and Bian (2019) reported high degradation rates of 96.5%, 60.8%, 95.9%, and 91.6% for the mycotoxins AFB1, DON, ZEA, and OTA, respectively in maize and aqueous solution. These authors also observed that the mycotoxins degradation was significantly affected by the ultrasonic intensity (2.2–11 W/cm^3) and sonication time range from 10 to 50 min. In other study, these authors observed an AFB1 degradation rate up to 85.1% in aqueous solution after 80 min of treatment under frequency of 20 kHz and power intensity of 6.6 W/cm^3 (Liu, Li, Liu, & Bian, 2019). In contrast, in another study, Mortazavia, Sania, and Mohsenib (2015) obtained lower reduction for AFs, about 41% after 10 min under frequency of 20 KHz.

Pesticide removal by USN. USN-assisted cleaning is considered as an environmental-friendly and effective process in eliminating pesticides comparing to conventional methods. It also enhances time and energy saving

(Azam et al., 2020). Several studies have reported USN cleaning as an efficient technique to remove pesticides from various vegetables itself or combined with other methods. For instance, Lozowicka, Jankowska, Hrynko, and Kaczynski (2016) observed that USN cleaning at a frequency of 40 kHz and power 2 × 240 W peak/period was effective in removing 16 pesticides from strawberries with reduction rates comprising between 45.1% and 91.2% after 5 min of treatment. USN cleaning contributed to a more efficient pesticide reduction than that obtained only by traditional cleaning.

In apple juices, Zhang et al. (2010) obtained degradation rates of 41.7% for malathion and 82.0% for chlorpyrifos pesticides after USN irradiation treatment with 25 kHz and 500 W during 120 min. In another study, these authors reported a degradation percentage of 51.3% for diazinon after USN treatment during 30 min (Zhang et al., 2010). Both initial concentration and USN power may influence the reductions rates obtained.

10.2.2 US Applications in Environmental Sustainability

10.2.2.1 USN and water treatment

The use of USN technology has received wide attention to be applied in wastewater treatment and the environmental remediation areas. For instance, USN has promising applications for the degradation of persistent organic contaminant compounds in wastewater. The USN destruction of pollutants in aqueous phase involves different reaction pathways such as pyrolysis and hydroxyl radical-mediated reactions (Matouq, Al-Anber, Susumu, Tagawa, & Karapanagioti, 2014).

In this line, Matouq et al. (2014) investigated USN high frequency treatment on the degradation of methyl orange dye, MOD, from industrial wastewater. Azo dyes are commonly used in textile, dyeing, printing, and cosmetic industries and produce toxic effects in ecosystem due its highly persistent in natural environments. These authors reported that USN technique successfully degraded azo dyes. The best condition for azo dyes degradation were initial concentration of 50 ppm, 15 mL of liquid volume and power of 25 V.

Over the last years, several studies have also investigated the use of USN to decontaminate pesticides from wastewater. For instance, Farooq, Shaukat, Khan, and Farooq (2008) reported that the combination of USN with H_2O_2 resulted in a good alternative in the decomposition of the methidathion pesticide. The decomposition observed was about 88% under USN treatment in combination with H_2O_2 at 120 μm, pH 3 and 90 min. In other study, Schieppati et al. (2019) investigated the application of USN-assisted photocatalytic in the degradation of the herbicide isoproturon in water. These authors reported that under power of 50 W cm^2, the degradation observed was about 100% after 1 h of treatment. Moreover, lower molecular weight of by-products was obtained compared to photocatalytic treatments alone. For diazinon insecticide, Wang and Shih (2016), observed that USN in combination with Fenton's and Fenton-like reagents facilitated its degradation and reduced its toxicity. After processing time of 60 min a 98% of diazinon reduction was reached with a mineralization efficiency of 30%. To sum up, the USN treatment in combination with other chemicals or oxidation processes is suggested to be an efficient tool to remove contaminant compounds from wastewater.

10.2.2.2 USN and recovery of high-added-value compounds from food wastes and side streams

A large amount of food wastes and side streams are produced during the food supply chain. The effective utilization of food waste and by-products has high potential to obtain high-added value compounds that can be employed after as functional ingredients. Valuable compounds such as polysaccharides, polyphenols, essential oils, pigments, proteins, flavor compounds, enzymes, and dietary fibers can be effectively recovered employing different extraction methods, such us ultrasounds (Barba, Zhu, Koubaa, Sant'Ana, & Orlien, 2016; Marić et al., 2018; Roselló-Soto et al., 2015).

For instance, Al Khawli et al. (2021) optimized the conditions based in a response surface methodology (RSM) to effective proteins and antioxidant compounds extraction from sea bass side streams (heads, skin, bones, and viscera) employing USN. The authors obtained a high percentage of proteins and antioxidant activity, especially in viscera side streams under time and temperature conditions of 30 min and 50°C, respectively.

Regarding bioactive compounds recovery from fruits and vegetable wastes, Jabbar et al. (2015) investigated the USN-assisted extraction technology (UAE) under operation parameters of extraction time (3–37 min), extraction temperature (10°C–60°C) and ethanol concentration (13%–97%) on total phenols, antioxidant capacity, chlorogenic acid, caffeic acid, catechin and epicatechin recovery from carrot pomace using response surface methodology. The results revealed that UAE was an eco-friendly and a time-saving process for obtain phenolic compounds from carrot pomace. In orange peel, Montero-Calderon, Cortes, Zulueta, Frigola, and Esteve (2019) studied the effect of USN on the antioxidant capacity, phenolic content, ascorbic acid, and total carotenoids obtained from orange peel under conditions of 400 W power, 30 min, and 50% ethanol in water. The results revealed total carotenoid concentration of 0.63 mg/100 g, 53.78 mg AA/100 g of vitamin C, phenolic concentration of 105.96 mg GAE/100 g, and antioxidant capacity (ORAC = 27.08 mM TE and TEAC = 3.97 mM TE). In olive pomace, Martínez-Patiño et al. (2019) studied the effect of US assisted extraction with ethanol-water mixtures on total phenolic compounds and flavonoids, as well as their antioxidant activity. The optimal conditions observed for the extraction were found to be 43.2% ethanol concentration, 70% amplitude, and 15 min of treatment that resulted in an extract with a phenolic and flavonoid content per gram of olive pomace of 57.5 mg gallic acid equivalent and 126.9 mg rutin equivalent, respectively. Moreover, the values for DPPH, ABTS, and FRAP assay were 56.7, 139.1, and 64.9 mg Trolox equivalent.

10.2.2.3 Ultrasound in biomass valorization

Biomass constitutes an alternative source to produce fuels, solvents, chemical building-blocks, and biopolymers. The biomass processing allows a reduction of petroleum dependence and consequently a decrease of environmental impact. Biomass consists in residues from food processing, wasted husks, leaves, and seeds. For biomass valorization, US has been applied in biomass pre-treatments, lignin degradation, cellulose conversion, and to increase sugar production by biochemical routes (Flores et al., 2021; Kuna, Behling, Valange, Chatel, & Colmenares, 2017; Ong & Wu, 2020).

Niglio, Procentese, Russo, Sannia, and Marzocchella (2020), studied a combined pre-treatment based on USN and mild alkaline solution to promote enzymatic hydrolysis of coffee silver skin. The maximum sugar yield was achieved after enzymatic hydrolysis of coffee silver skin pre-treated by sonication during 5 min, 11% w/v biomass loading, and 75 min autoclave in 5% w/v NaOH. In other study, Dong, Chen, Guan, Li, and Xin (2018) reported the efficiency of dual-frequency ultrasound combined with alkali pre-treatment method to facilitate efficient biogas production. These authors observed that the proposed methodology increased the cumulative biogas yield and gas production rate. Similarity, Hassan et al. (2020) reported that USN pre-treatment constitutes a promising technology for increased valorization of brewer's spent grain as a means of releasing fermentable sugars, and the subsequent production of ethanol from this lignocellulosic

source. The optimal conditions were found to be 20% of USN power, 60 min of treatment, temperature of 26.3°C, and 17.3% w/v of biomass in water. Under optimal conditions a 2.1-fold increase in reducing sugar yield was obtained. Fermentation studies using *S. cerevisiae* growing on the pre-treated brewer's spent grain resulted in a conversion of 66% of the total sugar content into ethanol.

10.2.3 USN and Economic Sustainability

Economic sustainability is focused in efficiently use of energy, time, and other resources, which can provide immense benefits from a reasonable reuse of resources.

10.2.3.1 USN application in clean extraction

Most methods used in industry are chemical based, involving corrosive or toxic chemicals. A significant advance would be the use of clean extraction technologies, such as ultrasound, which fits well with the growing market trend of used ingredients obtained by clean and environment-friendly extraction processes (Rutkowska, Namieśnik, & Konieczka, 2017). Several economic benefits can be derived from the use of USN technology and the reduction in the use of chemicals. USN system is a simple, efficient and cost-effective alternative method comparing to traditional techniques, that provides several benefits, enhancing extraction yield and extraction rates; giving the opportunity to use alternative generally recognized as safety solvents and enhancing the recovery of heat-sensitive components under conditions that employing traditional techniques would result in low yields (Roselló-Soto et al., 2015).

For instance, in olive oil industry, USN applications in extra-virgin olive oil extraction process can have many advantages compared to traditional techniques, such as efficiently and cost effective, the increase of the oil yield and the acceleration of enzymatic kinetics, better extraction of heathy minor compounds, low cost of equipment comparing to other emerging technologies, extraction efficiency with moderate increments of temperature, improvement of antioxidant contents and the preservation of composition and quality of the oil (Clodoveo, 2019). In microalgae production, ultrasound-assisted extraction of bioactive compounds from fresh microalgae is possible in a reduced time and in a simple and scalable pre-industrial system (Adam, Abert-Vian, Peltier, & Chemat, 2012).

10.2.3.2 USN application in drying

In industry the USN application in drying constitutes an interesting alternative to traditional drying reducing the time, which may involve an energy saving process.

In this sense, the effectiveness of ultrasound on thermal drying has been studied by different authors. Guoid and Lei (2020) investigated the application of USN to enhance the silt drying process and revealed that the length of drying time can be shortened by increasing temperature and ultrasound power. The combination of thermal drying at temperatures of 60°C and 100°C and 100 W power ultrasound was optimal in the length of drying time, shortened by 44.19% and 45.16%, respectively and in the energy consumption, saving up to 38.16%. In apple drying, Kowalski and Pawłowski (2015) also observed that USN makes the drying processes more effective enhancing the drying efficiency without increasing their temperature significantly. In this sense, the increase in drying efficiency is mainly due to the ultrasound vibration effect. In potato drying, Ozuna, Cárcel, García-Pérez, and Mulet (2011) evaluated the use of power ultrasound to improve drying and the influence in the water transport mechanisms. These authors observed that the higher is the applied power, the faster is the drying kinetic. Moreover, the ultrasonic effect in the water transport is based on mechanical phenomena with a low heating capacity, this fact allows to obtain dry products with high quality characteristics. Regarding the application of USN-assisted osmotic dehydration in drying of fruits. The USN-assisted osmotic dehydration promotes water loss and sugar gain during the process. The effect of this process on the effective water diffusivity depends on the degree of cells breakdown in fruit tissue. USN constitutes a cost-effective treatment when USN process allows to increase the effective diffusivity of water in the fruit reducing air-drying time (Fernandes & Rodrigues, 2008).

10.3 CONCLUSIONS

Ultrasound technology is a useful alternative in food processing to reach food safety and security sustainable goals, as well as environmental and economic sustainability objectives established in the UN's 2030 program. Some reports showed promising results in sustainable processing. Some of its principal contributions towards sustainability consist in allowing the reduction of pathogenic microorganisms and some food contaminants and in promoting or inhibiting enzyme activities. It also has shown promising applications in the degradation of organic contaminant compounds in wastewater, recovering valuable compounds from food wastes and by-products or improving biomass conversion into added-value products. Moreover, it has been shown to be an efficient and cost-effective alternative to

be applied in industry processes such as extraction and drying. This emerging technology can be a good alternative to respond to the society demand of eco-friendly processing without affecting sensorial and health food promoting attributes.

REFERENCES

Adam, F., Abert-Vian, M., Peltier, G., & Chemat, F. (2012). "Solvent-free" ultrasound-assisted extraction of lipids from fresh microalgae cells: A green, clean and scalable process. *Bioresource Technology, 114*, 457–465. https://doi.org/10.1016/j.biortech.2012.02.096.

Adewuyi, Y. G. (2005). Sonochemistry in environmental remediation. 1. Combinative and hybrid sonophotochemical oxidation processes for the treatment of pollutants in water. *Environmental Science & Technology, 39*(10), 3409–3420. https://doi.org/10.1021/es049138y.

Al Khawli, F., Pallarés, N., Martí-Quijal, F. J., Ferrer, E., Barba, F. J., Khawli, A., et al. (2021). Sea bass side streams valorization assisted by ultrasound. LC-MS/MS-IT determination of mycotoxins and evaluation of protein yield, molecular size distribution and antioxidant recovery. *Applied Sciences, 11*(5), 2160. https://doi.org/10.3390/app11052160.

Arshad, R. N., Abdul-Malek, Z., Roobab, U., Munir, M. A., Naderipour, A., Qureshi, M. I., et al. (2021). Pulsed electric field: A potential alternative towards a sustainable food processing. *Trends in Food Science and Technology, 111*, 43–54. https://doi.org/10.1016/j.tifs.2021.02.041.

Arvanitoyannis, I. S., Kotsanopoulos, K. V., & Savva, A. G. (2017). Use of ultrasounds in the food industry-methods and effects on quality, safety, and organoleptic characteristics of foods: A review. *Critical Reviews in Food Science and Nutrition, 57*(1), 109–128. https://doi.org/10.1080/10408398.2013.860514.

Azam, S. M. R., Ma, H., Xu, B., Devi, S., Siddique, M. A. B., Stanley, S. L., et al. (2020). Efficacy of ultrasound treatment in the and removal of pesticide residues from fresh vegetables: A review. *Trends in Food Science and Technology, 97*, 417–432. https://doi.org/10.1016/j.tifs.2020.01.028.

Barba, F. J., Koubaa, M., do Prado-Silva, L., Orlien, V., & Sant'Ana, A. D. S. (2017). Mild processing applied to the inactivation of the main foodborne bacterial pathogens: A review. *Trends in Food Science and Technology, 66*, 20–35. https://doi.org/10.1016/j.tifs.2017.05.011.

Barba, F. J., Zhu, Z., Koubaa, M., Sant'Ana, A. S., & Orlien, V. (2016). Green alternative methods for the extraction of antioxidant bioactive compounds from winery wastes and by-products: A review. *Trends in Food Science and Technology, 49*, 96–109. https://doi.org/10.1016/j.tifs.2016.01.006.

Bermúdez-Aguirre, D., Mawson, R., & Barbosa-Cánovas, G. V. (2011). Study of possible mechanisms of inactivation of *Listeria innocua* in thermo-sonicated milk using scanning electron microscopy and transmission electron microscopy. *Journal of Food Processing and Preservation, 35*, 767–777.

Chatel, G. (2018). How sonochemistry contributes to green chemistry? *Ultrasonics Sonochemistry, 40,* 117–122. https://doi.org/10.1016/j.ultsonch.2017.03.029.

Chemat, F., & Khan, M. K. (2011). Applications of ultrasound in food technology: Processing, preservation and extraction. *Ultrasonics Sonochemistry, 18*(4), 813–835. https://doi.org/10.1016/J.ULTSONCH.2010.11.023.

Cintas, P. (2016). Ultrasound and green chemistry—Further comments. *Ultrasonics Sonochemistry, 28,* 257–258. https://doi.org/10.1016/j.ultsonch.2015.07.024.

Clodoveo, M. L. (2019). Industrial ultrasound applications in the extra-virgin olive oil extraction process: History, approaches, and key questions. *Foods, 8*(4), 121. https://doi.org/10.3390/foods8040121.

D'amico, D. J., Silk, T. M., Wu, J., & Guo, M. (2006). Inactivation of microorganisms in milk and apple cider treated with ultrasound. *Journal of Food Protection, 69*(3), 556–563. https://doi.org/10.4315/0362-028X-69.3.556.

de São José, J. F. B., de Andrade, N. J., Ramos, A. M., Vanetti, M. C. D., Stringheta, P. C., & Chaves, J. B. P. (2014). Decontamination by ultrasound application in fresh fruits and vegetables. *Food Control, 45*(1), 36. https://doi.org/10.1016/j.foodcont.2014.04.015.

Desa, U. (2016). *Transforming our world: The 2030 agenda for sustainable development.* United Nations.

Dong, C., Chen, J., Guan, R., Li, X., & Xin, Y. (2018). Dual-frequency ultrasound combined with alkali pretreatment of corn stalk for enhanced biogas production. *Renewable Energy, 127,* 444–451. https://doi.org/10.1016/j.renene.2018.03.088.

Farooq, R., Shaukat, S. F., Khan, A. K., & Farooq, U. (2008). Ultrasonic induced decomposition of methidathion pesticide. *Journal of Applied Sciences, 8*(1), 140–145.

Fernandes, F. A. N., & Rodrigues, S. (2008). Application of ultrasound and ultrasound-assisted osmotic dehydration in drying of fruits. *Drying Technology, 26,* 1509–1516. https://doi.org/10.1080/07373930802412256.

Flores, E. M. M., Cravotto, G., Bizzi, C. A., Santos, D., & Iop, G. D. (2021). Ultrasound-assisted biomass valorization to industrial interesting products: State-of-the-art, perspectives and challenges. *Ultrasonics Sonochemistry, 72,* 105455. https://doi.org/10.1016/j.ultsonch.2020.105455.

García-Serna, J., Pérez-Barrigón, L., & Cocero, M. J. (2007). New trends for design towards sustainability in chemical engineering: Green engineering. *Chemical Engineering Journal, 133*(1–3), 7–30. https://doi.org/10.1016/j.cej.2007.02.028.

Gavahian, M., Chen, Y.-M., Mousavi Khaneghah, A., Barba, F. J., & Yang, B. B. (2018). In-pack sonication technique for edible emulsions: Understanding the impact of acacia gum and lecithin emulsifiers and ultrasound homogenization on salad dressing emulsions stability. *Food Hydrocolloids, 83,* 79–87. https://doi.org/10.1016/j.foodhyd.2018.04.039.

Gavahian, M., Pallares, N., Al Khawli, F., Ferrer, E., & Barba, F. J. (2020). Recent advances in the application of innovative food processing technologies for mycotoxins and pesticide reduction in foods. *Trends in Food Science and Technology, 106,* 209–218. https://doi.org/10.1016/j.tifs.2020.09.018.

Guoid, J., & Lei, G. (2020). Application of ultrasound to enhance the silt drying process: An experimental study. *PLoS One, 15*(7). https://doi.org/10.1371/journal.pone.0236492, e0236492.

Hashemi, S. M. B., Mousavi Khaneghah, A., Koubaa, M., Barba, F. J., Abedi, E., Niakousari, M., et al. (2018). Extraction of essential oil from Aloysia citriodora Palau leaves using continuous and pulsed ultrasound: Kinetics, antioxidant activity and antimicrobial properties. *Process Biochemistry, 65,* 197–204. https://doi.org/10.1016/j.procbio.2017.10.020.

Hassan, S. S., Ravindran, R., Jaiswal, S., Tiwari, B. K., Williams, G. A., & Jaiswal, A. K. (2020). An evaluation of sonication pretreatment for enhancing saccharification of brewers' spent grain. *Waste Management, 105,* 240–247. https://doi.org/10.1016/j.wasman.2020.02.012.

Haughton, P. N., Lyng, J. G., Morgan, D. J., Cronin, D. A., Noci, F., Fanning, S., et al. (2012). An evaluation of the potential of high-intensity ultrasound for improving the microbial safety of poultry. *Food and Bioprocess Technology, 5*(3), 992–998. https://doi.org/10.1007/s11947-010-0372-y.

Jabbar, S., Abid, M., Wu, T., Hashim, M. M., Saeeduddin, M., Hu, B., et al. (2015). Ultrasound-assisted extraction of bioactive compounds and antioxidants from carrot pomace: A response surface approach. *Journal of Food Processing and Preservation, 39*(6), 1878–1888. https://doi.org/10.1111/jfpp.12425.

Knorr, D., Froehling, A., Jaeger, H., Reineke, K., Schlueter, O., & Schoessler, K. (2011). Emerging technologies in food processing. *Annual Review of Food Science and Technology, 2,* 203–235.

Koubaa, M., Mhemdi, H., Barba, F. J., Roohinejad, S., Greiner, R., & Vorobiev, E. (2016). Oilseed treatment by ultrasounds and microwaves to improve oil yield and quality: An overview. *Food Research International, 85,* 59–66. https://doi.org/10.1016/j.foodres.2016.04.007.

Kowalski, S. J., & Pawłowski, A. (2015). Intensification of apple drying due to ultrasound enhancement. *Journal of Food Engineering, 156,* 1–9. https://doi.org/10.1016/j.jfoodeng.2015.01.023.

Kuna, E., Behling, R., Valange, S., Chatel, G., & Colmenares, J. C. (2017). Sonocatalysis: A potential sustainable pathway for the valorization of lignocellulosic biomass and derivatives. In *Chemistry and chemical technologies in waste valorization* (pp. 1–20). https://doi.org/10.1007/s41061-017-0122-y.

Liu, Y., Li, M., Bai, F., & Bian, K. (2019). Effects of pulsed ultrasound at 20 kHz on the sonochemical degradation of mycotoxins. *World Mycotoxin Journal, 12*(4), 1–10. https://doi.org/10.3920/WMJ2018.2431.

Liu, Y., Li, M., Liu, Y., & Bian, K. (2019). Structures of reaction products and degradation pathways of aflatoxin B 1 by ultrasound treatment. *Toxins, 11*(9), 526. https://doi.org/10.3390/toxins11090526.

Lozowicka, B., Jankowska, M., Hrynko, I., & Kaczynski, P. (2016). Removal of 16 pesticide residues from strawberries by washing with tap and ozone water, ultrasonic cleaning and boiling. *Environmental Monitoring* and Assessment, 188(1), 1–19. https://doi.org/10.1007/s10661-015-4850-6.

Marić, M., Grassino, A. N., Zhu, Z., Barba, F. J., Brnčić, M., & Rimac Brnčić, S. (2018). An overview of the traditional and innovative approaches for pectin extraction from plant food wastes and by-products: Ultrasound-, microwaves-, and enzyme-assisted extraction. Trends in Food Science and Technology, 76, 28–37. https://doi.org/10.1016/j.tifs.2018.03.022.

Martínez-Patiño, J. C., Gómez-Cruz, I., Romero, I., Gullón, B., Ruiz, E., Brnčić, M. B., et al. (2019). Ultrasound-assisted extraction as a first step in a biorefinery strategy for valorisation of extracted olive pomace. Energies, 12(14), 2679. https://doi.org/10.3390/en12142679.

Matouq, M., Al-Anber, Z., Susumu, N., Tagawa, T., & Karapanagioti, H. (2014). The kinetic of dyes degradation resulted from food industry in wastewater using high frequency of ultrasound. Separation and Purification Technology, 135, 42–47. https://doi.org/10.1016/j.seppur.2014.08.002.

Misra, N. N., Koubaa, M., Roohinejad, S., Juliano, P., Alpas, H., Inàcio, R. S., et al. (2017). Landmarks in the historical development of twenty first century food processing technologies. Food Research International, 97, 318–339. https://doi.org/10.1016/j.foodres.2017.05.001.

Misra, N. N., Martynenko, A., Chemat, F., Paniwnyk, L., Barba, F. J., & Jambrak, A. R. (2018). Thermodynamics, transport phenomena, and electrochemistry of external field-assisted nonthermal food technologies. Critical Reviews in Food Science and Nutrition, 58(11), 1832–1863. https://doi.org/10.1080/10408398.2017.1287660.

Montero-Calderon, A., Cortes, C., Zulueta, A., Frigola, A., & Esteve, M. J. (2019). Green solvents and ultrasound-assisted extraction of bioactive orange (Citrus sinensis) peel compounds. Scientific Reports, 9, 16120. https://doi.org/10.1038/s41598-019-52717-1.

Mortazavia, S. M., Sania, A. M., & Mohsenib, S. (2015). Destruction of AFT by ultrasound treatment. Journal of Applied Environmental and Biological Sciences, 4(11S), 198–202.

Niglio, S., Procentese, A., Russo, M. E., Sannia, G., & Marzocchella, A. (2020). Combined pretreatments of coffee silverskin to enhance fermentable sugar yield. Biomass Conversion and Biorefinery, 10(4), 1237–1249. https://doi.org/10.1007/s13399-019-00498-y.

Ong, V. Z., & Wu, T. Y. (2020). An application of ultrasonication in lignocellulosic biomass valorisation into bio-energy and bio-based products. Renewable and Sustainable Energy Reviews, 132. https://doi.org/10.1016/j.rser.2020.109924.

Ozuna, C., Cárcel, J. A., García-Pérez, J. V., & Mulet, A. (2011). Improvement of water transport mechanisms during potato drying by applying ultrasound. Journal of the Science of Food and Agriculture, 91(14), 2511–2517. https://doi.org/10.1002/jsfa.4344.

Picart-Palmade, L., Cunault, C., Chevalier-Lucia, D., Belleville, M.-P., & Marchesseau, S. (2019). Potentialities and limits of some non-thermal technologies to improve sustainability of food processing. Frontiers in Nutrition, 5, 130. https://doi.org/10.3389/fnut.2018.00130.

Rastogi, N. K. (2011). Opportunities and challenges in application of ultrasound in food processing. Critical Reviews in Food Science and Nutrition, 51(8), 705–722. https://doi.org/10.1080/10408391003770583.

Roohi, R., Abedi, E., Hashemi, S. M. B., Marszałek, K., Lorenzo, J. M., & Barba, F. J. (2019). Ultrasound-assisted bleaching: Mathematical and 3D computational fluid dynamics simulation of ultrasound parameters on microbubble formation and cavitation structures. Innovative Food Science and Emerging Technologies, 55, 66–79. https://doi.org/10.1016/j.ifset.2019.05.014.

Roselló-Soto, E., Galanakis, C. M., Brnčić, M., Orlien, V., Trujillo, F. J., Mawson, R., et al. (2015). Clean recovery of antioxidant compounds from plant foods, by-products and algae assisted by ultrasounds processing. Modeling approaches to optimize processing conditions. Trends in Food Science & Technology, 42(2), 134–149. https://doi.org/10.1016/j.tifs.2015.01.002.

Rudik, F. Y., Morgunova, N. L., & Krasnikova, E. S. (2020). Decontamination of grain by ultrasound. IOP Conference Series: Earth and Environmental Science, 421(2), 022022. https://doi.org/10.1088/1755-1315/421/2/022022.

Rutkowska, M., Namieśnik, J., & Konieczka, P. (2017). Ultrasound-assisted extraction. In The application of green solvents in separation processes (pp. 301–324). Elsevier Inc. https://doi.org/10.1016/B978-0-12-805297-6.00010-3.

Sachs, J. D. (2012). From millennium development goals to sustainable development goals. Lancet (London, England), 379(9832), 2206–2211. https://doi.org/10.1016/S0140-6736(12)60685-0.

Sagong, H. G., Lee, S. Y., Chang, P. S., Heu, S., Ryu, S., Choi, Y. J., et al. (2011). Combined effect of ultrasound and organic acids to reduce Escherichia coli O157:H7, Salmonella typhimurium, and Listeria monocytogenes on organic fresh lettuce. International Journal of Food Microbiology, 145(1), 287–292. https://doi.org/10.1016/j.ijfoodmicro.2011.01.010.

Sarkinas, A., Sakalauskiene, K., Raisutis, R., Zeime, J., Salaseviciene, A., Puidaite, E., et al. (2018). Inactivation of some pathogenic bacteria and phytoviruses by ultrasonic treatment. Microbial Pathogenesis, 123, 144–148. https://doi.org/10.1016/j.micpath.2018.07.004.

Schieppati, D., Galli, F., Peyot, M. L., Yargeau, V., Bianchi, C. L., & Boffito, D. C. (2019). An ultrasound-assisted photocatalytic treatment to remove an herbicidal pollutant from wastewaters. Ultrasonics Sonochemistry, 54, 302–310. https://doi.org/10.1016/j.ultsonch.2019.01.027.

Terefe, N. S., Gamage, M., Vilkhu, K., Simons, L., Mawson, R., & Versteeg, C. (2009). The kinetics of inactivation of pectin methylesterase and polygalacturonase in tomato juice by thermosonication. Food Chemistry, 117(1), 20–27. https://doi.org/10.1016/j.foodchem.2009.03.067.

United Nations. (2021). https://sdgs.un.org/es/goals. Accessed March 6, 2021.

Wang, C. K., & Shih, Y. H. (2016). Facilitated ultrasonic irradiation in the degradation of diazinon insecticide. Sustainable Environment Research, 26(3), 110–116. https://doi.org/10.1016/j.serj.2016.04.003.

World Health Organization. (2015). *Health in 2015: From MDGs, millennium development goals to SDGs, sustainable development goals.* World Health Organization.

Wu, J., Gamage, T. V., Vilkhu, K. S., Simons, L. K., & Mawson, R. (2008). Effect of thermosonication on quality improvement of tomato juice. *Innovative Food Science and Emerging Technologies, 9*(2), 186–195. https://doi.org/10.1016/j.ifset.2007.07.007.

Zhang, Y., Xiao, Z., Chen, F., Ge, Y., Wu, J., & Hu, X. (2010). Degradation behavior and products of malathion and chlorpyrifos spiked in apple juice by ultrasonic treatment. *Ultrasonics Sonochemistry, 17*(1), 72–77. https://doi.org/10.1016/j.ultsonch.2009.06.003.

Zhang, Y., Zhang, W., Liao, X., Zhang, J., Hou, Y., Xiao, Z., et al. (2010). Degradation of diazinon in apple juice by ultrasonic treatment. *Ultrasonics Sonochemistry, 17*(4), 662–668. https://doi.org/10.1016/j.ultsonch.2009.11.007.

Zinoviadou, K. G., Galanakis, C. M., Brncic, M., Grimi, N., Boussetta, N., Mota, M. J., et al. (2015). Fruit juice sonication: Implications on food safety and physicochemical and nutritional properties. *Food Research International, 77*(4), 743–752. https://doi.org/10.1016/j.foodres.2015.05.032.

CHAPTER 11

Innovative Technologies in Sustainable Food Production: Cold Plasma Processing

IWONA NIEDŹWIEDŹ • MAGDALENA POLAK-BERECKA

Department of Microbiology, Biotechnology and Human Nutrition, University of Life Sciences in Lublin, Lublin, Poland

11.1 INTRODUCTION

Food consumption is increasing worldwide and, as predicted by the World Health Organization (WHO), the food industry will have to provide food for around 9.7 milliard people in 2050 (WHO, 2015). According to the Food and Agriculture Organization of the United Nations (FAO), around 800 million people in the world are currently undernourished (FAO, 2017). Sustainable Development Goals (SDG) have been set up to eliminate hunger and provide healthy and safe food for all by 2030. The need to produce more food generates environmental risks, such as air and water pollution, soil degradation, deforestation, loss of animal habitats, loss of biodiversity, and depletion of natural resources (Whitmee et al., 2015). Data shows that about a quarter of greenhouse gases are emitted to the atmosphere as a result of food industry activities, and agricultural production consumes about 70% of fresh water in the world. In addition, pesticides used in agriculture, which are intended to contribute to higher yields, are unfortunately often released into the environment, e.g., into water bodies, creating a risk to human and animal health (Lindgren et al., 2018; Vermeulen, Campbell, & Ingram, 2012). For these reasons, there is an increasing emphasis on sustainable environmentally friendly food production. To apply sustainable production, the food industry aims to reduce water and energy consumption and product waste. Attention is also drawn to the problem of the amount of pesticides used in agriculture and food-borne diseases. In addition, food manufacturers are aware that consumers are increasingly looking for minimally processed products. For this reason, scientists are increasingly looking for new technologies to ensure the highest possible quality of food products while maintaining their safety (Misra & Roopesh, 2019).

Cold plasma (CP), considered by the scientific community to be the fourth state of matter, is an ionized gas containing many reactive compounds, excited and basic particles, or UV photons. Its reactive composition has attracted the attention of researchers, who carry out numerous studies to check its potential use in agriculture, medicine, or food production. CP shows a capability of inactivation of microorganisms, which makes it a potential new technique used for decontamination. However, despite numerous studies, the exact mechanism of CP action on microbial cells is still unknown. Its effectiveness is determined by many factors, e.g., the type and power of the generator used, the duration of the process, the type of sample to be treated, and the individual characteristics of the microorganism (Liao et al., 2017; Niedźwiedź, Waśko, Pawłat, & Polak-Berecka, 2019). Studies confirming the effectiveness of cold plasma in the inactivation of biological agents have prompted scientists to carry out more research in food samples or in agriculture. It has thus been proven that cold plasma can extend the shelf life of food by reducing the number of microorganisms (Misra, Keener, Bourke, & Cullen, 2015) and degrading pesticides (Jiang, Zheng, & Wu, 2016) and mycotoxins produced by fungi (Hojnik, Cvelbar, Tavčar-Kalcher, Walsh, & Križaj, 2017). In addition, it has been shown that cold plasma can be used to treat wastewater from food production (Tampieri et al., 2018) and, importantly, can contribute to faster germination of seeds by increasing production efficiency (Jiayun et al., 2014; Ling et al., 2014). Additionally, the CP technology is characterized by low process temperatures, energy efficiency, and use of low amounts of water (Misra & Roopesh, 2019). All this makes it an interesting technology for sustainable food production. On the other hand, the insufficient information about the interaction of reactive compounds

Sustainable Production Technology in Food. https://doi.org/10.1016/B978-0-12-821233-2.00007-1
Copyright © 2021 Elsevier Inc. All rights reserved.

with food ingredients, water, or other sterilized samples makes it difficult to introduce this technology on an industrial scale. There is still little information on the potential toxicity of the resulting by-products during the process and their impact on human and animal health and the environment (Ekezie, Sun, & Cheng, 2017). This aspect therefore needs to be thoroughly investigated in the future. The aim of this review is to characterize the cold plasma technology to provide the latest information on its application in food technology and to describe potential directions for its future use in sustainable production.

11.2 COLD PLASMA TECHNOLOGY

11.2.1 Definition

The concept of plasma was introduced for the first time in 1928 by American physicochemist Irving Langmuir (Niedźwiedź et al., 2019). Plasma, which is considered as the fourth state of matter, is a partially or fully ionized gas consisting of charged particles, reactive compounds, excited and basic state particles, and UV photons (Bruggeman et al., 2016). The main criteria for plasma division include atmospheric pressure (low-pressure, high-pressure plasma), composition of the plasma gas (single component, multi-component plasma), and temperature (low-temperature, high-temperature plasma) (Bourke, Ziuzina, Han, Cullen, & Gilmore, 2017). In the potential application of plasma in sustainable food production, the division into temperature-related types is most important. It is based on the temperature of the electrons (T_e) contained in the plasma stream. The high-temperature plasma is characterized by $T_e = 10^6$–10^8 K, while low-temperature plasma has a temperature of $T_e = 10^4$–10^5 K (Fridman, Chirokov, & Gutsol, 2005). In terms of its thermodynamic balance, low-temperature plasma can be divided into equilibrium plasma (thermal) and non-thermal plasma (non-thermal). Non-thermal plasma, called cold plasma, is used in food sterilization. Thermodynamic imbalance means that electrons have higher energy than other plasma components; hence, the temperature of the process itself is close to room temperature and should not exceed 60°C (Liao et al., 2017). In the context of using cold plasma in food technology, this feature is extremely important, because the process temperature affects the quality of the final food product.

11.2.2 Plasma Source

Cold plasma for technological purposes is produced by means of various electrical discharges, i.e., barrier, microwave, and glow discharge. As a result, most of the energy from the electric field is collected by electrons as a consequence of their collision, while some is transferred to neutral particles. This creates a Te ≥ Tn state characterizing non-thermal plasma. The type of discharge applied affects the composition and abundance of reactive compounds in the resulting plasma. In the food industry and agriculture, dielectric barrier discharges and plasma jets are most commonly used to obtain cold plasma (Bourke et al., 2017). The diagram of a dielectric barrier discharge (DBD) generator, called silent discharges, is based on two electrodes with different potential, in which the dielectric material is usually located on one of the electrodes (rarely on both electrodes). The system is powered by a high voltage generator with a frequency usually around 100 Hz. The plasma formed in the space between the electrodes is usually non-thermal and non-equilibrium. The material to be sterilized is placed directly on the insulator electrode (Miao & Yun, 2012). Due to the atmospheric pressure level, such processes require high levels of energy. The second most commonly used plasma source is the plasma jet generator. In practice, it is most often shaped like a cone and is connected to a current generator of about 30 kHz. Ionized gas (plasma) escapes through a nozzle and is directed to the sample. These generators are simple designs that can easily be modified and adapted to the technology.

11.2.3 Plasma Chemistry

Plasma chemistry is crucial due to the essence of the sterilization mechanism. The type of plasma-generating gas and the type of electrical discharge applied determine the subsequent chemical composition of the plasma stream, and thus influence its sterilization efficiency (Zhang et al., 2019). Free radicals and reactive plasma species are responsible for the modification of the sterilized surface, the oxidation of biomolecules, and the maintenance of products. Active plasma compounds are formed in many reactions as a result of ion neutralization, ion reactions, Penning ionization, or electronic collisions such as dissociation, excitation, and vibration. Depending on the type of gas used, plasma species differ. Reactive oxygen species (ROS) and reactive nitrogen species (RNS) are formed in plasma in which the working gas is oxygen, nitrogen, or a mixture of these gases. ROS formed in plasma can include ozone, hydrogen peroxide, singlet oxygen, hydroperoxyl, hydroxyl radical, alkoxyl, or peroxyl anion. RNS present in plasma are nitrogen oxide, peroxynitrite, nitrogen dioxide radical, alkyl peroxynitrite, or peroxynitric acid (Arjunan, Sharma,

& Ptasinska, 2015). The determination of plasma chemistry is difficult and complex. Reactions leading to the formation of these compounds take place over a wide time range from a nanosecond to even an hour. Nowadays, optical absorption or emission spectroscopy, laser-induced fluorescence, and chemical methods are used to identify plasma components (Misra & Jo, 2017; Zhang et al., 2019).

11.2.4 Antimicrobial Mechanism

The sterilizing effect of cold plasma was confirmed by many authors in their studies (Fernandez, Noriega, & Thompson, 2013; Laroussi, Karakas, & Hynes, 2011; Ziuzina, Patil, Cullen, Keener, & Bourke, 2014). However, the exact mechanism of action is still unclear but depends on many factors (Niedźwiedź et al., 2019). The action of reactive compounds, ionized atoms and particles, high-energy electrons, and UV radiation seems to be crucial in the elimination of undesirable microorganisms. The components contained in cold plasma mainly affect the external structures of microorganisms and DNA causing their damage (Liao et al., 2017; Niedźwiedź et al., 2019). The loss of cell membrane integrity may result from the formation of unsaturated fatty acid peroxides and membrane protein oxidation due to the presence of ROS and RNS. In addition, electrostatic forces created by the accumulation of charged compounds on the outer side of the membrane cause its fractures. Another possible mechanism that can occur after using cold plasma is electroporation. This is the result of production of a pulsating electric field, causing surface structures to break due to the formation or growth of cellular micropores (Liao et al., 2017). In addition to the damage to the external structures of microorganisms that contribute to the leakage of internal cell components, cold plasma, or more precisely UV radiation, can cause changes in the structure of the genetic material. Photons present in the plasma can lead to the formation of nitrogen-based dimers, thus destroying the DNA replication capacity (Beggs, 2002). Due to these properties, low-temperature plasma can be used to eliminate spoilage microorganisms from food products.

11.2.5 Effectiveness of Cold Plasma

Cold plasma can be successfully used as sterilizing agent in sustainable food production, but it must be pointed out that the effectiveness of sterilization is influenced by many factors. The most important are process parameters (reactor type and power, frequency and voltage, chemical composition of the working gas,

process duration), environmental conditions (matrix type, pH, humidity), and the specific properties of the eliminated microorganisms (morphological characteristics, physiological state). The influence of particular parameters on CP efficiency based on the available literature is presented in Table 11.1 (Liao et al., 2017; Niedźwiedź et al., 2019).

TABLE 11.1

Examples of the Influence of Process Conditions on Cold Plasma Sterilization Efficiency.

Parameter Type	Results	References
PROCESSING PARAMETERS		
Gas type	He/O$_2$ ↑ than He	Eto, Ono, Ogino, and Nagatsu (2008)
Treatment time	↑ Longer treatment time	Jayasena et al. (2015)
Flow rate	↑ Higher flow rate	Niemira and Sites (2008)
Input power	↑ Higher input power	Yun et al. (2010)
ENVIRONMENTAL FACTORS		
Humidity	↑ Higher humidity	Hähnel, von Woedtke, and Weltmann (2010)
Acidity	↑ Lower pH	Kayes et al. (2007)
Matrix	Various degrees of inactivation on the surface of lettuce, strawberry, and potato	Fernandez et al. (2013)
PROPERTIES OF MICROORGANISMS		
Physiological state	Spores more resistant than vegetative cells	Tseng, Abramzon, Jackson, and Lin (2012)
Initial inoculum	↑ Less inoculum	Laroussi et al. (2011)
Type of microorganisms	Gram-positive bacteria more resistant than Gram-negative bacteria	Ziuzina et al. (2014)

* ↑ higher inactivation efficiency.

11.3 COLD PLASMA IN FOOD PROCESSING

11.3.1 Food Preservation

Microbiological contamination of food is a major problem for the food industry around the world (López et al., 2019; Sarangapani, Patange, Bourke, Keener, & Cullen, 2018). Consumers are increasingly looking for minimally processed products; however, despite their beneficial health effects, such foodstuffs may be potential carriers of bacterial, parasitic, and viral pathogens (Abadias, Usall, Anguera, Solsona, & Viñas, 2008). The most common microorganisms contaminating food and causing human diseases include bacteria of the genera *Salmonella* sp. *Listeria, Escherichia* (mainly *E. coli*), yeasts mainly of the genus *Saccharomyces* spp., *Rhodothorula* sp. and *Pichia* sp., and molds such as *Penicillium* sp., *Aspergillus* sp., *Alternaria* sp., *Cladosporium* spp., and *Botrytis* sp. (Fernandez et al., 2013; Hygreeva, Pandey, & Radhakrishna, 2014; Olaimat & Holley, 2012; Raybaudi-Massilia, Mosqueda-Melgar, Soliva-Fortuny, & Martín-Belloso, 2009). A huge challenge in ensuring food safety is the branch of the food industry of fresh products such as spinach, tomato lettuce, pepper, and strawberries. Diseases caused by pathogens transmitted via this route are widely reported worldwide. In 2010, 10% of reported foodborne outbreaks in the European Union were caused by pathogens contaminating fresh vegetables and fruits (Van Boxstael et al., 2013). Due to their rich nutrient pool, food products of animal origin, i.e., meat, dairy products, seafood, and fish, also provide an ideal environment for the development of undesirable pathogenic microflora (Martín et al., 2014). Unwanted microflora inhabits various environments in contact with food (industrial equipment, water distribution systems), usually growing as a biofilm rather than as planktonic cells. As a result, their elimination is difficult as they become more resistant to sterilizing agents (Sharma et al., 2014). The currently used methods of food preservation and surface disinfection, i.e., the use of disinfectants, addition of biopreservatives to products, heat treatment, ozone technology, etc., usually give good results, but can have a negative impact on the physicochemical properties (Scholtz, Pazlarova, Souskova, Khun, & Julak, 2015). Therefore, scientists from all over the world are looking for new alternative techniques with high decontamination efficiency and minimal impact on product quality. Research on the use of cold plasma in food technology has started relatively recently, making plasma the first-generation technology, which means that it is at its early stages of development (Bermúdez-Aguirre & Barbosa-Cánovas, 2010). The studies carried out on the

TABLE 11.2
Summary of the Effect of Cold Plasma Processing on the Elimination of Unwanted Microorganisms From Food Products.

Product	Microorganisms	References
FRESH PRODUCT		
Apples	*E. coli* ~3.6 *Salmonella* ~3.7	Niemira and Sites (2008)
Spinach	*E. coli* ~5	Klockow and Keener (2009)
Lettuce	*Aeromonas hydrophila* ~5	Jahid, Han, and Ha (2014)
Red chicory	*L. monocytogenes* ~>4	Trevisani et al. (2017)
FRUIT JUICE		
Orange juice	*S. aureus, E. coli* and *Candida albicans* ~>5	Shi et al. (2011)
Grape juice	*Saccharomyces cerevisiae* ~7.4	Pankaj, Wan, Colonna, and Keener (2017)
ANIMAL ORIGIN PRODUCT		
Pork	*Listeria* ~2.0 *E. coli* ~2.5 *Salmonella* ~2.6	Jayasena et al. (2015)
Beef	*Listeria* ~1.9 *E. coli* ~2.5 *Salmonella* ~2.5	Jayasena et al. (2015)
Chicken breast/poultry	*Salmonella typhimurium* ~1.2	Kim et al. (2013)
Milk	*E. coli Salmonella typhimurium* ~2.4 *Listeria monocytogenes*	Kim et al. (2015)

efficacy of this method in the elimination of undesirable microorganisms from fresh products, juices, and products of animal origin are presented in Table 11.2.

As mentioned in the previous section, the decontamination efficiency of cold plasma in the context of food preservation depends on process parameters that need to be optimized at the stage of the experiment design (Misra, Tiwari, Raghavarao, & Cullen, 2011; Niemira & Sites, 2008). Exposure of the food product for cold plasma decontamination may be direct or indirect. In the case of the direct exposure, the food product, as the name suggests, will be directly exposed to the plasma stream. This approach maximizes the interaction of food ingredients with reactive gas-produced compounds. The other approach consists of indirect exposure where the sample is at a certain distance from the plasma source and only relatively stable reactive species interact with food. This technique reduces the potential negative effects of the direct exposure on product's sensitive tissues (Sarangapani et al., 2018). In addition, an interesting issue that requires careful investigations is the effectiveness of plasma-activated water (PAW) in the process of decontamination of fresh products. In this technique, water is exposed to plasma discharge, which leads to formation of long-lived reactive compounds such as hydrogen peroxide, nitrates, and nitrites. Due to this high amount of RONS (reactive oxygen and nitrogen species) in the composition, PAW is very effective in the elimination of microorganisms with little impact on the physicochemical characteristics of the product. For these reasons, PAW may become an interesting alternative decontamination method in the future (Ma et al., 2015; Sarangapani et al., 2018). Studies on the effectiveness of PAW were carried out by Ma et al. (2015), who dipped strawberries for 5, 10, and 15 min in activated water. They showed that the contamination of *Staphylococcus aureus*-inoculated fruit was reduced at the level of 1.6 to 2.3 log; moreover, after 4 days of storage, the elimination level of the unwanted microorganism increased by 1 log (Ma et al., 2015). Another interesting application for cold plasma in food technology is the in-package cold plasma technology (Misra, Pankaj, et al., 2014; Misra, Patil, et al., 2014; Misra et al., 2015). This technique involves placing the food product in a sealed rigid or flexible package whose space is filled with air or modified gas (Misra et al., 2014). Then, the package with the product is subjected to a strong electric field, resulting in the formation of an ionized gas - plasma. By spreading in the package, the decomposition products, i.e., RONS, eliminate unwanted microorganisms that can potentially spoil food. Over time, the unstable RONS extinguish to form the output gas (Misra et al., 2014). In food technology, cold plasma has found application not only in the decontamination of food products but also in the reduction of the use of pesticides or extension of the shelf life of products (Fig. 11.1). These issues will be addressed in detail in the next section on the use of cold plasma in sustainable food production (López et al., 2019; Sarangapani et al., 2018).

11.3.1.1 Physiochemical quality of food

Besides effective inactivation of microorganisms, food decontamination methods should have a minimum effect on the physicochemical properties of the product

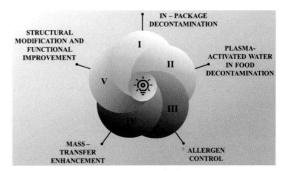

FIG. 11.1 New trends for cold plasma application in food industry.

and should additionally extend its shelf life. The most important food quality parameters can be divided into physical (color, size, structure), chemical (nutritional value, pH), and microbiological (spoilage microorganisms) traits. In cold plasma, reactive compounds, which are mainly responsible for the elimination of unwanted microflora, can also interact with food ingredients (Pankaj, Wan, & Keener, 2018). Color is one of the basic physical parameters of food quality discerned by the consumer first. Available literature data on the effects of CP on the color of fresh products indicate that the intensity of the impact depends on the type of sample, the type of generator used, and other process parameters such as voltage, gas type, and duration of the process. In addition, the storage time of a plasma-treated product is important, since the color of products may change during storage as a result of inactivation of some enzymes and undesirable chemical reactions (Misra et al., 2011; Pankaj et al., 2018). Baier, Ehlbeck, Knorr, Herppich, and Schlüter (2015) observed increased browning in carrots during storage after plasma treatment. On the other hand, a study of cut apples treated with CP for 30 min and then stored for 4 h showed that the darkening of the fruit after the plasma treatment was less intense than in the control sample (Tappi et al., 2014). This effect was explained by inhibition of enzymatic browning. Changes in color brightness after cold plasma application can also be associated with moisture loss (Wang et al., 2012). In the case of fruit juices, a change in color may occur due to changes in pigments caused by polymerization of phenolic compounds (Kovačević et al., 2016). Generally, available literature data suggest that a short CP processing time has a minimal impact on the color of food products (Misra, Pankaj, et al., 2014; Misra, Patil, et al., 2014; Niemira & Sites, 2008). As regards changes in the firmness of the products tested, most of the available studies did not show any structural changes after plasma treatment (Misra, Pankaj, et al., 2014; Misra, Patil, et al., 2014; Tappi et al., 2016; Ziuzina et al., 2016). In turn, in the case of strawberries immersed in PAC, no change in hardness occurred even after 4 days of storage (Ma et al., 2015). Moreover, some authors have indicated that the CP treatment of some product categories such as grains may induce desired changes in the structure, i.e., reduction of hardness, which is associated with a shorter processing time before consumption (Sarangapani et al., 2016; Thirumdas, Saragapani, Ajinkya, Deshmukh, & Annapure, 2016). Another feature of a food product is its chemical properties, i.e., pH or nutrients. Available studies indicate that CP can change the acidity of a product, which is related to the interaction of plasma components with moisture. The degree of acidity change depends on the type of product (solid, liquid) and the buffering capacity or physiological activity of the living tissues in the product (Misra, 2016; Oehmigen et al., 2010). A major challenge in the food industry is the degradation of vitamins contained in food products. Vitamins, which are characterized by low stability (A, C, E), can pose a problem for conventional thermal decontamination methods (Dionísio, Gomes, & Oetterer, 2009). In most experiments assessing the effect of CP on vitamin stability in a product, the degree of degradation of ascorbic acid was checked. The majority of the results obtained did not show a significant decrease in vitamin C (Ramazzina et al., 2015; Song et al., 2015). On the other hand, cut fruit showed a decrease in vitamin C content down to 4% after plasma treatment (Wang et al., 2012). In summary, cold plasma has great potential as a new non-thermal food preservation method. However, the contradictory results concerning the influence of plasma on the physicochemical quality of products necessitate further experiments to investigate the mechanisms of the interaction of plasma components with food ingredients. Appropriate optimization of the process (adapted to the type of product) will minimize the adverse effects on the product and thus allow the commercial application of this technique.

11.3.2 Food Packaging

Besides the elimination of undesirable microorganisms from food products, cold plasma has found application in sterilization and modification of product packaging materials. The materials protect the product from contact with the external environment during transport and distribution, thus fulfilling its basic function of ensuring food safety. Before filling, polymer bottles are usually sterilized chemically with hydrogen peroxide or peracetic acid. Despite their high efficiency, these methods generate large amounts of wastewater, which in turn affects the cost of the product (Misra, Ziuzina, Cullen, & Keener, 2013). Currently, the low-temperature plasma technology is mainly used to modify certain properties of packaging materials, e.g., to reduce the permeability of oxygen or carbon dioxide. The use of CP for sterilization of packaging materials, i.e., plastic bottles, foils, or lids, facilitates rapid elimination of microorganisms. However, when choosing the sterilization parameters, it is important to consider possible changes in the properties of the packaging material. An experiment conducted by Ferrante, Iannace, and Monetta (1999) showed that 200 s of low-temperature plasma action on the polymer reduced its strength by about 20% (Ferrante et al., 1999).

11.4   COLD PLASMA IN SUSTAINABLE FOOD PRODUCTION

Agriculture, food industry, and medicine each day face the huge challenge of meeting society's needs while ensuring an adequate level of safety. Every year, the human population is growing, and by 2050, the global total estimated figure is expected to be around 9.7 milliard people. For the food industry, this means that food production must increase by 50% by 2050 compared to 2012 (Garcia, Osburn, & Jay-Russell, 2020). The negative effects of intensified food production on the environment in the form of lost biodiversity, fresh water shortages, and soil erosion are already being observed worldwide. For this reason, there is a trend towards development of solutions for sustainable safe food production (Bourke, Ziuzina, Boehm, Cullen, & Keener, 2018; Garcia et al., 2020). An important role in providing safe products is played by preservation methods, which are designed to eliminate unwanted microorganisms that can potentially endanger the health of consumers by causing diseases. With its advantages such as low process temperature, short processing time, energy efficiency, water efficiency, and high antimicrobial efficacy with a minimal impact on the food quality and the environment, cold plasma has become the main focus of research as a potential technique for sustainable food production (Misra & Roopesh, 2019). Food production can generally be divided into two stages. The first is primary processing, which includes crop or animal husbandry, harvesting or slaughtering of animals, and stages of preparation of products for sale, e.g., transport. The second stage involves secondary processing, i.e., creation of a new food product for sale (Bourke et al., 2018). Many scientific reports indicate that cold plasma or plasma-activated water can

FIG. 11.2 Cold plasma in sustainable food production.

contribute positively at each stage of food production. The treatments can accelerate seed germination (Jiayun et al., 2014; Ling et al., 2014), effectively decompose pesticides and mycotoxins (Sarangapani et al., 2016; Shi, Ileleji, Stroshine, Keener, & Jensen, 2017; Ten Bosch et al., 2017), and eliminate undesirable microorganisms (Fernandez et al., 2013; Laroussi et al., 2011; Ziuzina et al., 2014). This section presents the possibilities of using cold plasma in sustainable food production (Fig. 11.2.).

11.4.1 Germination and Plant Growth Enhancement

Due to the rapid development of industrialization and urbanization, the population is steadily increasing; hence, the problem of hunger around the world is becoming increasingly serious. Unfortunately, it is anticipated that such natural resources as water may be exhausted. Therefore, it is necessary to obtain the largest possible quantity of crops in a safe and sustainable manner. Traditional methods for crop production, including fertilization and irrigation, lead to environmental imbalance. The cold plasma technology can be used to ensure sustainable food production already at the agricultural level. There are scientific reports that indicate improvement in seed germination after CP application (Jiayun et al., 2014; Ling et al., 2014), control of undesirable microorganisms and insects in the seed (El-Aziz, Mahmoud, & Elaragi, 2014; Selcuk, Oksuz, & Basaran, 2008), or enhanced plant growth (Sera, Spatenka, Sery, Vrchotova, & Hruskova, 2010). Basic research on the use of cold plasma in agriculture included the elimination of microorganisms from seeds. One of the first experiments on this topic was carried out by Selcuk et al. (2008), who studied the effect of cold plasma on the decontamination of the seed surfaces of tomato, beans, soy, oats, and rye contaminated with fungi of the genus *Aspergillus* and *Penicillium*. The

authors obtained a 3-log reduction after 15 min of exposure to cold plasma using SF6 (Selcuk et al., 2008). Subsequent studies confirmed the sterilizing effect of cold plasma (Hashizume et al., 2014; Schnabel et al., 2012). In addition to microbial contamination, insects destroying agricultural material should be taken into account in ensuring seed safety. CP has been shown to lead to increased mortality in the larval and pupil stage of Indian meal moths (*Plodia interpunctella*) (El-Aziz et al., 2014). In addition to protecting the seed from biological hazards, cold plasma also contributes to improving seed germination by increasing the speed and efficiency of the process (Jiayun et al., 2014; Ling et al., 2014). One of the most likely causes of this phenomenon is a change in the wetting properties of the surfaces and a decrease in the apparent contact angle (Bormashenko et al., 2015). In fact, some studies indicate that plants whose seeds have been treated with cold plasma may have different growth parameters (root and shoot length, dry matter) or different nutrients such as phenols (Sera et al., 2010). The effect of CP on the growth of plant cells results from a change in their antioxidant activity, which is the result of ROS generated in the plasma, and this in turn causes the production of growth factors in plants (Dobrin, Magureanu, Mandache, & Ionita, 2015). Plant diseases caused by fungal pathogens are another problem contributing to reduction of yields. Besides elimination of undesirable microorganisms, as confirmed by numerous studies (Hashizume et al., 2014; Schnabel et al., 2012; Selcuk et al., 2008), cold plasma has the potential to increase the resistance of exposed plants to fungal pathogens (Siddique, Hardy, & Bayliss, 2018). Plants (wheat, lupine, maize) whose seeds were exposed to CP had a lower incidence of diseases caused by *Fusarium* spp. and *Ustilago maydis* (Filatova et al., 2016) In sustainable food production, an increase in germination efficiency is beneficial for the malting or brewing industry, as technologies that accelerate the germination process can significantly reduce energy consumption. In addition, increased resistance to fungal pathogens can reduce the use of chemical plant protection products, which have a negative impact on both human health and the environment.

11.4.2 Degradation of Mycotoxins and Pesticides

According to FAO data, as many as ¼ of the world's crops are contaminated with fungal mycotoxins during plant growth or during harvest storage. Aflatoxin, ochratoxin, and fumonisins are considered to be the most toxic to human and animal health.

These compounds are characterized by resistance to high temperatures up to 306°C (aflatoxin), which makes elimination thereof extremely difficult. The action of cold plasma has been demonstrated to allow full degradation and reduction of the toxicity of some mycotoxins, i.e., aflatoxin, deoxynivalenol, and nivalenol (Hojnik et al., 2017). Various studies indicate that the reduction of toxicity is probably associated with changes in the structure of these compounds induced by the exposure to cold plasma (Wang et al., 2015). Most studies have tested plasma efficiency on synthetic compounds (Hojnik et al., 2017; Wang et al., 2015). In turn, available scientific reports indicate that also mycotoxins in food products can be degraded under the influence of plasma; however, this depends on the type of food and mycotoxin and the level of moisture in the samples (Shi et al., 2017; Ten Bosch et al., 2017).

Besides mycotoxins, which are toxic secondary metabolites of fungi, pesticides that are widely used in the agricultural sector to increase yields and prevent plant diseases are equally dangerous to the environment and humans. Studies conducted so far indicate that, as a result of the activity of reactive compounds, cold plasma can eliminate residues of organochlorine and organophosphate pesticides in various types of products (solid, liquid) (Jiang et al., 2016). For example, Dorraki, Mahdavi, Ghomi, and Ghasempour (2016) achieved 88% elimination of diazinon from the surface of cucumbers, compared to the initial concentration, after 10 min of exposure to cold plasma (Dorraki et al., 2016). On the surface of apples, the degradation of paroxone after plasma-air treatment was as high as 95.5% (Heo et al., 2014). In addition to the elimination of pesticides from the surface of fresh fruit and vegetables, cold plasma can be used to reduce these compounds from sewage. Sarangapani et al. (2016) used an air plasma jet as a working gas generated by a DBD reactor to degrade pesticides from water. After 8 min of the process, significant reduction in dichlorvos, malathion, and endosulfan was achieved (Sarangapani et al., 2016). Additionally, the analysis of samples showed that the degradation products were less toxic (Misra, Pankaj, et al., 2014; Misra, Patil, et al., 2014; Sarangapani et al., 2016). However, complete elimination of these compounds from the environment may be hindered, as some insects show resistance to pesticides used in plant protection, forcing farmers to apply higher doses. Consequently, bigger amounts of pesticides enter the soil and wastewater, posing a serious risk to the environment (Bourke et al., 2018).

11.4.3 Soil Remediation

The use of pesticides and the lack of sustainable plant cultivation and animal husbandry have led to the erosion of many soils. Decontamination of contaminated land and reclamation is a difficult and expensive process. Available methods such as phytostabilization or electrokinetic reclamation are not fully satisfactory due to their costs. Therefore, scientists have focused their attention on new technologies, including cold plasma. Available studies indicate that the CP technology can effectively influence the degradation of soil contaminants (Aggelopoulos et al., 2016; Lou, Lu, Li, Wang, & Wu, 2012). Lou et al. (2012) investigated the level of remediation of soil contaminated by chloramphenicol exposed to cold plasma. They achieved 81% reduction in soil contamination, compared to the control, after 20 min of the process and at 10% soil moisture. In addition, they found that the efficiency of this process depends on the voltage, type of gas, and soil moisture Lou et al. (2012). Another experiment indicated effective glyphosate soil degradation (about 94%) with no negative effect on seed germination and seedling growth (Aggelopoulos et al., 2016). On the other hand, the use of cold plasma to cleanse the soil may result in a reduction in the number of microorganisms inhabiting it. This is beneficial for the elimination of soil microorganisms that cause plant diseases. On the other hand, there are drawbacks of such application of CP, e.g., elimination of beneficial nitrogen bacteria (Stryczewska, Ebihara, Takayama, Gyoutoku, & Tachibana, 2005).

11.4.4 Wastewater Treatment

The food industry uses huge amounts of water, which results in generation of wastewater. Water with chlorine is most often used for cleaning fresh food or disinfecting rooms and equipment. Chlorine is a strong disinfectant with a harmful effect on the environment and health, due to the possible presence of its carcinogenic derivatives. In addition, given the concerns about depletion of fresh water resources, solutions should be sought to reduce the use of natural water resources and recycle wastewater. Cold plasma has been used for many years alone or in combination with other technologies to remove water pollution (Ölmez & Kretzschmar, 2009). Reactive compounds contained in the plasma stream are responsible for the degradation of harmful compounds. The effectiveness of new generators in water decontamination is usually checked using redox dyes (Misra et al., 2015). Recently, a prototype reactor has been constructed, which proved to be effective in the degradation of phenol, rhodamine B, and metallochlorine (Tampieri et al., 2018). As mentioned in the subsection on pesticides, the elimination of e.g., dichlophosphorus or endosulfan (Sarangapani et al., 2016) from water samples was demonstrated. Important research on the effective decomposition of organic matter in washing water and wastewater from the food industry was conducted by Sarangapani et al. (2017), who tested the effectiveness of plasma generated by dielectric discharges with air as a working gas in the decomposition of fats from milk and meat. They noted substantial degradation of these organic compounds from sewage produced by the dairy and meat industries. Therefore, it can be concluded from the available studies that cold plasma can be used as a potential method to eliminate pesticide residues as well as other organic compounds contained in wastewater from the food industry.

11.4.5 Extension of Shelf Life

A long shelf life of food products is desirable on the globalized market, where products are often exported to other countries. In connection with the problem of hunger in the world, there is social opposition against wasting food. Fresh and meat products that easily become microbiologically contaminated are a particular challenge. An FAO report presented in 2011 revealed that the waste of some fresh products from harvest to sale could be as high as 30% of the total quantity (Pan, Cheng, & Sun, 2019). Due to its widely proven inactivation effect on different groups of microorganisms (Fernandez et al., 2013; Laroussi et al., 2011; Ziuzina et al., 2014), and the ability to modify storage packaging, cold plasma is becoming an interesting alternative technique to extend the shelf-life of products (Misra et al., 2015). A study carried out by Tappi et al. (2016) on freshly cut melon is an example confirming the delayed spoilage of products after exposure to CP. The fruit was exposed to cold plasma obtained via dielectric discharge and then it was stored and subjected to microbiological analysis and qualitative parameters change. The authors found a significant increase in the shelf life of fruits, because the growth of mesophilic and psychophilic microorganisms was delayed (Tappi et al., 2016). More examples of the use of cold plasma in the decontamination of food products including its effects on their physicochemical properties are discussed in section 3.

11.5 FUTURE PERSPECTIVE

Before the industrial-scale application of cold plasma as a new non-thermal technique for sustainable food production, it is necessary to examine its safety.

It is important to check whether it generates toxic by-products as a result of RONS interactions with plasma-treated samples. Unfortunately, this issue is still not fully elucidated and there are relatively few scientific reports investigating this aspect (Ekezie et al., 2017). Some of them checked the influence of feeding rats (males and females) with edible foil previously subjected to plasma treatment on their organism. After 2 weeks, the authors of this experiment noticed small changes in the blood results of the examined animals, but they did not exceed the norm. Moreover, the authors did not observe any changes in the appearance of the liver (Han, Suh, Hong, Kim, & Min, 2016). In turn, other experiments indicated that some CP-treated samples showed cytotoxic activity (Keidar, 2015). It is therefore necessary to establish the persistence of potentially formed cytotoxic substances, their concentration, and the dose that can be considered safe for human consumption. Nitrates and nitrites generated in solutions subjected to the plasma process should be considered as well. In water subjected to CP sterilization, the concentrations of these compounds may even exceed the WHO limits. For this reason, the process should be monitored and appropriate process conditions should be selected to make it environmentally safe (Bourke et al., 2018). In addition to the production of harmful products, cold plasma can cause negative chemical changes in the product composition depending on the type of product (Thirumdas et al., 2016). Therefore, it is possible that the cold plasma technology will be limited to specific product types or areas where there is no risk to health and the environment. In sustainable food production, it is essential that every technology used is energy-efficient, uses small amounts of water, and is relatively cheap. For CPs generated under atmospheric pressure, as reported by Misra and Roopesh (2019), the energy consumption is low, as the maximum power draw at 80 kV does not exceed 130 W (Misra & Roopesh, 2019). Similarly, small amounts of water are used, and the method itself can be called environmentally friendly in this respect. As far as the costs of this technology are concerned, the cost of the equipment and the type of gases used must be taken into account. In economic terms, the use of air as the working gas is the best solution to generate the lowest costs. However, its effectiveness in biological inactivation may be lower than that of the helium/oxygen gas mixture. On the other hand, the use of a working gas that does not contain oxygen in food processing can be useful for products that are sensitive to oxidation. Therefore, it will be difficult to harmonize the process conditions and it will be necessary to determine them individually for certain groups of compounds (Kodama, Thawatchaipracha, & Sekiguchi, 2014).

Finally, every new technology strives to move from pre-laboratory research to commercial applications. In legal terms, international agencies are responsible for bringing new technologies to market. Although this may vary from country to country, each of them has the task of careful verification that the new technology does not pose a threat to human safety and the environment. In addition, they must assess its costs and possible economic benefits. In the United States, three agencies are responsible for assessment: Environmental Protection Agency (EPA), Food and Drug Administration (FDA), and United States Department of Agriculture (USDA). In Europe, the conditions for the approval of new technologies are established in MEMO-15-5875, and the European Food Safety Authority (EFSA) is responsible for control.

11.6 SUMMARY

This review was intended to provide the latest scientific information on the potential use of cold plasma in sustainable food production. The technique has great potential in this respect, as it is energy efficient, requires little water, and is highly effective in inactivating different groups of microorganisms. This new innovative non-thermal sterilization technology can be used for decontamination of food products, degradation of mycotoxins and pesticides, modification of food packaging materials, decontamination of aqueous wastewater, acceleration of germination processes, and modification of the properties of food. However, although many studies confirm the effectiveness of cold plasma, not all interactions of plasma active ingredients with food components are still clear. There is little information on the risk of toxic by-products and their impact on health and the environment. Further research is needed to address the safety concerns raised by this new technology in order to transfer laboratory research results to commercial industrial applications. This will increase the chance of future approval for commercial introduction of this non-thermal technology from international organizations.

REFERENCES

Abadias, M., Usall, J., Anguera, M., Solsona, C., & Viñas, I. (2008). Microbiological quality of fresh, minimally-processed fruit and vegetables, and sprouts from retail establishments. *International Journal of Food Microbiology*, 123(1–2), 121–129.

Aggelopoulos, C. A., Gkelios, A., Klapa, M. I., Kaltsonoudis, C., Svarnas, P., & Tsakiroglou, C. D. (2016). Parametric analysis of the operation of a non-thermal plasma reactor for the remediation of NAPL-polluted soils. *Chemical Engineering Journal*, 301, 353–361.

Arjunan, K. P., Sharma, V. K., & Ptasinska, S. (2015). Effects of atmospheric pressure plasmas on isolated and cellular DNA—A review. *International Journal of Molecular Sciences*, 16(2), 2971–3016.

Baier, M., Ehlbeck, J., Knorr, D., Herppich, W. B., & Schlüter, O. (2015). Impact of plasma processed air (PPA) on quality parameters of fresh produce. *Postharvest Biology and Technology*, 100, 120–126.

Beggs, C. B. (2002). A quantitative method for evaluating the photoreactivation of ultraviolet damaged microorganisms. *Photochemical & Photobiological Sciences*, 1, 431–437.

Bermúdez-Aguirre, D., & Barbosa-Cánovas, G. V. (2010). Recent advances in emerging nonthermal technologies. In *Food engineering interfaces* (pp. 285–323). New York, NY: Springer.

Bormashenko, E., Shapira, Y., Grynyov, R., Whyman, G., Bormashenko, Y., & Drori, E. (2015). Interaction of cold radiofrequency plasma with seeds of beans (Phaseolus vulgaris). *Journal of Experimental Botany*, 66(13), 4013–4021.

Bourke, P., Ziuzina, D., Boehm, D., Cullen, P. J., & Keener, K. (2018). The potential of cold plasma for safe and sustainable food production. *Trends in Biotechnology*, 36(6), 615–626.

Bourke, P., Ziuzina, D., Han, L., Cullen, P. J., & Gilmore, B. F. (2017). Microbiological interactions with cold plasma. *Journal of Applied Microbiology*, 123(2), 308–324.

Bruggeman, P. J, Kushner, M. J., Locke, B. R., Gardeniers, G. E., Graham, W. G., Graves, D. B., et al..... (2016). Plasma–liquid interactions: A review and roadmap. *Plasma Sources Science and Technology*, 25, 053002.

Dionísio, A. P., Gomes, R. T., & Oetterer, M. (2009). Ionizing radiation effects on food vitamins: A review. *Brazilian Archives of Biology and Technology*, 52(5), 1267–1278.

Dobrin, D., Magureanu, M., Mandache, N. B., & Ionita, M. D. (2015). The effect of non-thermal plasma treatment on wheat germination and early growth. *Innovative Food Science & Emerging Technologies*, 29, 255–260.

Dorraki, N., Mahdavi, V., Ghomi, H., & Ghasempour, A. (2016). Elimination of diazinon insecticide from cucumber surface by atmospheric pressure air-dielectric barrier discharge plasma. *Biointerphases*, 11(4), 041007.

Ekezie, F. G. C., Sun, D. W., & Cheng, J. H. (2017). A review on recent advances in cold plasma technology for the food industry: Current applications and future trends. *Trends in Food Science & Technology*, 69, 46–58.

El-Aziz, M. F. A., Mahmoud, E. A., & Elaragi, G. M. (2014). Non thermal plasma for control of the Indian meal moth, Plodia interpunctella (Lepidoptera: Pyralidae). *Journal of Stored Products Research*, 59, 215–221.

Eto, H., Ono, Y., Ogino, A., & Nagatsu, M. (2008). Low-temperature sterilization of wrapped materials using flexible sheet-type dielectric barrier discharge. *Applied Physics Letters*, 93(22), 221502.

FAO. (2017). *The future of food and agriculture—Trends and challenges*. Rome: Food and Agriculture Organization of the United Nations. Available online at: http://www.fao.org/3/a-i6583e.pdf.

Fernandez, A., Noriega, E., & Thompson, A. (2013). Inactivation of *Salmonella enterica* serovar *Typhimurium* on fresh produce by cold atmospheric gas plasma technology. *Food Microbiology, 33*(1), 24–29.

Ferrante, D., Iannace, S., & Monetta, T. (1999). Mechanical strength of cold plasma treated PET fibers. *Journal of Materials Science, 34*(1), 175–179.

Filatova, I., Azharonok, V., Lyushkevich, V., Zhukovsky, A. G., Mildažienė, V., Paužaitė, G., et al. (2016). The effect of pre-sowing plasma seeds treatment on germination, plants resistance to pathogens and crop capacity. In *IWOPA 2016: 1st international workshop on plasma agriculture, May 15th–20th 2016, AJ Drexel plasma institute [electronic resource]: Scientific program*. Camden, New Jersey: AJ Drexel Plasma Institute, 2016.

Fridman, A., Chirokov, A., & Gutsol, A. (2005). Non-thermal atmospheric pressure discharges. *Journal of Physics D: Applied Physics, 38*(2), R1.

Garcia, S. N., Osburn, B. I., & Jay-Russell, M. T. (2020). One health for food safety, food security, and sustainable food production. *Frontiers in Sustainable Food Systems, 4*, 1.

Hähnel, M., von Woedtke, T., & Weltmann, K. D. (2010). Influence of the air humidity on the reduction of Bacillus spores in a defined environment at atmospheric pressure using a dielectric barrier surface discharge. *Plasma Processes and Polymers, 7*(3-4), 244–249.

Han, S. H., Suh, H. J., Hong, K. B., Kim, S. Y., & Min, S. C. (2016). Oral toxicity of cold plasma-treated edible films for food coating. *Journal of Food Science, 81*(12), T3052–T3057.

Hashizume, H., Ohta, T., Takeda, K., Ishikawa, K., Hori, M., & Ito, M. (2014). Quantitative clarification of inactivation mechanism of Penicillium digitatum spores treated with neutral oxygen radicals. *Japanese Journal of Applied Physics, 54*(1S), 01AG05.

Heo, N. S., Lee, M. K., Kim, G. W., Lee, S. J., Park, J. Y., & Park, T. J. (2014). Microbial inactivation and pesticide removal by remote exposure of atmospheric air plasma in confined environments. *Journal of Bioscience and Bioengineering, 117*(1), 81–85.

Hojnik, N., Cvelbar, U., Tavčar-Kalcher, G., Walsh, J. L., & Križaj, I. (2017). Mycotoxin decontamination of food: Cold atmospheric pressure plasma versus "classic" decontamination. *Toxins, 9*(5), 151.

Hygreeva, D., Pandey, M. C., & Radhakrishna, K. (2014). Potential applications of plant based derivatives as fat replacers, antioxidants and antimicrobials in fresh and processed meat products. *Meat Science, 98*(1), 47–57.

Jahid, I. K., Han, N., & Ha, S. D. (2014). Inactivation kinetics of cold oxygen plasma depend on incubation conditions of Aeromonas hydrophila biofilm on lettuce. *Food Research International, 55*, 181–189.

Jayasena, D. D., Kim, H. J., Yong, H. I., Park, S., Kim, K., Choe, W., et al. (2015). Flexible thin-layer dielectric barrier discharge plasma treatment of pork butt and beef loin: Effects on pathogen inactivation and meat-quality attributes. *Food Microbiology, 46*, 51–57.

Jiang, B., Zheng, J., & Wu, M. (2016). Nonthermal plasma for effluent and waste treatment. In *Cold plasma in food and agriculture* (pp. 309–342). Academic Press.

Jiayun, T., Rui, H. E., Xiaoli, Z., Ruoting, Z., Weiwen, C., & Size, Y. (2014). Effects of atmospheric pressure air plasma pretreatment on the seed germination and early growth of Andrographis paniculata. *Plasma Science and Technology, 16*(3), 260.

Kayes, M. M., Critzer, F. J., Kelly-Wintenberg, K., Roth, J. R., Montie, T. C., & Golden, D. A. (2007). Inactivation of foodborne pathogens using a one atmosphere uniform glow discharge plasma. *Foodborne Pathogens and Disease, 4*(1), 50–59.

Keidar, M. (2015). Plasma for cancer treatment. *Plasma Sources Science and Technology, 24*(3), 033001.

Kim, H. J., Yong, H. I., Park, S., Kim, K., Bae, Y. S., Choe, W., et al. (2013). Effect of inactivating *Salmonella Typhimurium* in raw chicken breast and pork loin using an atmospheric pressure plasma jet. *Journal of Animal Science and Technology, 55*(6), 545–549.

Kim, H. J., Yong, H. I., Park, S., Kim, K., Choe, W., & Jo, C. (2015). Microbial safety and quality attributes of milk following treatment with atmospheric pressure encapsulated dielectric barrier discharge plasma. *Food Control, 47*, 451–456.

Klockow, P. A., & Keener, K. M. (2009). Safety and quality assessment of packaged spinach treated with a novel ozone-generation system. *LWT- Food Science and Technology, 42*(6), 1047–1053.

Kodama, S., Thawatchaipracha, B., & Sekiguchi, H. (2014). Enhancement of essential oil extraction for steam distillation by DBD surface treatment. *Plasma Processes and Polymers, 11*(2), 126–132.

Kovačević, D. B., Putnik, P., Dragović-Uzelac, V., Pedisić, S., Jambrak, A. R., & Herceg, Z. (2016). Effects of cold atmospheric gas phase plasma on anthocyanins and color in pomegranate juice. *Food Chemistry, 190*, 317–323.

Laroussi, M., Karakas, E., & Hynes, W. (2011). Influence of cell type, initial concentration, and medium on the inactivation efficiency of low-temperature plasma. *IEEE Transactions on Plasma Science, 39*(11), 2960–2961.

Liao, X., Liu, D., Xiang, Q., Ahn, J., Chen, S., Ye, X., et al. (2017). Inactivation mechanisms of non-thermal plasma on microbes: A review. *Food Control, 75*, 83–91.

Lindgren, E., Harris, F., Dangour, A. D., Gasparatos, A., Hiramatsu, M., Javadi, F., et al. (2018). Sustainable food systems—A health perspective. *Sustainability Science, 13*(6), 1505–1517.

Ling, L., Jiafeng, J., Jiangang, L., Minchong, S., Xin, H., Hanliang, S., et al. (2014). Effects of cold plasma treatment on seed germination and seedling growth of soybean. *Scientific Reports, 4*, 5859.

López, M., Calvo, T., Prieto, M., Múgica-Vidal, R., Muro-Fraguas, I., Alba-Elías, F., et al. (2019). A review on non-thermal atmospheric plasma for food preservation: Mode of action, determinants of effectiveness and applications. *Frontiers in Microbiology, 10*, 622.

Lou, J., Lu, N., Li, J., Wang, T., & Wu, Y. (2012). Remediation of chloramphenicol-contaminated soil by atmospheric pressure dielectric barrier discharge. *Chemical Engineering Journal, 180*, 99–105.

Ma, R., Wang, G., Tian, Y., Wang, K., Zhang, J., & Fang, J. (2015). Non-thermal plasma-activated water inactivation of food-borne pathogen on fresh produce. *Journal of Hazardous Materials, 300*, 643–651.

Martín, B., Perich, A., Gómez, D., Yangüela, J., Rodríguez, A., Garriga, M., et al. (2014). Diversity and distribution of Listeria monocytogenes in meat processing plants. *Food Microbiology, 44*, 119–127.

Miao, H., & Yun, G. (2012). The effect of air plasma on sterilization of *Escherichia coli* in dielectric barrier discharge. *Plasma Science and Technology, 14*(8), 735.

Misra, N. N. (2016). Quality of cold plasma treated plant foods. In *Cold plasma in food and agriculture* (pp. 253–271). Academic Press.

Misra, N. N., & Jo, C. (2017). Applications of cold plasma technology for microbiological safety in meat industry. *Trends in Food Science & Technology, 64*, 74–86.

Misra, N. N., Keener, K. M., Bourke, P., & Cullen, P. J. (2015). Generation of in-package cold plasma and efficacy assessment using methylene blue. *Plasma Chemistry and Plasma Processing, 35*(6), 1043–1056.

Misra, N. N., Pankaj, S. K., Walsh, T., O'Regan, F., Bourke, P., & Cullen, P. J. (2014). In-package nonthermal plasma degradation of pesticides on fresh produce. *Journal of Hazardous Materials, 271*, 33–40.

Misra, N. N., Patil, S., Moiseev, T., Bourke, P., Mosnier, J. P., Keener, K. M., et al. (2014). In-package atmospheric pressure cold plasma treatment of strawberries. *Journal of Food Engineering, 125*, 131–138.

Misra, N. N., & Roopesh, M. S. (2019). Cold plasma for sustainable food production and processing. In *Green food processing techniques: Preservation, transformation and extraction* (p. 431). Academic Press.

Misra, N. N., Tiwari, B. K., Raghavarao, K. S. M. S., & Cullen, P. J. (2011). Nonthermal plasma inactivation of food-borne pathogens. *Food Engineering Reviews, 3*(3–4), 159–170.

Misra, N. N., Ziuzina, D., Cullen, P. J., & Keener, K. M. (2013). Characterization of a novel atmospheric air cold plasma system for treatment of packaged biomaterials. *Transactions of the ASABE, 56*, 1011–1016.

Niedźwiedź, I., Waśko, A., Pawłat, J., & Polak-Berecka, M. (2019). The state of research on antimicrobial activity of cold plasma. *Polish Journal of Microbiology, 68*(2), 153–164.

Niemira, B. A., & Sites, J. (2008). Cold plasma inactivates Salmonella Stanley and Escherichia coli O157: H7 inoculated on golden delicious apples. *Journal of Food Protection, 71*(7), 1357–1365.

Oehmigen, K., Hähnel, M., Brandenburg, R., Wilke, C., Weltmann, K. D., & Von Woedtke, T. (2010). The role of acidification for antimicrobial activity of atmospheric pressure plasma in liquids. *Plasma Processes and Polymers, 7*(3–4), 250–257.

Olaimat, A. N., & Holley, R. A. (2012). Factors influencing the microbial safety of fresh produce: A review. *Food Microbiology, 32*(1), 1–19.

Ölmez, H., & Kretzschmar, U. (2009). Potential alternative disinfection methods for organic fresh-cut industry for minimizing water consumption and environmental impact. *LWT- Food Science and Technology, 42*(3), 686–693.

Pan, Y., Cheng, J. H., & Sun, D. W. (2019). Cold plasma-mediated treatments for shelf life extension of fresh produce: A review of recent research developments. *Comprehensive Reviews in Food Science and Food Safety, 18*(5), 1312–1326.

Pankaj, S. K., Wan, Z., Colonna, W., & Keener, K. M. (2017). Effect of high voltage atmospheric cold plasma on white grape juice quality. *Journal of the Science of Food and Agriculture, 97*(12), 4016–4021.

Pankaj, S. K., Wan, Z., & Keener, K. M. (2018). Effects of cold plasma on food quality: A review. *Foods, 7*(1), 4.

Ramazzina, I., Berardinelli, A., Rizzi, F., Tappi, S., Ragni, L., Sacchetti, G., et al. (2015). Effect of cold plasma treatment on physico-chemical parameters and antioxidant activity of minimally processed kiwifruit. *Postharvest Biology and Technology, 107*, 55–65.

Raybaudi-Massilia, R. M., Mosqueda-Melgar, J., Soliva-Fortuny, R., & Martín-Belloso, O. (2009). Control of pathogenic and spoilage microorganisms in fresh-cut fruits and fruit juices by traditional and alternative natural antimicrobials. *Comprehensive Reviews in Food Science and Food Safety, 8*(3), 157–180.

Sarangapani, C., Devi, R. Y., Thirumdas, R., Trimukhe, A. M., Deshmukh, R. R., & Annapure, U. S. (2017). Physico-chemical properties of low-pressure plasma treated black gram. *LWT- Food Science and Technology, 79*, 102–110.

Sarangapani, C., Misra, N. N., Milosavljevic, V., Bourke, P., O'Regan, F., & Cullen, P. J. (2016). Pesticide degradation in water using atmospheric air cold plasma. *Journal of Water Process Engineering, 9*, 225–232.

Sarangapani, C., Patange, A., Bourke, P., Keener, K., & Cullen, P. J. (2018). Recent advances in the application of cold plasma technology in foods. *Annual Review of Food Science and Technology, 9*, 609–629.

Schnabel, U., Niquet, R., Krohmann, U., Winter, J., Schlüter, O., Weltmann, K. D., et al. (2012). Decontamination of microbiologically contaminated specimen by direct and indirect plasma treatment. *Plasma Processes and Polymers, 9*(6), 569–575.

Scholtz, V., Pazlarova, J., Souskova, H., Khun, J., & Julak, J. (2015). Nonthermal plasma—A tool for decontamination and disinfection. *Biotechnology Advances, 33*(6), 1108–1119.

Selcuk, M., Oksuz, L., & Basaran, P. (2008). Decontamination of grains and legumes infected with Aspergillus spp. and Penicillum spp. by cold plasma treatment. *Bioresource Technology, 99*(11), 5104–5109.

Sera, B., Spatenka, P., Sery, M., Vrchotova, N., & Hruskova, I. (2010). Influence of plasma treatment on wheat and oat germination and early growth. *IEEE Transactions on Plasma Science, 38*(10), 2963–2968.

Sharma, G., Rao, S., Bansal, A., Dang, S., Gupta, S., & Gabrani, R. (2014). *Pseudomonas aeruginosa* biofilm: Potential therapeutic targets. *Biologicals, 42*(1), 1–7.

Shi, H., Ileleji, K., Stroshine, R. L., Keener, K., & Jensen, J. L. (2017). Reduction of aflatoxin in corn by high voltage atmospheric cold plasma. *Food and Bioprocess Technology, 10*(6), 1042–1052.

Shi, X. M., Zhang, G. J., Wu, X. L., Li, Y. X., Ma, Y., & Shao, X. J. (2011). Effect of low-temperature plasma on microorganism inactivation and quality of freshly squeezed orange juice. *IEEE Transactions on Plasma Science, 39*(7), 1591–1597.

Siddique, S. S., Hardy, G. S. J., & Bayliss, K. L. (2018). Cold plasma: A potential new method to manage postharvest diseases caused by fungal plant pathogens. *Plant Pathology, 67*, 1011–1021.

Song, A. Y., Oh, Y. J., Kim, J. E., Song, K. B., Oh, D. H., & Min, S. C. (2015). Cold plasma treatment for microbial safety and preservation of fresh lettuce. *Food Science and Biotechnology, 24*(5), 1717–1724.

Stryczewska, H. D., Ebihara, K., Takayama, M., Gyoutoku, Y., & Tachibana, M. (2005). Non-thermal plasma-based technology for soil treatment. *Plasma Processes and Polymers, 2*, 238–245.

Tampieri, F., Giardina, A., Bosi, F. J., Pavanello, A., Marotta, E., Zaniol, B., et al. (2018). Removal of persistent organic pollutants from water using a newly developed atmospheric plasma reactor. *Plasma Processes and Polymers, 15*(6), 1700207.

Tappi, S., Berardinelli, A., Ragni, L., Dalla Rosa, M., Guarnieri, A., & Rocculi, P. (2014). Atmospheric gas plasma treatment of fresh-cut apples. *Innovative Food Science & Emerging Technologies, 21*, 114–122.

Tappi, S., Gozzi, G., Vannini, L., Berardinelli, A., Romani, S., Ragni, L., et al. (2016). Cold plasma treatment for fresh-cut melon stabilization. *Innovative Food Science & Emerging Technologies, 33*, 225–233.

Ten Bosch, L., Pfohl, K., Avramidis, G., Wieneke, S., Viöl, W., & Karlovsky, P. (2017). Plasma-based degradation of mycotoxins produced by Fusarium, Aspergillus and Alternaria species. *Toxins, 9*(3), 97.

Thirumdas, R., Saragapani, C., Ajinkya, M. T., Deshmukh, R. R., & Annapure, U. S. (2016). Influence of low pressure cold plasma on cooking and textural properties of brown rice. *Innovative Food Science & Emerging Technologies, 37*, 53–60.

Trevisani, M., Berardinelli, A., Cevoli, C., Cecchini, M., Ragni, L., & Pasquali, F. (2017). Effects of sanitizing treatments with atmospheric cold plasma, SDS and lactic acid on verotoxin-producing *Escherichia coli* and *Listeria monocytogenes* in red chicory (radicchio). *Food Control, 78*, 138–143.

Tseng, S., Abramzon, N., Jackson, J. O., & Lin, W. J. (2012). Gas discharge plasmas are effective in inactivating *Bacillus* and *Clostridium* spores. *Applied Microbiology and Biotechnology, 93*(6), 2563–2570.

Van Boxstael, S., Habib, I., Jacxsens, L., De Vocht, M., Baert, L., Van de Perre, E., et al. (2013). Food safety issues in fresh produce: Bacterial pathogens, viruses and pesticide residues indicated as major concerns by stakeholders in the fresh produce chain. *Food Control, 32*(1), 190–197.

Wang, S. Q., Huang, G. Q., Li, Y. P., Xiao, J. X., Zhang, Y., & Jiang, W. L. (2015). Degradation of aflatoxin B1 by low-temperature radio frequency plasma and degradation product elucidation. *European Food Research and Technology, 241*, 103–113.

Vermeulen, S. J., Campbell, B. M, & Ingram, J. S. (2012). Climate change and food systems. *Annual Review of Environment and Resources, 37*, 195–222. https://doi.org/10.1146/annurev-environ-020411-130608.

Wang, R. X., Nian, W. F., Wu, H. Y., Feng, H. Q., Zhang, K., Zhang, J., et al. (2012). Atmospheric-pressure cold plasma treatment of contaminated fresh fruit and vegetable slices: Inactivation and physiochemical properties evaluation. *The European Physical Journal D, 66*(10), 276.

Whitmee, S., Haines, A., Beyrer, C., Boltz, F., Capon, A. G., de Souza Dias, B. F., … et al. (2015). Safeguarding human health in the Anthropocene epoch: Report of The Rockefeller Foundation–Lancet Commission on planetary health. *The Lancet, 386*, 1973–2028.

World Health Organization. World Health Day. (2015). *Food safety—The global view* (p. 2105). http://www.who.int/campaigns/world-health-day/2015/en/.

Yun, H., Kim, B., Jung, S., Kruk, Z. A., Kim, D. B., Choe, W., et al. (2010). Inactivation of Listeria monocytogenes inoculated on disposable plastic tray, aluminum foil, and paper cup by atmospheric pressure plasma. *Food Control, 21*(8), 1182–1186.

Zhang, K., Perussello, C. A., Milosavljević, V., Cullen, P. J., Sun, D. W., & Tiwari, B. K. (2019). Diagnostics of plasma reactive species and induced chemistry of plasma treated foods. *Critical Reviews in Food Science and Nutrition, 59*(5), 812–825.

Ziuzina, D., Misra, N. N., Cullen, P. J., Keener, K. M., Mosnier, J. P., Vilaró, I., et al. (2016). Demonstrating the potential of industrial scale in-package atmospheric cold plasma for decontamination of cherry tomatoes. *Plasma Medicine, 6*, 3–4.

Ziuzina, D., Patil, S., Cullen, P. J., Keener, K. M., & Bourke, P. (2014). Atmospheric cold plasma inactivation of *Escherichia coli, Salmonella enterica* serovar typhimurium and *Listeria monocytogenes* inoculated on fresh produce. *Food Microbiology, 42*, 109–116.

CHAPTER 12

Nanotechnology

PAULO EDUARDO SICHETTI MUNEKATA[A] • MIRIAN PATEIRO[A] •
RUBÉN DOMÍNGUEZ[A] • MOHAMED A. FARAG[B,C] •
THEODOROS VARZAKAS[D] • JOSÉ MANUEL LORENZO[A,E]

[a]Galician Meat Technology Center, Galicia Technology Park, Ourense, Spain, [b]Department of Chemistry, School of Sciences and Engineering, The American University in Cairo, New Cairo, Egypt, [c]Pharmacognosy Department, College of Pharmacy, Cairo University, New Cairo, Egypt, [d]Department of Food Science and Technology, Faculty of Agriculture and Food, University of the Peloponnese, Kalamata, Greece, [e]Food Technology Area, Faculty of Sciences of Ourense, Vigo University, Ourense, Spain

12.1 INTRODUCTION

Sustainability plays a major role in the development of the current food production system. However, carrying out activities and changes to improve the sustainable level of the current food system is a complex task with dynamic and multicomponent factors (Ben-Eli, 2018). An important aspect related to this scenario is the definition of sustainability that has not been comprehensively defined yet. Moreover, the definition of sustainability comprises many aspects that leads to different perspectives within its general concept due to its application to the difference areas of knowledge (Ben-Eli, 2018; Waseem & Kota, 2017). For instance, possible definitions are: *the harmonious improvements made in the use of resources, investments, technological development, and regulations to improve the potential to meet the necessities and objectives of the current and future generations* (Koltun, 2010) and *a well-balanced interaction between a population and the carrying capacity of the environment where this population develops and express its potential without generating irreversible effects in the carrying capacity of their environment* (Ben-Eli, 2018). Moreover, different definitions exists: aiming for the limits of consumption, centering in the three pillars (economic, social, and environment domains), preservation and continuation of activities, ensuring human welfare, and development and advance ensuring sustainable practices (Waseem & Kota, 2017).

In terms of food production, the view of the Food Agriculture Organization of the United Nations indicates a relevant adaptation of the definition of sustainability: production of healthy, safe, and nutritious foods in a system that is profitable for all partners, has broad social benefits and reduced/neutral environmental impact. Moreover, achieving sustainability is a gradual and ever-growing process that require strategic changes in the current system (FAO/WHO, 2018). Current changes that show the progression of the food system are the development of food industries in modern society, especially by increasing the options beyond staple foods in order to follow the demand for new products with consumer-tailored characteristics (quality, composition, and sensory attributes, for instance). However, important challenges remain as a consequence of the development of society: energy-intense processes associated with an important ecological footprint, recurrent food outbreaks, continuous food losses and waste generation, limited access of small-scale producers and processors to more profitable markets, and wide production and consumption of low nutritional quality foods (FAO/WHO, 2018; McKenzie & Williams, 2015).

In order to address these issues and progress to a more sustainable food production system, technology plays a center role by modifying the supplies, processing conditions, food components, shelf life, and improving the interaction with consumers. In this sense a multidisciplinary and transcendent strategic use of technology can effectively address these issues due to the complexity of food production system at local, regional, national, and global levels (FAO/WHO, 2018; Herrero et al., 2020). The importance of a multidisciplinary tool to improve the sustainability of food production system is justified by the effect of producers, professionals of the food industry, and consumers have in the environmental performance of a food product. It is also relevant to mention that the progression of society, trends, current populational increase, technological development, modifications in the policy of

Sustainable Production Technology in Food. https://doi.org/10.1016/B978-0-12-821233-2.00012-5
Copyright © 2021 Elsevier Inc. All rights reserved.

food and related areas are also important factors that can affect the sustainable performance of food system (FAO/WHO, 2018).

Nanotechnology can be seen as a multidiscipline filed of research and application by integrating the knowledge of chemistry, physics, biology, and engineering to produce and built materials and devices in the nanoscale (at least of its dimensions is smaller than 100 nm) (Bajpai et al., 2018; He, Deng, & Hwang, 2019; Yu et al., 2018). These novel materials can be used to perform traditional and innovative uses in several research areas, which also include the production of food. Particularly for food production area, the use of nanotechnology has been associated with major advances in several research areas from crop and animal production to preservation and consumption (Singh, Shukla, Kumar, Wahla, & Bajpai, 2017; Yu et al., 2018). The insertion of nanotechnology in the sustainable food production is an important advance with specific aims and effects that can provide solutions, at different levels, to several challenges of the whole food production chain (Fig. 12.1). This chapter aims to present recent applications and solutions of nanotechnology in the production of food and discuss the relevance of this technology to develop the current food system to a more sustainable.

12.2 CROP PRODUCTION

12.2.1 Nanofertilizers

Fertilizers are essential components in modern agriculture. This condition is derived from at least one of the following factors: the pursuit of maximum production yield, overcome competitors, the reduction of working force in the agricultural sector, to provide regular production under uncertain environmental conditions, and the absence or difficulties to access technical information to manage crop production (Zheng, Luo, & Hu, 2020).

However, the excessive use of fertilizers (in the face of the increasing demand for food and feeding) causes important environmental concerns associated to emission of greenhouse gases, deterioration of cultivated areas and contamination of water bodies (Wang, Zhu, Zhang, & Wang, 2018). In this sense, exploring environmental friendly alternatives to improve the production of crops became a major topic among researchers and farmers.

A relevant example of the use of nanotechnology in the production of biofertilizers is the use of onion silver nanoparticles to improve the growth of tomato and brinjal (Gosavi et al., 2020). The authors observed that using the 15 mL/L every 2 days for 7 days improved the vigor and biomass of plants in both crops in comparison to lower concentrations and control. In a similar way, a recent experiment with a biofertilizer produced

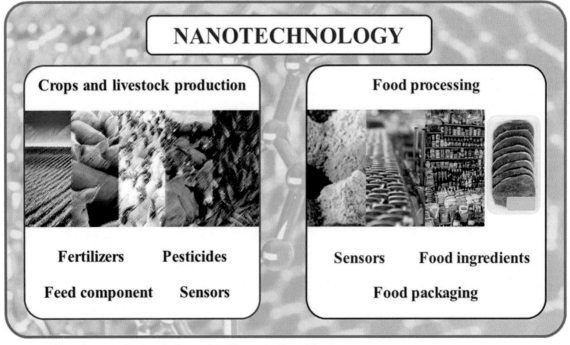

FIG. 12.1 Use of nanotechnology in the sustainable production of food.

with seaweed *Turbinaria ornata* and ZnO nanoparticles (Itroutwar et al., 2020). This new fertilizer improved the characteristics of germinated seeds of rice (grain weight, seed length, seed thickness, and seed width).

A related experiment with magnesium nanoparticles improved the content of magnesium and the nitrate reductase in green beans (Salcido-Martínez et al., 2020). The doses of 50 and 100 ppm of fertilizer produced was the most appropriate to improve the germination of these seeds. Similarly, the use of a commercial nanofertilizer (Biozar®) composed of Fe, Zn, and Mn increased the yield of wheat production (Mardalipour, Zahedi, & Sharghi, 2014). Specifically, this fertilizer improved the length and number of spikes as well as the amount, amount per spike, and weight of wheat seeds. An interesting approach was explored by (Bettencourt, Degenhardt, Torres, Tanobe, & Soccol, 2020), who studied the effect of nanoparticles containing iron and manganese. According to the authors, the germination of maize seeds as well as root development, and fresh weight were enhanced.

A relevant approach is the combined use of natural fertilizers and nanoparticles to improve the production yield of economically relevant crops. For instance, a recent study explored the combined use of microbial strains (*Azotobacter chrocoocum*, *Azosperilium lipoferum*, and *Pseudomonas putida*) and nano zinc-iron oxide particles in wheat under water deficit conditions (Sharifi, Khalilzadeh, Pirzad, & Anwar, 2020). In this case, the combination of Azotobacter and bimetallic nanoparticles composed of Zn–Fe oxide produced the higher increase in the grain yield (88%). Differently, the production of maize was improved by microbial strains (*Pseudomonas putida* strain 146 and mycorrhiza) but non-significant effect was reported for iron oxide under drought stress (Eliaspour, Sharifi, Shirkhani, & Farzaneh, 2020). Moreover, the authors also indicated that the combined strains also reduce the deleterious effects of drought stress.

12.2.2 Nanopesticides

Controlling pests in food and feeding crops is a primary action to reduce the losses associated with invasive plants, insects, microorganisms, for example (Zhang, 2018). The use of these substance is controlled and follow strict protocols for approval and application that varies from one country to another. For instance, the Europe Union has an online database that can be used to check whether a substance is approved for use, pending approval or not approved as well as other information related to its toxicity for humans and animals, maximum residue levels and the acceptable daily intake (European Commission, 2020a, 2020b).

The concern related to the use of pesticides, especially among consumers, is related to uncertain and unknown effects of synthetic pesticides in health (Devi, Duraimurugan, Chandrika, Gayatri, & Prasad, 2019). Scientific evidence supports the potential risk to develop asthma cancers and leukemia that can occur by different routes such as oral, dermal, respiratory, and ocular in agricultural workers (Kim, Kabir, & Jahan, 2017). Moreover, the residues generated by application of pesticides in agricultural practices can also affect the nonhuman biota and the environment by persisting for longer periods or generate other toxic products (Carvalho, 2017).

Exploring sustainable options to protect crops using nanotechnology is an interesting approach. In this sense, nanopesticides have been studied to protect plant crops and have reduced impact in humans, nonhuman biota, and the environment. The nanomaterials produce to achieve this goal have active pesticidal components or tailored structure to cause a pesticidal effect. For instance, the use of α-pinene and linalool in silica nanoparticles deterred the feeding of castor semilooper (*Achaea janata* L.) and tobacco cutworm (*Spodoptera litura* F.) pests (Rani, Madhusudhanamurthy, & Sreedhar, 2014). Likewise, another experiment evaluated the effect of the essential oil of *Xylopia aromatica* nanoencapsulated in poly-ε-caprolactone against silverleaf whitefly (*Bemisia tabaci*) (Peres et al., 2020). Another relevant experiment with natural pesticides was carried out using neem oil encapsulated in zein nanoparticles against bean weevil (*Acanthoscelides obtectus*), silverleaf whitefly (*Bemisia tabaci*), and red spider mite (*Tetranychus urticae*) (Pascoli et al., 2020). The authors observed a that both pests were killed after being exposed to this nanopesticide.

Another related experiment evaluated the effect of *Lippia citriodora* essential oil in polyvinyl alcohol nanofibers (Mahdavi, Rafiee-Dastjerdi, Asadi, Razmjou, & Achachlouei, 2020). The results of this study indicated that nanofibers were more lethal than non-encapsulated oil against potato tuber moth (*Phthorimaea operculella*). Similarly, the nanoencapsulation of chili oil, cinnamon oil, and neem oils was effective against nematodes (Nguyen et al., 2020). Another study with a similar objective explored the influence of nickel-oxide nanoparticles obtained by green synthesis method in pulse beetle (*Callosobruchus maculatus* F.) in black gram (*Vigna mungo* L.). According to the authors, the use of these nanoparticles was efficient in reducing fecundity and extended the developmental period of this pest (Rahman et al., 2020).

Likewise, the essential oil of black pepper (*Piper nigrum*) was tested against rice weevil (*Sitophilus oryzae*) and red flour beetle (*Tribolium castaneum*) (Rajkumar et al.,

2020). Similar to aforementioned studies, nanoencapsulation improve the pesticidal effect of the active component. Another relevant outcome associated with nanopesticide was reported in a recent study with nanoemulsions loaded with neem and citronella oils (Ali et al., 2017).

A recent study carried out with kaolin (aluminosilicate mineral) nanoflakes reduced the adhesion of Southern green stink bug (*Nezara viridula*) and Mediterranean fruit fly (*Ceratitis capitata*) in the leaves of cherry (*Prunus laurocerasus* L. "caucasica") and sunflower (*Helianthus annuus* L.) (Salerno, Rebora, Kovalev, Gorb, & Gorb, 2020). A similar experiment was carried out with surface-functionalized silica nanoparticle that caused 90% mortality rate in rice weevil (*Sitophilus oryzae*) (Debnath et al., 2011). Collectively, nanotechnology is key tool to produce biopesticides against several pests from different classes (microorganisms, insects, and nematodes, for instance) in both in vitro and in situ experiments.

12.2.3 Nanosensors

The use of sensors is another important aspect related to sustainable production of food by improving the information and control during food production, especially using nanocomponents in the assembling of these devices (Kuswandi, Futra, & Heng, 2017). A sensor is a device, machine, or a system that detect quantitative or semiquantitative changes in a selected variable. A transducer converts this change in a signal that is further analyzed by an electronic device (usually a computer) where the information can be interpreted (Giraldo, Wu, Newkirk, & Kruss, 2019; Lorenzo et al., 2019). In the case of nanosensors, these devices convey information occurring in nanoscale to a macroscopic scale (Kuswandi et al., 2017). Some relevant studies exploring the use of nanosensors in food crops are displayed in Table 12.1.

TABLE 12.1
Nanosensors Applied in Food Crops.

Crop- and Food-Related Applications	Sensor Components	Characteristics	Ref.
Response to stressors (UV-B, light, mechanical wound, and pathogenic microorganism peptide) *Arabidopsis thaliana*	Hermin, single-walled carbon nanotubes and ss DNA	Sensible to UV-B, light, and pathogenic microorganisms	Wu et al. (2020)
Response to stressors (chitin, a pathogen peptide, extreme temperature, salinity, and osmotic stress) *Oryza sativa* L. spp. japonica cv. Zhonghua11	Carbon nanotubes and DNA fragments (pFLIPglu–2 μΔ13 and pFLIPglu–600 μΔ13)	FLIPglu–2 μΔ13 rice was suitable to detect glucose variations as a response to stressors	Zhu et al. (2017)
Response to pathogenic microorganism metabolite (*p*-ethylguaiacol) In vitro	SnO_2 and TiO_2 nanoparticles and screen-printed carbon	Low detection limit and high sensitivity for both nanoparticles	Fang, Umasankar, & Ramasamy (2014)
Metabolic response to pathogenic microorganism exposure (salicylic acid) *Brassica napus* L.	Copper nanoparticles-modified gold electrode in a glass tube	Nanosesor had similar temporal response to HPLC method	Wang et al. (2010)
Response to DNA of pathogenic microorganism Soil samples	Gold nanoparticles in an oligonucleotide probe	Fast nanosensor with low detection limit	Khaledian, Nikkhah, Shams-bakhsh, & Hoseinzadeh (2017)
Response to DNA of pathogenic microorganism *Arabidopsis thaliana*	Gold nanoparticles and screen-printed carbon electrodes	Low detection limit	Lau et al. (2017)
Response to toxic metal (arsenic) *Spinacia oleracea* and *Oryza sativa*	single-walled carbon nanotube, $(GT)_5$ sequence and TO-PRO-1	Low sensitivity	Lew, Park, Cui, and Strano (2021)

In terms of food production, the biological changes in selected variables of a given plant is monitored in real time and facilitates the decision-making processing to use precise amounts of fertilizer or pesticides and any other action necessary (such as irrigation) to ensure the development of crop up to harvesting (Giraldo et al., 2019). This broad and generic definition supports a crucial characteristic of sensors: their flexibility of applications, especially using nanotechnology. A relevant example of the use of nanosensors in crop production is the monitoring of stress response in *Arabidopsis thaliana* (a model species used in research) using single-walled carbon nanotubes to evaluate the release of H_2O_2, a molecule involved in plant stress response (Wu et al., 2020). The authors produced a hemin complexed aptamer DNA single-walled carbon nanotubes sensor that operated in the physiologic range of H_2O_2 (10–100 μM) for the following stressors: UV-B, common light, and pathogen peptide exposure. However, this nanosensor was not adequate to indicate mechanical leaf damage.

Another relevant application of nanosensors in plants is the monitoring of physiological response to external stressors. For instance, a recent experiment reported the development of genetically encoded Förster resonance energy transfer nanosensor (Zhu et al., 2017). The authors observed significant changes in sensor response to the exposure to chitin, a pathogen peptide, extreme temperature, salinity, and osmotic stress in rice plant.

Monitoring of alterations caused by fungi and bacteria infections in plants is another interesting application of nanosensors to improve the control of crop production and preserve food production. In this sense, (Fang et al., 2014) developed a screen-printed carbon nanosensor with TiO_2 or SnO_2 nanoparticles to monitor the formation of *p*-ethylguaiacol a volatile compounds produced from the infection of *Phytophthora cactorum*, a pathogenic fungus in strawberries. This nanosensor displayed detection limits between 35 and 62 nM. Similarly, a related experiment with oilseed rape explored the use of a sensor composed of gold electrode modified with copper nanoparticles for the quantification of salicylic acid (an important signaling molecule in plants) in oilseed rape infected with *Sclerotinia sclerotiorum* (Wang et al., 2010). The authors observed a similar temporal behavior in the quantification of salicylic acid in plant leaves between the sensor and quantification using a HPLC method.

In the case of bacterial infections, Khaledian et al. (2017) developed a sensor with gold nanoparticles to detect the potato bacterial wilt (*Ralstonia solanacearum*). The detection limit for this nanosensor was 7.5 ng of fungal DNA. Moreover, the short time required to perform the assay (15 min) and the fact that soil samples can be tested with this sensor are relevant aspects cited by the authors. In a similar way, a gold nanoparticle-based electrochemical sensor was develop by Lau et al. (2017) to detect the presence of *Pseudomonas syringae* in *Arabidopsis thaliana*. Interestingly, the authors reported reliable detection with samples contaminated with 13 pg of DNA.

The monitoring of contaminants acquired from soil is another important aspect to be considered in the production of crops. In this sense, the use of nanobionics has an important role in the development of sensors for real time monitoring of contaminants in food crops. This strategy consists in adding a non-native response to a plant (using nanomaterials containing active factors) that will respond to a target compound or stressor (Lew, Koman, Gordiichuk, Park, & Strano, 2020). With this in mind, a recent study with single-walled carbon nanotube sensor containing $(GT)_5$ sequence and TO-PRO-1 (a cyanine dye intercalated with DNA) incorporated in spinach (*Spinacia oleracea*) and rice (*Oryza sativa*) (Lew et al., 2021). However, the authors reported low sensitivity to arsenic (as soil contaminant).

12.3 ANIMAL PRODUCTION

Another relevant aspect in food production consists on farming animals. Traditionally, animals have been used to obtain several food products such as meat, milk, and eggs but the concern around farming animals been raised in modern and intensive practices have been changing. Animal well-fare became a central topic of discussion and investigation due to the complexity of the factors involving the physical and mental state of animals during their life and slaughter (Keeling et al., 2019; Velarde, Fàbrega, Blanco-Penedo, & Dalmau, 2015). Consequently, factors directly related to health, comfort, nourishment, safety, prevention of unpleasant states, and the possibility to express its normal behavior gained even more importance in animal production (Velarde et al., 2015).

Particularly in animal health, the main actions are associated to use of nanotechnology in preservation of homeostasis or improving health by increasing the antioxidant response (endogenous antioxidant enzymes such as glutathione peroxidase and superoxide dismutase, GPx and SOD, respectively), reducing serum lipids (such as low-density lipoprotein cholesterol (LDL) and free fatty acids), favoring the development, and reducing incidence of infections and related death rate can be cited as relevant applications (Table 12.2). It is important to notice that these effects are associated

TABLE 12.2
Nanotechnology in Animal Production.

Animal (Breed or Scientific Name)	Nanocomponent (Concentration in Feeding)	Effect	Refs.
Rabbit (Californian)	Selenium nanoparticles (400 µg/kg)	Increased digestibility, body weight gain; improved liver and kidney functions; reduced feed conversion ratio	Abdel-Wareth et al. (2019)
Rabbit (n.i.)	Selenium nanoparticles (25, and 50 mg/kg)	Increased body weight, feed intake, carcass percentage; kidney and liver functions, GSH, and CAT; reduced feed conversion ratio and IL-4	Sheiha et al. (2020)
Sheep (Shal)	Zinc oxide nanoparticles (28 mg/kg)	Increased Zn absorption, rumen and blood antioxidant status, IgG; reduced blood urea level	Alijani, Rezaei, and Rouzbehan (2020)
Lamb (Jalauni)	Zinc oxide nanoparticles (60 ppm)	Increased Zn retention; no effect in blood metabolites	Singh, Maity, and Maity (2018)
Lambs (Moghani)	Selenium nanoparticles (1 and 2 mg/kg)	Increased serum GPx1 and SOD levels, expression of GPx1 and SEPW1 in liver	Ghaderzadeh, Aghjehgheshlagh, Nikbin, and Navidshad (2020)
Pig (Landrace × Large White × Duroc)	Zinc oxide nanoparticles (150, 300, or 450 mg kg)	Increased weight gain, feed intake, villus height to crypt depth ratio in the duodenum and jejunum, zinc level in serum, heart, liver, spleen and kidney, serum IgA, IL-6, and TNF-α levels; reduced diarrhea incidence, zinc excretion, *Escherichia coli* counts in cecum, colon, and rectum, serum IgM level	Pei et al. (2019)
Pig (Duroc × Landrace × Yorkshire)	Zinc oxide nanoparticles (0.3, 0.4, 0.5 or 0.6 g/kg)	Increased weight gain, reduced diarrhea incidence, serum alkaline phosphatase, IgG, IgM, SOD and insulin levels, liver SOD and metallothionein levels, lactic acid bacteria and total anaerobic bacteria counts, reduced serum, and liver malondialdehyde, *Escherichia coli* counts in cecal digesta	Sun et al. (2019)
Pig (Landrace × Yorkshire × Duroc)	Chromium nanoparticles (200 µg/kg)	Increased absorption of chromium, feed intake and weight gain, chromium content in meat; reduced serum free fatty acids and back fat thickness	Li, Fu, and Lien (2017)
Pig (Landrace × Yorkshire × Duroc)	Selenium nanoparticles (0.30 ppm)	Increased serum selenium and GPx levels; expression of Selenoprotein W, GPx1, GPx3, GPx4, and GPx5	Lee et al. (2020)
Calves (Holstein)	Zinc oxide nanoparticles (82–85 mg/kg)	Increased dry matter intake, weight gain, nutrient digestibility, leukocyte, hematocrit and serum total antioxidant capacity and zinc level	Abdollahi, Rezaei, and Fazaeli (2020)
Calves (Holstein)	Zinc oxide nanoparticles (30 and 60 mg/kg)	Increased weight gain and final weight and SOD	Seifdavati et al. (2018)

TABLE 12.2
Nanotechnology in Animal Production—cont'd

Animal (Breed or Scientific Name)	Nanocomponent (Concentration in Feeding)	Effect	Refs.
Cows (Holstein)	Selenium nanoparticles (0.30 mg/kg)	Increased the expression of GPx1, GPx2, GPx4, TrxR2, TrxR3, selenoproteins F, K, T, and W in mammary tissue; selenium content and GPx activity in milk	Han, Pang, Fu, Phillips, and Gao (2021)
Broiler (Ross 308)	Selenium nanoparticles (0.1–0.5 mg/kg)	Increased weight gain and feed conversion ratio, energy and protein utilization, and breast and drumsticks percentages; reduced abdominal fat; improved immune system	Ahmadi, Ahmadian, and Seidavi (2018)
Broiler (n.i.)	Zinc oxide nanoparticles (0.03, 0.06, and 0.3 ppm)	Increased body weight; reduced feed conversion ratio; no effect in breast, thighs, and drumsticks yield	Sahoo et al. (2016)
Broiler (CARIBRO Vishal)	Green and market zinc oxide nanoparticles (40, 60, and 80 ppm)	Increased serum SOD, GPx, CAT, zinc, calcium, and phosphorus levels; enhanced meat antioxidant potential and reduced fat and cholesterol content	Dukare et al. (2020)
Broiler (n.i.)	Selenium nanoparticles (0.5 mg/kg)	Increased weight gain, breast and thigh weight, fat content, expression of SOD, GPx, IR, IGF-I and glucose transporters in muscle; reduced serum total cholesterol and triglyceride level; no effect in lipid oxidation in muscles, serum aspartate aminotransferase HDL-cholesterol and triiodothyronine levels	Saleh and Ebeid (2019)
Broiler (Arbor Acres)	Chromium nanoparticles (1200 µg/kg)	Increased serum chromium levels; reduced LDL-cholesterol	Lin, Huang, Li, Cheng, and Lien, (2015)
Broiler (Cobb 500)	Nano-composite adsorbents (2.5 and 5.0 g/kg; magnetic graphene oxide with chitosan)	Increased feed conversion ratio; reduced aflatoxin absorption and organ lesions	Saminathan, Selamat, Pirouz, Abdullah, and Zulkifli (2018)
Nile tilapia (*Oreochromis niloticus*)	Selenium nanoparticles (1 mg/kg)	Increased weight gain, intestinal villus length and width and number of goblet cells, SOD, GPx, CAT, phagocytic, and lysozyme activities, phagocytic index, NBT level, expression of IL-1β and TNF-α in liver; reduced feed conversion ratio and malonaldehyde level	Dawood, Zommara, Eweedah, Helal, and Aboel-Darag (2020)
Nile tilapia (*Oreochromis niloticus*)	Selenium nanoparticles (0.7 mg/kg)	Increased weight gain, final weight, red blood cells, GPx, SOD, and CAT activity, expression and content of IgM, nitric oxide, lysozyme; reduced feed conversion ratio, MCV, alanine and aspartate aminotransferase, alkaline phosphatase, and lactate dehydrogenase	Neamat-Allah, Mahmoud, and Abd El Hakim (2019)

Continued

TABLE 12.2
Nanotechnology in Animal Production—cont'd

Animal (Breed or Scientific Name)	Nanocomponent (Concentration in Feeding)	Effect	Refs.
Nile tilapia (*Oreochromis niloticus*)	Chitosan nanoparticle (5 g/kg)	Increased weight gain, final weight, feed efficiency, intestinal villus length, red blood cells count, GSH, CAT, amylase, and lipase activity; reduced total anaerobic and aerobic intestinal bacterial counts, red blood cells volume, liver creatinine, creatinine, and malonaldehyde levels	Abd El-Naby, Al-Sagheer, Negm, and Naiel (2020)
Nile tilapia (*Oreochromis niloticus*)	Nano-zeolite (5 and 10 g/kg)	Increased weight gain, final weight, feed conversion ratio, amylase and chymotrypsin activity, red and white blood cells; reduced serum alanine and aspartate aminotransferase and alkaline phosphatase, malondialdehyde level, MCV, and DNA fragmentation	Hassaan et al. (2020)
Rohu (*Labeo rohita*)	Zinc oxide nanoparticles (10, 20 and 30 mg/kg)	Increased weight gain, amylase, protease, and lipase activity, G6P, FBP, and G6PD activity, serum glucose, protein, lipid, cholesterol, GOT and GPT, phagocytic, peroxidise activity, superoxide anion; reduced feed conversion ratio	Mondal et al. (2020)
Rainbow trout (*Onchorhynchus mykiss*)	Selenium nanoparticles (1 mg/kg)	Increased liver selenium level, serum globulin, GPx, SOD, and CAT activity; reduced hematocrit, malondialdehyde level; no effect in muscle chemical composition	Naderi, Keyvanshokooh, Ghaedi, and Salati (2019))
Crab (*Eriocheir sinensis*)	Cerium oxide nanoparticles (0.2, 0.4, 0.8, 1.6, 3.2, 6.4, or 12.8 mg/kg)	Increased weight gain, immunological response, hemocyte counts, SOD, CAT, expression of cathepsin L; reduced feed conversion ratio, mortality rate (exposed to stressor and infection), and malondialdehyde level	Qin et al. (2019)

CAT, catalase; *FBP*, fructose-1, 6-bis phosphatase; *G6P*, glucose-6-phosphatase; *G6PD*, glucose-6-phosphate dehydrogenase; *GOT*, glutamate oxaloacetic transaminase; *GPT*, glutamic pyruvate transaminase; *GPx1*, glutathione peroxidase 1; *GPx2*, glutathione peroxidase 2; *GPx3*, glutathione peroxidase 3; *GPx4*, glutathione peroxidase 4; *GPx5*, glutathione peroxidase 5; *GSH*, reduced glutathione; *IgA*, immunoglobulin A; *IGFJI*, insulin-like growth factor-I; *IgG*, immunoglobulin G; *IgM*, immunoglobulin M; *IL-1β*, interleukin 1β; *IL-4*, interleukin 4; *IL-6*, interleukin 6; *IR*, insulin receptor; *LDL*, low-density lipoprotein; *MCV*, mean corpuscular volume; *NBT*, nitro-blue tetrazolium; *n.i.*, not indicated; *SEPW1*, selenoprotein W1; *SOD*, superoxide dismutase; *TNF-α*, tumor necrosis factor alpha; *TrxR*, thioredoxin reductase.

with nutritional effects such as the better absorption and utilization of nutrients (selenium, zinc, and chromium, for instance) with the eventual increase in growth performance of animals.

An interesting example of the role of nanotechnology in the production of rabbit by including nanonutrients in the feeding (Abdel-Wareth et al., 2019). In this case, the use of Selenium nanoparticles improved the digestibility of nutrients, body weight gain, and also the liver and kidney functions. The feed conversion ratio of animals supplemented with nano-selenium particles was also reduced according to the authors. A related experiment with selenium nanoparticles indicated a similar improvement in terms of animal production and animal health (Sheiha et al., 2020). In this case, the authors evaluated the influence of source of selenium nanoparticles: biological (lactic fermentation) vs synthetic at two concentrations (25 and 50 mg/kg). Supplementing the animals with 50 mg/kg from biological source in the feeding reduced the feed conversion ratio and also increased the body weight, feed intake, and carcass percentage. The animals in this treatment also displayed improved kidney and liver functions, glutathione (GSH), and catalase (CAT) and as well as reduced feed conversion ratio and interleukin 4 (IL-4) levels.

Sheep and lamb production can also be improved by nanotechnology. The incorporation of zinc oxide nanoparticles was recently evaluated in the production of sheep (Alijani et al., 2020). This mineral used in nanosize improve the antioxidant status of rumen and blood of supplemented animals. Additionally, the supplementation also improved the absorption of Zn and induce the production of immunoglobulin G (IgG). In a similar study, the inclusion of nanoparticles of this mineral in the diet of lambs improved the Zn absorption and did not influenced other blood serum metabolites (Singh et al., 2018). Likewise, the use of selenium nanoparticles was also explored in lamb production and lead to significant expression and activity of the endogenous antioxidant system in the liver of supplemented animals (Ghaderzadeh et al., 2020).

In the case of pig production, nanoparticles were studied to improve the growth of piglets. For instance, zinc oxide nanoparticles were associated with bigger villus height to crypt depth ratio in the duodenum and jejunum as well as with improved feed intake and weight gain. Zinc deposition in several tissues and immunological response (in terms of serum immunoglobulin A, interleukin 6, and tumor necrosis factor alpha levels) were increased. Moreover, the counts of *Escherichia coli* count in cecum, colon, and rectum were reduced along with the incidence of diarrhea (Pei et al., 2019; Sun et al., 2019). The use of chromium nanoparticles was another relevant advance in the production of pigs (Li et al., 2017). The supplemented animals had higher feed intake and weight gain as well as lower back fat thickness than animals consuming chromium of regular size. In the incorporation of nanoselenium in the feeding of pigs significantly increased the expression and serum levels of GPx in relation to selenium of conventional size (Lee et al., 2020).

The use of nanotechnology can also improve the production of cattle. Particularly for zinc, the inclusion of nanoparticles of this mineral improved the feed intake and weight gain of calves. In terms of metabolic indicators, the digestibility of nutrients and serum antioxidant capacity and zinc levels were improved (Abdollahi et al., 2020; Seifdavati et al., 2018). The effect of selenium nanoparticles was also studied in the production of calves (Han et al., 2021). In this case, the expression of GPx1, GPx2, GPx4, TrxR2, and selenoproteins F, K, T, and W was bigger than obtained from animals fed with selenium of conventional size.

Several studies reported the effect of nanonutrients in poultry production. For instance, a recent experiment explored the use of zinc oxide nanoparticles in broiler production and reported significant effect related to its concentration in the diet (Sahoo et al., 2016). Using 0.06 ppm produced the most meaningful effect in terms of animal production but no effect cut yield was indicated by the authors. A related experiment explored the effect of the source of zinc nanoparticles: organic vs commercial and their concentration in the diet of broilers (Dukare et al., 2020). The authors of the study observed an increase in the serum levels of antioxidant enzymes and mineral, especially with 80 ppm regardless of the source.

Selenium nanoparticle can also influence the growth and health of broilers (Saleh & Ebeid, 2019). In terms of growth performance, supplemented animals had higher weight gain and breast and thigh weights in comparison to non-supplemented group. The expression of SOD, GPx, insulin receptor, and glucose transporter were also induced by nanoparticles whereas the serum levels of total cholesterol and triglycerides reduced. Another relevant component in broiler production is chromium that is better absorbed when nanoparticles are used and also reduce LDL-cholesterol in relation to conventional size chromium (Lin et al., 2015).

Nanotechnology has also a relevant role in the production of fish. In a similar way reported for other farmed animals, selenium nanoparticles can influence the endogenous antioxidant defense by increasing the activity of SOD, GPx, GSH, and CAT activity as well as inducing the immunological system (by means of IL-1β) and improving the antioxidant status (reducing malonaldehyde level) and improving weight gain and final weight (Dawood et al., 2020; Neamat-Allah et al., 2019). Another interesting nanocomponent included in the feeding of Nile tilapia is chitosan (Abd El-Naby et al., 2020). In this case, the benefits were higher weight gain, final weight, feed efficiency, better intestinal development, antioxidant defenses during the development of Nile tilapia. Additionally, this nanocomponent also reduced total anaerobic and aerobic bacterial counts in intestine and malonaldehyde levels.

The production of rohu (*Labeo rohita*) can be influenced by nanotechnology. For this species, the inclusion of zinc oxide nanoparticles (20 mg/kg) increase the growth along with a reduction in the feed conversion ratio (Mondal et al., 2020). Moreover, the activity of digestive and metabolic enzymes was also increased by supplementing animals with this nanocomponent. Likewise, the growth performance, antioxidant defense, and status could be positively modulated by selenium nanoparticles in rainbow trout (*Onchorhynchus mykiss*) (Naderi et al., 2019). An experiment with cerium oxide nanoparticles in the feeding of crab (*Eriocheir sinensis*) revealed that its use as feeding component (0.8 mg/kg feed) improved weight gain, immunological and antioxidant response in comparison to non-supplemented animals (Qin et al., 2019).

Interestingly, nanomaterials and nanocomponents can also improve the defense of farm animals against stressors and pathogenic microorganisms. A relevant example of this effect was the incorporation of nano-composite adsorbent (composed of magnetic graphene oxide with chitosan; 5.0 g/kg in feed) in the diet of broilers reduced the absorption of aflatoxin and also lesions associated with the ingestion of this toxin (Saminathan et al., 2018). A similar experiment with Nile tilapia exposed to aflatoxin indicated that the use of nano-zeolite (natural clay mineral) in the diet reduced the metabolic alterations associated with this toxin (Hassaan et al., 2020). Another interesting use of nanoparticles was reported for the incorporation of selenium in Nile tilapia exposed to *Streptococcus iniae* (Neamat-Allah et al., 2019). According to the authors, the impact of bacterial infection in supplemented group was reduced in relation to non-supplemented animals. In production of crabs, the incorporation of cerium oxide nanoparticles (0.8 mg/kg) successfully reduced the mortality rate of animals exposed to ammonia nitrogen and *Aeromonas hydrophila* infection (Qin et al., 2019).

The foods produced with animals feed with nanoparticles was also considered in the studies of animal production (Table 12.2). in terms of mineral composition, the addition of nanoparticle containing chromium and selenium can increase the deposition of these minerals in pork meat and cow milk, respectively (Han et al., 2021; Li et al., 2017). In case of zinc, an increase in the antioxidant potential along with a reduction in fat and cholesterol content in broiler meat was reported (Dukare et al., 2020). In terms of lipid oxidation, no effect was associated with selenium nanoparticles supplementation in broiler meat (Saleh & Ebeid, 2019). Finally, this mineral also did not affect the chemical composition of fish muscles (Naderi et al., 2019).

12.4 FOOD PROCESSING AND PACKAGING

12.4.1 Nanosensors

The use of nanosensors in food processing follows the same approach as discussed for crop: indicate the presence or accumulation of a target compounds that can reduce the quality and lead to eventual loss (Kuswandi et al., 2017). From the information acquired from the nanosensors it is possible to evaluate the quality and safety of food products before and after processing (Table 12.3).

Identifying the presence and quantifying food contaminants is another relevant aspect related to nanosensor in the context of sustainability by increasing the control in the food production in terms of food safety and security (Neethirajan, Ragavan, Weng, & Chand, 2018). In order to provide safe food and monitoring microbial, biological, and chemical contaminants in foods is indispensable. The use of nanosensors has a crucial role in this scenario by indicating the presence and quantifying these contaminants. In terms of quality, monitoring the freshness in food products is one of the strategies to ensure safety of food products. Facilitating the evaluation of quality and safety is an important factor in terms of sustainability in order to improve the control in the management of food chain, waste and the actions related to its eventual proper management/disposal. An example of the use of nanosensors in the study carried out by Aghaei et al. (2020) with fresh fish with a protein-based halochromic electrospun. In this study, the authors explored the nanosensor that indicated the spoilage of samples by changing its color (from yellow to magenta) as the content of TVB-N (compounds formed during fish spoilage) increased during storage. Once sensor displayed a purplish color, the fish must be consumed soon and in the case of magenta color, the sample is considered inadequate for consumption.

Another aspect related to food safety is the presence of pathogenic microorganisms (such as *Escherichia coli*, *Salmonella* spp., and *Staphylococcus* spp.) and their toxins. In this case, nanosensors aim to produce a signal as a specific bacteria or toxin contaminate food. For instance, a recent study reported the development of a sensor to detect *Escherichia coli* (STEC) using gold nanoparticles in a electrochemiluminescence sensor (Liu et al., 2018). According to the authors, the detection of limit for this sensor was 0.3 pM with effective detection in the 1 pM-5 nM of DNA. Similarly, another nanosensor was developed to detected the presence of *Salmonella* typhi (Sanjay et al., 2016). In this case, the sensor contained TiO_2-PANI, gold and an antibody to

TABLE 12.3

Nanosensors for Detection of Freshness in Food, Pathogenic Microorganisms, Microbial Toxins, and Toxic Contaminants.

Indication of freshness In trout fillet	Zein, alizarin, and glycerin	Suitable correlation between color change and the accumulation of TVB-N below safe limits	Aghaei, Ghorani, Emadzadeh, Kadkhodaee, and Tucker (2020)
Response to pathogenic microorganism (*Escherichia coli* (STEC)) In vitro	BN quantum dots, carbono quantum dots, and gold nanoparticles	Low detection limit and reliable measurements in the range of pM and nM	Liu, Chen, Wang, and Ma (2018)
Response to pathogenic microorganism (*Salmonella typhi*) In vitro	TiO_2 nanoparticles, PANI, gold, SiO_2, and antibodies	Sensor was sensible to different concentrations of the pathogen	Sanjay, Gayithri, Naveen Kumar, Krishna, and Jagadeesh Prasad (2016)
Response to pathogenic microorganism (*Staphylococcus aureus*) In vitro	Carbon nanotubes with vancomycin	Low time required for detection and detection limit of 1.0 CFU/mL	Shen et al. (2020)
Response to pathogenic microorganism toxin (aflatoxin) In milk	Silver nanoparticles, α-cyclodextrin and graphene quantum dots	Low detection limit and reliable range of quantification between 0.015 and 25 mM	Shadjou, Hasanzadeh, Heidar-poor, and Shadjou (2018)
Response to pathogenic microorganism toxin (aflatoxin) In vitro	Gold nanoparticles, allyl mercaptan, hydroxyethyl-methacrylate, and glass slide	Detection limit and linear range of detention of 1 pg/mL and 0.0001 ng/mL and 10.0 ng/mL	Akgönüllü, Yavuz, and Denizli (2020)
Response to organic contaminants (benzo[a] pyrene, fluoranthene, and naphthalene) In vitro	4-dodecyl benzenediazonium-tetrafluoroborate and gold nanoparticles	Low detection limits for all tested compounds individually or collectively evaluated	Tijunelyte et al. (2017)
Response to organic contaminants (benzo[a] pyrene) In vitro	Gold nanoparticles, 5,5′- dithiobis(succinimidyl-2-nitrobenzoate) and monoclonal antibodies	Low detection limit (2 nM)	Dribek et al. (2017)
Response to heavy metals (lead and copper) In vitro	Gold nanoparticles and dansyl fluorophore	Linear quantification between 0.001 and 10 ppm	Nath, Arun, and Chanda (2015)
Response to heavy metal (lead) In vitro	Graphene quantum dots-aptamer probe and graphene oxide	Low detection limit and quantification range	Qian, Shan, Chai, Chen, and Feng (2015)
Response to heavy metal (mercury) In vitro	Superparamagnetic Fe_2O_3 nanoparticles, silica, cysteamine, and CdTe quantum dots	Detection limit of 0.49 nM	Satapathi et al. (2018)
Response to heavy metal (mercury) In vitro	Gold nanoparticles and oligonucleotide aptamer	Low linear quantification (between 1 and 32 ng/mL) and detection limit (0.28 ng/mL)	Xiao et al. (2016)

detect this pathogen. A related nanosensor was developed by Shen et al. (2020) to detect *Staphylococcus aureus*. The fluorescent sensor was composed of carbon nanotubes with vancomycin and a dual-functionalized aptamer and detected this pathogen (10^8 CFU/mL) in 30 min.

In terms of microbial toxins, the detection of aflatoxin and its level is a relevant use of nanosensors. Aiming to detect this toxin in milk samples, a recent study reported the development of sensor containing silver nanoparticles and α-cyclodextrin (Shadjou et al., 2018). This sensor was able to indicate the presence of aflatoxin in a concentration of 2 μM and suitable detection range between 0.015 and 25 mM. Another sensor developed for the detection of afllaxoin displayed a lower detection limit and operational range (Akgönüllü et al., 2020). In this case, the nanosensor surface plasmon resonance gold chip and was successfully applied to detect aflatoxin between 0.0001 ng/mL and 10.0 ng/mL with detection limit of 1 pg/mL.

The consumption of contaminants in water and food is associated with increased risk of developing severe diseases. With this concern in mind, Tijunelyte et al. (2017) assembled an sensor using 4-dodecyl benzenediazonium-tetrafluoroborate to detect benzo[*a*]pyrene, fluoranthene, and naphthalene. This sensor displayed limit of detection of 0.026, 0.064 and 3.94 mg/mL for benzo[*a*]pyrene, fluoranthene, and naphthalene, respectively. Moreover, the sensor was able to simultaneously detect the three contaminants in a mix solution. A related experiment was carried out to detected benzo[*a*]pyrene using gold nanoparticles, 5,5′- dithiobis(succinimidyl-2-nitrobenzoate), and monoclonal antibodies to construct the sensor (Dribek et al., 2017). According to the authors, a detection limit of 2 nM was obtained using this sensor.

Heavy metals are also important contaminants imposing important health risk to consumers. In order to identify the presence of lead and copper, Nath et al. (2015) constructed a sensor containing gold nanoparticles and dansyl fluorophore. The detection of both metals was possible at concentration in the range of 0.001–10 ppm. Another study explored the use of graphene quantum dots-aptamer probe and graphene oxide was assembled to detect lead (Qian et al., 2015). In this case, the detection of lead was possible with concentrations below 400 nM with a detection limit of 0.6 nM.

The development of nanosensor has also been applied to detect mercury, another toxic heavy metal. For instance, a recent study reported the development of a sensor composed of superparamagnetic Fe$_2$O$_3$ nanoparticles coated with silica and attached to cysteamine capped CdTe quantum dots (Satapathi et al., 2018). The authors reported a detection limit of 0.49 nM. Likewise, Xiao et al. (2016) developed a sensor to detect mercury using gold nanoparticles and oligonucleotide aptamer in water samples. This sensor provided reliable measurements 1–32 ng/mL with a detection limit of 0.28 ng/mL, according to the authors. Collectively, the use of nanosensors is an important part in the sustainable production of food by providing reliable information about the presence of specific compounds and microorganisms (in both crops and foods) that can drastically impact the food chain production and also affect consumer health.

12.4.2 Nano-Ingredients

Nanoencapsulation is a strategy that ultimately leads to entrapment of active compounds (core component) into protective capsule (wall material). The process can be carried out by several strategies. Some of the technologies used in the production of nanoencapsulated materials are: anti-solvent precipitation, electrospinning, electrospraying, emulsion-evaporation method, high pressure micro-fluidization, ionic gelation, lyophilization, nanoemulsification, spray-drying, and ultrasonication-assisted emulsification (Kurozawa & Hubinger, 2017; Liu et al., 2020; Modarres-Gheisari, Gavagsaz-Ghoachani, Malaki, Safarpour, & Zandi, 2019; Walia, Dasgupta, Ranjan, Ramalingam, & Gandhi, 2019). The selection of method is dependent of many factors such as core component and wall material properties as well as the interaction between them and the intended application.

Relevant studies that support the role of nanoencapsulation in the context of sustainable food production are found in Table 12.4. For instance, the incorporation of hydrophobic compounds in food products is a challenge due to their reduced miscibility among food components. In this sense, Horuz and Belibağlı (2018) explored the use of gelatin nanofibers to improve the solubilization of lycopene and obtained a significant increase in the solubility in aqueous solution by encapsulating with gelatin nanofibers. Moreover, the authors also indicated that nanoencapsulated lycopene was stable for up to 14 days of storage. Medeiros et al. (2019) carried out a similar experiment with carotenoids extracted from cantaloupe melons using different protein sources (porcine gelatin, whey protein isolate and whey concentrate). Among all emulsifiers, porcine gelatin produced the most stable emulsion in aqueous system. Additionally, the authors also explored the effect of encapsulated carotenoids in yogurt and reported a

TABLE 12.4
Nano Encapsulation of Bioactive Compounds and Their Applications in Sustainable Food Production.

Encapsulation Process	Source and Bioactive Compounds	Wall Material(s)	Effect	Ref.
Electrospinning	Tomato peel carotenoids	Gelatin nanofibers	Improved water solubility and stability of lycopene during storage	Horuz and Belibağlı (2018)
Lyophilization	Cantaloupe melon carotenoids	Porcine gelatin, whey protein isolate, and whey concentrate	Porcine gelatin increased water solubility of carotenoids and the stability of yogurt color	Medeiros et al. (2019)
High pressure micro-fluidization	*Thymus capitatus* essential oil	Sodium dodecylsulfate	Increased antimicrobial activity (*Staphylococcus aureus*, *Bacillus licheniformis*, *Enterococcus hirae*, *Escherichia coli*, and *Pseudomonas aeruginosa*); and reduced antioxidant activity	Jemaa et al. (2018)
Spray-drying	*Eucalyptus staigeriana* essential oil	Cashew gum	Particles with antimicrobial activity against *Listeria monocytogenes* and *Salmonella* Enteritidis; storage stability was influenced by core and wall material ratio	Herculano, de Paula, de Figueiredo, Dias, and Pereira (2015)
Emulsion-evaporation method	Guabiroba fruit polyphenols	Poly (DL-lactide-co-glycolide)	Increased antibacterial (*Listeria innocua*) and antioxidant activity	Pereira et al. (2018)
Ionic gelation	Jujube pulp and seed polyphenols	Chitosan	Encapsulation increased the antioxidant activity of polyphenols	Han, Lee, Park, Ahn, and Lee (2015)
Nanoemulsification	Olive leaf polyphenols	Whey protein concentrate with pectin	Reduced the lipid oxidation in soybean oil and thermal stability	Mohammadi, Jafari, Esfanjani, and Akhavan (2016)
Nanoemulsification	Grape and apple pomace polyphenols	Chitosan and soy protein isolate	Soy protein isolate had higher encapsulation efficiency than chitosan; encapsulation increased antioxidant activity; controlled release and protection of polyphenols	Ahmed et al. (2020)
Electrospraying	Green tea polyphenols	Zein	Improved gastrointestinal stability and absorption of polyphenols; ratio zein/polyphenols influenced these variables	Bhushani, Kurrey, and Anandharamakrishnan (2017)
Anti-solvent precipitation	Cinnamon bark polyphenols	Xanthan gum	Protected against the gastric stage of digestion and had easy release in the intestine; and increased thermal stability of polyphenols	Muhammad et al. (2020)

Continued

TABLE 12.4
Nano Encapsulation of Bioactive Compounds and Their Applications in Sustainable Food Production—cont'd

Encapsulation Process	Source and Bioactive Compounds	Wall Material(s)	Effect	Ref.
Ultrasonication-assisted emulsification	Vitamin D	Tween-20	Increased bioaccessibility in simulated gastric digestion	Walia, Dasgupta, Ranjan, Chen, and Ramalingam (2017)
Nanoemulsification	Olive leaf polyphenols	Whey protein concentrate with pectin	The combined use of both emulsifiers delayed the release of polyphenols in comparison to whey protein single emulsion	Mohammadi, Jafari, Assadpour, and Faridi Esfanjani (2016)

significant increase in color stability in comparison to non-encapsulated yogurt.

Another relevant aspect related to the nanoencapsulation of bioactive compounds is their technological role in food quality and storage stability. The growth of spoilage and pathogenic microorganisms play a central role in the production of food by reducing shelf life and leading to dispose as waste. In this sense, including natural antimicrobials in foods is necessary action to improve shelf life. For instance, nanoencapsulation was proven to increase the antimicrobial activity of *Thymus capitatus* essential oil against *Staphylococcus aureus*, *Bacillus licheniformis*, *Enterococcus hirae*, *Escherichia coli*, and *Pseudomonas aeruginosa* (Jemaa et al., 2018). Herculano et al. (2015) reported a similar outcome using nanoencapsulated *Eucalyptus staigeriana* essential oil in cashew gum in *Listeria monocytogenes* and *Salmonella* Enteritidis. The authors also highlighted the importance of exploring the effect of core component-wall material ratio due to the influence of this variable has in the stability of nanocapsules during storage. Likewise, the nanocapsulated produced with poly ($_{DL}$-lactide-*co*-glycolide) improved the antibacterial activity of guabiroba fruit polyphenols *Listeria innocua* and also increased its antioxidant activity (Pereira et al., 2018).

The antioxidant activity of natural compounds is another relevant aspect found in foods and their processing by-products to improve the stability of foods (Echegaray et al., 2018). In this context, Han et al. (2015) indicated that nanoencapsulated jujube pulp and seed polyphenols had higher antioxidant activity than free polyphenols counter-part. A related experiment carried out with soybean oil indicated an antioxidant effect in the preservation of soybean oil with nanoencapsulated olive leaf polyphenols (Mohammadi, Jafari, Esfanjani, & Akhavan, 2016).

The nanoencapsulation of grape and apple pomace polyphenols was associated with increased antioxidant activity in relation to non-encapsulated, especially using soy protein isolate (Ahmed et al., 2020). In this experiment, the authors also explored an important characteristic of nanoencapsulated material: controlled release. In this case, the encapsulated material displayed a slower release over time and also improved the stability of polyphenols during storage.

Other studies also explored this characteristic of nanoencapsulated materials in order to prevent losses during gastric digestion and improve bioaccessibility during intestinal passage. For instance, Bhushani et al. (2017) produced green tea polyphenols (mainly composed of catechin derivatives) nanocapsules using zein as wall material. The authors observed a significant improvement in the stability of green tea catechins during simulated digestion and facilitated the release of these compounds once the capsulated in Caco-2 cells. a related experiment reported a similar outcome for cinnamon bark polyphenols encapsulated in xanthan gum (Muhammad et al., 2020). Additionally, the authors of this study also indicated that encapsulated polyphenols had higher thermal stability than free cinnamon bark polyphenols. Likewise, Walia et al. (2017) reported a higher bioaccessibility of Vitamin D nanoencapsulated with Tween-20.

The use of complex (double layer) nanoencapsulation was also explored in order to control the release of olive leaf polyphenols (Mohammadi, Jafari, Assadpour, & Faridi Esfanjani, 2016). In this case, the authors observed that whey protein concentrate and pectin (one component per layer) lead to higher stability and controlled release of these polyphenols in comparison to a single layer nanocapsule with whey protein concentrate as wall material. From all the information discussed above, the role of nanoencapsulation in sustainable food production is sustained by facilitating the incorporation, improving storage stability, and also enhancing the fraction of bioaccessible compounds.

One of the main concerns in public health is the reduction of sodium ingestion due to its relation with increased risk of developing cardiovascular diseases. Although the simplest solution is the reduction in the amount of salt added in processing of food, this strategy causes important modification in food characteristics and potential loss of quality (Inguglia, Zhang, Tiwari, Kerry, & Burgess, 2017; Pateiro, Munekata, Cittadini, Domínguez, & Lorenzo, 2021; Vidal, Lorenzo, Munekata, & Pollonio, 2020). In this sense, one of the possible solutions to reduce the sodium content in food products consist in reducing sodium chloride particle size. This strategy facilitates the dissolution of sodium chloride in saliva an increasing the perception of sodium by receptors in the mouth and makes a low-sodium been perceived as a product with "conventional" salt content (Inguglia et al., 2017).

This approach was explored in a study with cheese crackers (Moncada et al., 2015). The nanoparticles of sodium chloride were produced by nano spray dryer and applied in the surface of samples (1%, 1.5% and 2% of sodium chloride) and received higher scores for saltness than control (2% of sodium chloride) in fresh samples. The authors also highlighted that saltness in samples produced with nanoparticles had higher scores than control after 4 months of storage. A related experiment was carried out with potato chips (Vinitha, Leena, Moses, & Anandharamakrishnan, 2021). In this case, electrohydrodynamic atomized drying method was used to generate the sodium chloride nanoparticles. According to the authors, the use of nanoparticles increased the perception of saltness in relation to control chips with commercial size sodium chloride. Moreover, the authors also indicated that a reduction of 65% in sodium chloride could be achieved by using this nanoadditive.

Interestingly, other solutions to reduce content with nanotechnology consist in the use of nanochitin as active ingredient to increase the perception of saltness in food. The mechanism associated with this effect can be explained by the lowering in pH that protonates the chitin chains and favors the interaction with ions such Cl^-. An apparent increase in the proportion of Na^+ occurs (distribute over the diffuse layer), which also favors the exposure of Na^+ to receptors in the mouth and lead to an enhanced perception of saltness using lower sodium chloride content (Jiang, Tsai, & Liu, 2017). The effect of this strategy was evaluated in a recent study with tilapia fillets (Hsueh, Tsai, & Liu, 2017). The experiment was carried out using chitin nanofibers combined with citric and malic acids in the curing solution and lead to an increased perception of saltness in the fish fillets. Moreover, the authors also indicated that using 3 g/L citric acid or 4 g/L malic acid, no additional sourness was perceived.

12.4.3 Nano-Packaging

Food packaging is a main concern in terms of sustainable food production due the widely use of plastic from non-renewable sources (oil). Regardless of the way the plastic packaging ended its purposed as protective barrier (after food consumption or food discarded due to microbial or chemical contamination), millions of tons of plastic materials are discarded in soil every year leading a major environmental, social, and economic burden (Guillard et al., 2018; Licciardello, 2017). In this sense, some of the relevant uses of nanotechnology in the food production for the development of sustainable food packages consist on improving package composition by adding natural active nanoencapsulated components to extend food shelf life and improving safety and also exploring the development of biodegradable materials composed of natural nanocomponents. Two main approaches have been used in the production of films and coating materials with natural and biodegradable materials: casting and coating (Umaraw et al., 2020). Both approaches share common steps in the preparation of film/coating solutions (addition of matrix material, solvent, plasticizer, and active ingredient) whereas the differences are centered in the application method on foods. Films are produced prior to application in foods whereas the coating consist in dipping the food for a short period and letting the coating dry.

In terms of extension of increasing food shelf life with nano-structural components or nanoencapsulated in the packaging material. For instance. a recent study explored the incorporation of chitosan nanoparticles as coating solution to preserve the cherry fruits during refrigerated storage (Arabpoor, Yousefi, Weisany, & Ghasemlou, 2021). During the trial period, the quality indicators of coated samples were superior to uncoated samples, which was improved with the incorporation of encapsulated *Eryngium campestre* essential oil.

A similar outcome has been reported for the use of nanoencapsulated active components. This approach was explored in a study with nanoencapsulated tetraorthosilicate (a compounds with low toxicity and easily degraded in water) in a coating solution to improve the preservation of Loquat fruit (Song et al., 2016). The coated samples had better quality than uncoated fruits at the end of storage period (40 days at 5°C). Additionally, this system also induced the antioxidant endogenous defenses (superoxide dismutase and catalase, for instance) in the fruit for a longer period than observed for control samples and prevented the action of enzymes related to degradation of quality (such as polyphenoloxidase). Likewise, the preservation of lamb steaks was improved with a coating solution with nanoencapsulated *Satureja khuzestanica* essential oil during 20 days at 4°C (Pabast, Shariatifar, Beikzadeh, & Jahed, 2018). According to the authors, the coated samples displayed lower microbial counts, pH change, lipid oxidation, and sensory quality (red color, discoloration, and off-odor) loss than uncoated samples.

In terms of safety, an active mat with thymol-loaded nanofiber (composed of whey protein and polyvinyl alcohol) inhibited the growth of aflatoxin-producing *Aspergillus parasiticus* in kashar cheese during 7 days at 25°C (Tatlisu, Yilmaz, & Arici, 2019). A related experiment with nanoencapsulated β-carotene in a cassava starch and glycerol film improved the oxidative stability of sunflower oil during accelerated storage conditions (30 days at 30°C) (Assis, Pagno, Costa, Flôres, & Rios, 2018). Interestingly, this study also characterized the biodegradability of the innovative film and the same degradation rate during 15 days was observed in films regardless of the addition of nanoencapsulated β-carotene. The development of this type of package is one of the essential roles of nanotechnology in the development of sustainable options in food production.

In this sense, major efforts have been made in the last decades to develop new materials known as nanobiocomposites. Nanocomposites are materials composed of a matrix as continuous phase and a discontinuous phase also known as filler where at least one of the dimensions of at least one of the components is in the nanoscale (Honarvar, Hadian, & Mashayekh, 2016; Sandri et al., 2016). Producing nanocomposite using biodegradable materials permits the use of "nanobiocomposite" term, which is a major advance in production of sustainable food packaging materials (Kumar, Kaur, & Bhatia, 2017). In the context of food science, the use of nanobiocomposites (biopolymers filled with nanoparticles) in the production of packaging materials has been studied in recent decades (Table 12.5), wherein the expected

effect is the production of films and coatings with enhanced properties. For instance, chitosan nanoparticles (modified with sodium tripolyphosphate) reduced the permeability of water of a fish gelatin film (Hosseini et al., 2015). Moreover, the film also displayed protection against UV light. A related experiment with bamboo nanofibers indicated a significant increase in the elastic modulus in both chitosan and cassava films (Llanos & Tadini, 2018). However, the authors observed a significant increase in the permeability of oxygen and reductions in tensile strength and elongation at break.

Cellulose nanoparticles are interesting filler materials to improve the properties of biopolymers. One of the experiments that support this statement was carried out in a carboxymethyl cellulose and starch film incorporated with cellulose nanocrystals (El Miri et al., 2015). This nanobiocomposite displayed reduced water vapor permeability and increased elastic modulus and tensile strength. A recent experiment with bacterial cellulose nanowhiskers (rod-like crystalline structures) produced by *Komagataeibacter xylinus* improved the strength and barrier properties of bovine gelatin film (Haghighi et al., 2021). Additionally, the authors also observed that no significant effect in transparency was obtained from the addition of cellulose nanowhiskers in this colorless film. A similar outcome was obtained from a polyvinyl alcohol and chitosan film with rice straw cellulose nanocrystals in terms of tensile strength and thermal stability (Perumal et al., 2018). Furthermore, the film with cellulose nanocrystals also displayed fast degradability rate and antimicrobial activity against plant (*Colletotrichum gloeosporioides* and *Lasiodiplodia theobromae*) and human (*Staphylococcus aureus*, *Streptococcus mutans*, *Escherichia coli*, and *Pseudomonas aeruginosa*) pathogenic microorganisms in vitro.

Another interesting experiment with nanobiocomposites in films with antimicrobial activity (in vitro) was reported in a chitosan and polyvinyl alcohol film with the organoclay thiabendazoluim-montmorillonite (El Bourakadi et al., 2019). In this case, the microorganisms inhibited by the film were *Escherichia coli*, *Pseudomonas aeruginosa*, and *Staphylococcus aureus*. Additionally, the authors also indicated that this filler increased the Young's modulus and tensile strength of the film. Likewise, the incorporation of chitin nanoparticles in a chitosan film inhibited the growth of *Aspergillus niger* and also increased film strength (thermal stability, ultimate tensile strength, Young's modulus, and elongation at break) (Sahraee et al., 2017).

TABLE 12.5
Development and Application of Films and Coatings With Nanobiocomposites for Food Packaging.

Film Components	Effect of Nanoparticles	Ref.
Fish gelatin, glycerol, sodium tripolyphosphate, and chitosan nanoparticles	Increased tensile strength and elastic modulus; decreased the elongation at break, water vapor permeability; protection against UV light	Hosseini, Rezaei, Zandi, and Farahmandghavi (2015)
Chitosan, acetic acid, glycerol, and bamboo nanofibers	Increased elastic modulus and oxygen permeability; reduced tensile strength, elongation at break; no effect in water vapor permeability	Llanos and Tadini (2018)
Cassava starch, glycerol, and bamboo nanofibers	Increased tensile strength, elongation at break and oxygen permeability; no effect elastic modulus, water vapor permeability	Llanos and Tadini (2018)
Carboxymethyl cellulose, starch, glycerol, and cellulose nanocrystals	Preservation of transparency, reduced water vapor permeability, increased elastic modulus and tensile strength	El Miri et al. (2015)
Bovine gelatin, polyvinyl alcohol, glycerol, and bacterial cellulose nanowhiskers	Increased elastic modulus, elongation at break and tensile strength; reduced water vapor transmission rate and water vapor permeability; no effect in films transparency	Haghighi et al. (2021)
Polyvinyl alcohol, chitosan, polyethylene glycol, and rice straw cellulose nanocrystals	Increased tensile strength, thermal stability, and biodegradability rate; no effect in transparency; antifungal and antibacterial activity	Perumal, Sellamuthu, Nambiar, and Sadiku (2018)
Chitosan, polyvinyl alcohol, acetic acid, glycerol, and thiabendazoluim-montmorillonite nanoparticles	Increased Young's modulus and tensile strength; antimicrobial activity	El Bourakadi et al. (2019)
Gelatin, glycerol, glutaraldehyde, and chitin nanoparticles	Increased thermal stability, ultimate tensile strength, Young's modulus, and elongation at break; antimicrobial activity	Sahraee, Milani, Ghanbarzadeh, and Hamishehkar (2017)
Corn starch, chitosan, acetic acid, glycerol, and wood cellulose nanofibrils (evaluated in corn oil for 7 days at 70°C and in beef for 8 days at 4°C)	Increased the rigidity, light barrier, oxygen barrier and water vapor barrier, and opacity; reduced oxidation in corn oil and microbial growth rate in beef	Yu et al. (2017)
Tilapia gelatin, glycerol, and Cloisite Na$^+$ nanoclay (evaluated in mackerel meat powder stored for 30 days at 28–30°C)	Preservation of moisture; reduced lipid oxidation, total volatile basic-N, a* and b* increase, total color variation, and loss of sensorial quality; no effect in pH and L*	Nagarajan, Benjakul, Prodpran, and Songtipya (2015)
Squid gelatin, glycerol, and Cloisite Na$^+$ nanoclay (evaluated in mackerel meat powder stored for 30 days at 28–30°C)	Preservation of moisture; reduced lipid oxidation, total volatile basic-N, a* and b* increase, total color variation, volatile compounds, and loss of sensorial quality; no effect in pH and L*	Nagarajan et al. (2015)
Chitosan, acetic acid, and nano-cellulose (evaluated in cherries stored for 6 weeks at 1°C)	Reduced weight loss; slowed anthocyanin accumulation; no effect in acidity and malic acid content	Nabifarkhani, Sharifani, Garmakhany, Moghadam, and Shakeri (2015)

L*, luminosity; a*, redness; b*, yellowness.

The advances made in terms of film properties are part of the solution provided by nanotechnology in terms of sustainable food packaging. The protective effect during food storage is another crucial outcome to support the advances made in the field. In this sense, a film composed of corn starch, chitosan, and wood cellulose nanofibrils was developed to improve the preservation of corn oil and fresh beef (Yu et al., 2017). In terms of film properties, the addition of this nanocomponent improved the barrier properties but increased the opacity of the film. Reading the preservation of both foods, a significant reduction in the formation of lipid oxidation products (peroxide value reduced by 23%) was reported for corn oil and an inhibition of total microbial growth (4 log reduction) was indicated after refrigerated storage of fresh beef with the film with nanoparticles in relation to control film (without cellulose nanofibrils).

In order to preserve mackerel meat powder, the effect of Cloisite Na^+ nanoclay as modifier in films using either tilapia gelatin or squid gelatin as biopolymers was studied (Nagarajan et al., 2015). Both films slowed the progression of lipid oxidation, accumulation of total volatile basic-nitrogen, changes in color, and also the loss of quality (particularly for overall likeness) in relation to samples wrapped with polyethylene and control treatment (unwrapped). Finally, a chitosan film with nano-cellulose as modifier reduced the weight loss and did not affect the acidity and malic acid content of cherries during 6 weeks at 1°C (Nabifarkhani et al., 2015).

12.5 SUSTAINABLE ASSESSMENT OF NANOTECHNOLOGY

Nanotechnology has several applications in the production of foods, which can be also seen as an important tool to progress towards a more sustainable system to produce food for the current and future generations, as indicated in previous sections but considerations from the sustainable perspective are also necessary. The assessment of sustainable impact considering each domain has been a topic of major discussion and some indicators have been proposed to understand the value of proposed strategies. In terms of environmental domain, important indicators are carbon foot print, soil and water acidification, chemical pollution and insertion of new harmful agents, fossil and mineral resources, food loss and waste generation, and food and agricultural resources (Dong & Hauschild, 2017).

In the case of economic domain, financial health or internal financial stability, economic performance in terms of perceived value by shareholders, potential financial benefits beyond profit, and potential to create and sustain trading opportunities can be cited as relevant indicators (Labuschagne, Brent, & Van Erck, 2005). Social sustainability can be measured by diversity and equal opportunities, employment benefits, basic human rights practice, community funding and support, consumer health and safety, and product management aligned with consumer satisfaction (Popovic, Barbosa-Póvoa, Kraslawski, & Carvalho, 2018).

Moreover, it is also important to consider the interconnections between two domains when sustainable development is carried out in more than one domain: inclusive growth (economic with social domains), eco-social progress (social and environmental domains), and green growth (environmental and economic domains). Ultimately, the combination of actions in all domains leads to the concept of sustainability: economic, social and environmental-oriented development (FAO/WHO, 2018).

Alternatively, the environment assessment can be carried out by the Life Cycle Assessment (LCA) method (Dong & Hauschild, 2017). LCA consist in the comparative evaluation of environmental impact of a specific activity. For instance, a recent study compared the environmental impact of a polylactic acid-based nanocomposite active packaging with conventional plastic packaging where the entire life of the materials was considered (from cradle to grave) (Lorite et al., 2017). In this study, the packaging film containing the nanoparticles had bigger impact in climate change due to the higher energy consumption (production and mixing of film components) in relation to conventional packaging material (polyethylene terephthalate—PET). However, considering the increase of shelf life of 30%, the film containing the nanoparticles was equally or more environmentally friendly than PET. According to the authors, this outcome could be explained by the reduction in the waste generated (increased shelf life and reduced consumption of packaging materials) and fewer trips to the store. In terms of waste disposal, the burden associated with this novel nano-packaging was lower than obtained for PET due to the minimal contribution of decomposition to climate change.

However, more studies about the use of nanotechnology are necessary, especially to obtain LCA studies. LCA is dependent of technical information that has been increasing in the last decades but does not provide the same amount of information about the impact in the environment and humans (Salieri, Turner, Nowack, & Hischier, 2018).

The LCA of nanotechnology is dependent of technical information that is currently scarce.

Another relevant aspect to be considered in the sustainability of nanotechnology is the connection with Circular Economy. This integration consists on the waste valorization by incorporating waste materials from by-products of food industry as source of high-added value compounds in another processing line, which leads to a reduction in total waste generation (Lowry, Avellan, & Gilbertson, 2019).

Nutritional, safe, and healthier foods for consumers are the main aspects related to social domain for food production with the use of nanotechnology. Providing nutritious and safe food products to the population is one of the vital aspects for any food producer and industry. However, the supplying foods with tailored characteristics for specific shares of the consumer markets linked with health-related necessities is an important advance in the production of food. The main justifications for this advance are the increasing scientific evidence supporting a healthy diet and the consumption of functional foods, increasing level of knowledge among consumers and the consequent demand for these foods. in other words, providing tailored foods for consumers is an important aspect to allow consumer to decide (from their own experience and knowledge as well as the information in the label of foods) the most cost-effect food choices in order to maintain or improve health (Sibbel, 2007). The impact of nanotechnology in other indicators of the social domain, in the context of food production, requires additional studies.

The economic viability is major factor that can define the fate of any activity. In this sense, the economic evaluation can provide some valuable data. For instance, the combined use of nano-zeolite-loaded nitrogen and biofertilizers in the production of caraway (*Carum carvi* L.) plants reduced production cost per hectare and also increased the gross income in comparison to chemical fertilizers (El-Attar, Mahmoud, & Mahmoud, 2017). In a similar way to observed for social dimension, the economic data regarding the use of nanotechnology in the context of sustainability are scarce.

12.6 CONCLUSION

The large amount and density of information that the potential use of nanotechnology in food production provides view of the complexity of this technology, its applications, and solutions in each stage of food processing. Considering the three domains of sustainability, the production of food using nanotechnology can be seen as major tool to improve the current food system. However,

the information to strengthen the technological evidence and selection of the most cost, social, and environmental effective approaches within the wide variety of solutions provided by nanotechnology is scarce and require more efforts in this crucial area of human development.

ACKNOWLEDGMENTS

Thanks to GAIN (Axencia Galega de Innovación) for supporting this study (grant number IN607A2019/01). Paulo E. S. Munekata acknowledges postdoctoral fellowship support from the Ministry of Economy and Competitiveness (MINECO, Spain) "Juan de la Cierva" program (FJCI-2016-29486).

REFERENCES

Abd El-Naby, A. S., Al-Sagheer, A. A., Negm, S. S., & Naiel, M. A. E. (2020). Dietary combination of chitosan nanoparticle and thymol affects feed utilization, digestive enzymes, antioxidant status, and intestinal morphology of Oreochromis niloticus. *Aquaculture, 515*, 734577.

Abdel-Wareth, A. A. A., Ahmed, A. E., Hassan, H. A., Abd El-Sadek, M. S., Ghazalah, A. A., & Lohakare, J. (2019). Nutritional impact of nano-selenium, garlic oil, and their combination on growth and reproductive performance of male Californian rabbits. *Animal Feed Science and Technology, 249*, 37–45.

Abdollahi, M., Rezaei, J., & Fazaeli, H. (2020). Performance, rumen fermentation, blood minerals, leukocyte and antioxidant capacity of young Holstein calves receiving high-surface ZnO instead of common ZnO. *Archives of Animal Nutrition, 74*(3), 189–205.

Aghaei, Z., Ghorani, B., Emadzadeh, B., Kadkhodaee, R., & Tucker, N. (2020). Protein-based halochromic electrospun nanosensor for monitoring trout fish freshness. *Food Control, 111*, 107065.

Ahmadi, M., Ahmadian, A., & Seidavi, A. R. (2018). Effect of different levels of nano-selenium on performance, blood parameters, immunity and carcass characteristics of broiler chickens. *Poultry Science Journal, 6*(1), 99–108.

Ahmed, G. H. G., Fernández-González, A., García, M. E. D., Gaber Ahmed, G. H., Fernández-González, A., & Díaz García, M. E. (2020). Nano-encapsulation of grape and apple pomace phenolic extract in chitosan and soy protein via nanoemulsification. *Food Hydrocolloids, 108*, 105806.

Akgönüllü, S., Yavuz, H., & Denizli, A. (2020). SPR nanosensor based on molecularly imprinted polymer film with gold nanoparticles for sensitive detection of aflatoxin B1. *Talanta, 219*, 121219.

Ali, E. O. M., Shakil, N. A., Rana, V. S., Sarkar, D. J., Majumder, S., Kaushik, P., et al. (2017). Antifungal activity of nano emulsions of neem and citronella oils against phytopathogenic fungi, Rhizoctonia solani and Sclerotium rolfsii. *Industrial Crops and Products, 108*, 379–387.

Alijani, K., Rezaei, J., & Rouzbehan, Y. (2020). Effect of nano-ZnO, compared to ZnO and Zn-methionine, on performance, nutrient status, rumen fermentation, blood enzymes, ferric reducing antioxidant power and immunoglobulin G in sheep. *Animal Feed Science and Technology*, *267*, 114532.

Arabpoor, B., Yousefi, S., Weisany, W., & Ghasemlou, M. (2021). Multifunctional coating composed of Eryngium campestre L. essential oil encapsulated in nano-chitosan to prolong the shelf-life of fresh cherry fruits. *Food Hydrocolloids*, *111*, 106394.

Assis, R. Q., Pagno, C. H., Costa, T. M. H., Flôres, S. H., & Rios, A. D. O. (2018). Synthesis of biodegradable films based on cassava starch containing free and nanoencapsulated β-carotene. *Packaging Technology and Science*, *31*(3), 157–166.

Bajpai, V. K., Kamle, M., Shukla, S., Mahato, D. K., Chandra, P., Hwang, S. K., et al. (2018). Prospects of using nanotechnology for food preservation, safety, and security. *Journal of Food and Drug Analysis*, *26*(4), 1201–1214.

Ben-Eli, M. U. (2018). Sustainability: Definition and five core principles, a systems perspective. *Sustainability Science*, *13*(5), 1337–1343.

Bettencourt, G. M. D. F., Degenhardt, J., Torres, L. A. Z., Tanobe, V. O. D. A., & Soccol, C. R. (2020). Green biosynthesis of single and bimetallic nanoparticles of iron and manganese using bacterial auxin complex to act as plant bio-fertilizer. *Biocatalysis and Agricultural Biotechnology*, *30*, 101822.

Bhushani, J. A., Kurrey, N. K., & Anandharamakrishnan, C. (2017). Nanoencapsulation of green tea catechins by electrospraying technique and its effect on controlled release and in-vitro permeability. *Journal of Food Engineering*, *199*, 82–92.

Carvalho, F. P. (2017). Pesticides, environment, and food safety. *Food and Energy Security*, *6*(2), 48–60.

Dawood, M. A. O., Zommara, M., Eweedah, N. M., Helal, A. I., & Aboel-Darag, M. A. (2020). The potential role of nano-selenium and vitamin C on the performances of Nile tilapia (*Oreochromis niloticus*). *Environmental Science and Pollution Research*, *27*(9), 9843–9852.

Debnath, N., Das, S., Seth, D., Chandra, R., Bhattacharya, S. C., & Goswami, A. (2011). Entomotoxic effect of silica nanoparticles against *Sitophilus oryzae* (L.). *Journal of Pest Science*, *84*(1), 99–105.

Devi, P. S. V., Duraimurugan, P., Chandrika, K. S. V. P., Gayatri, B., & Prasad, R. D. (2019). Nanobiopesticides for Crop Protection. In K. A. Abd-Elsalam, & R. Prasad (Eds.), *Nanobiotechnology applications in plant protection* (pp. 145–168). Springer.

Dong, Y., & Hauschild, M. Z. (2017). Indicators for environmental sustainability. *Procedia CIRP*, *61*, 697–702.

Dribek, M., Rinnert, E., Colas, F., Crassous, M. P., Thioune, N., David, C., et al. (2017). Organometallic nanoprobe to enhance optical response on the polycyclic aromatic hydrocarbon benzo[a]pyrene immunoassay using SERS technology. *Environmental Science and Pollution Research*, *24*(35), 27070–27076.

Dukare, S., Mir, N. A., Mandal, A. B., Dev, K., Begum, J., Rokade, J. J., et al. (2020). A comparative study on the antioxidant status, meat quality, and mineral deposition in broiler chicken fed dietary nano zinc viz-a-viz inorganic zinc. *Journal of Food Science and Technology*, *58*(3), 834–843.

Echegaray, N., Gómez, B., Barba, F. J., Franco, D., Estévez, M., Carballo, J., et al. (2018). Chestnuts and by-products as source of natural antioxidants in meat and meat products: A review. *Trends in Food Science and Technology*, *82*, 110–121.

El Bourakadi, K., Merghoub, N., Fardioui, M., Mekhzoum, M. E. M., Kadmiri, I. M., Essassi, E. M., et al. (2019). Chitosan/polyvinyl alcohol/thiabendazoluim-montmorillonite bio-nanocomposite films: Mechanical, morphological and antimicrobial properties. *Composites Part B: Engineering*, *172*, 103–110.

El Miri, N., Abdelouahdi, K., Barakat, A., Zahouily, M., Fihri, A., Solhy, A., et al. (2015). Bio-nanocomposite films reinforced with cellulose nanocrystals: Rheology of film-forming solutions, transparency, water vapor barrier and tensile properties of films. *Carbohydrate Polymers*, *129*, 156–167.

El-Attar, A. B., Mahmoud, A. W. M., & Mahmoud, A. A. (2017). Economic evaluation of nano and organic fertilisers as an alternative source to chemical fertilisers on *Carum carvi* l. plant yield and components. *Agriculture*, *63*(1), 33–49.

Eliaspour, S., Sharifi, R. S., Shirkhani, A., & Farzaneh, S. (2020). Effects of biofertilizers and iron nano-oxide on maize yield and physiological properties under optimal irrigation and drought stress conditions. *Food Science and Nutrition*, *8*(11), 5985–5998.

European Commission. (2020a). *EU Pesticides Database: Active substances, safeners and synergists*. Retrieved January 8, 2021, from https://ec.europa.eu/food/plant/pesticides/eu-pesticides-database/active-substances/index.cfm?event=search.as&a_from=&a_to=&e_from=&e_to=&additionalfilter__class_p1=&additionalfilter__class_p2.

European Commission. (2020b). *EU pesticides database: Maximum residual levels*. Retrieved January 8, 2021, from http://ec.europa.eu/sanco/pesticides/public/.

Fang, Y., Umasankar, Y., & Ramasamy, R. P. (2014). Electrochemical detection of p-ethylguaiacol, a fungi infected fruit volatile using metal oxide nanoparticles. *Analyst*, *139*(15), 3804–3810.

FAO/WHO. (2018). *Sustainable food systems concept and framework*.

Ghaderzadeh, S., Aghjehgheshlagh, F. M., Nikbin, S., & Navidshad, B. (2020). Stimulatory effects of nano-selenium and conjugated linoleic acid on antioxidant activity, trace minerals, and gene expression response of growing male moghani lambs. *Veterinary Research Forum*, *11*(4), 385–391.

Giraldo, J. P., Wu, H., Newkirk, G. M., & Kruss, S. (2019). Nanobiotechnology approaches for engineering smart plant sensors. *Nature Nanotechnology*, *14*(6), 541–553.

Gosavi, V. C., Daspute, A. A., Patil, A., Gangurde, A., Wagh, S. G., Sherkhane, A., et al. (2020). Synthesis of green nanobiofertilizer using silver nanoparticles of *Allium cepa* extract short title: Green nanofertilizer from *Allium cepa*. *International Journal of Chemical Studies*, *8*(4), 1690–1694.

Guillard, V., Gaucel, S., Fornaciari, C., Angellier-Coussy, H., Buche, P., & Gontard, N. (2018). The next generation of sustainable food packaging to preserve our environment in a circular economy context. *Frontiers in Nutrition, 5*, 121.

Haghighi, H., Gullo, M., La China, S., Pfeifer, F., Siesler, H. W., Licciardello, F., et al. (2021). Characterization of bio-nanocomposite films based on gelatin/polyvinyl alcohol blend reinforced with bacterial cellulose nanowhiskers for food packaging applications. *Food Hydrocolloids, 113*, 106454.

Han, H. J., Lee, J. S., Park, S. A., Ahn, J. B., & Lee, H. G. (2015). Extraction optimization and nanoencapsulation of jujube pulp and seed for enhancing antioxidant activity. *Colloids and Surfaces B: Biointerfaces, 130*, 93–100.

Han, L., Pang, K., Fu, T., Phillips, C. J. C., & Gao, T. (2021). Nano-selenium supplementation increases selenoprotein (Sel) gene expression profiles and milk selenium concentration in lactating dairy cows. *Biological Trace Element Research, 199*(1), 113–119.

Hassaan, M. S., Nssar, K. M., Mohammady, E. Y., Amin, A., Tayel, S. I., & El-Haroun, E. R. (2020). Nano-zeolite efficiency to mitigate the aflatoxin B1 (AFB1) toxicity: Effects on growth, digestive enzymes, antioxidant, DNA damage and bioaccumulation of AFB1 residues in Nile tilapia (*Oreochromis niloticus*). *Aquaculture, 523*, 735123.

He, X., Deng, H., & Hwang, H.m. (2019). The current application of nanotechnology in food and agriculture. *Journal of Food and Drug Analysis, 27*(1), 1–21.

Herculano, E. D., de Paula, H. C. B., de Figueiredo, E. A. T., Dias, F. G. B., & Pereira, V. de A. (2015). Physicochemical and antimicrobial properties of nanoencapsulated Eucalyptus staigeriana essential oil. *LWT - Food Science and Technology, 61*(2), 484–491.

Herrero, M., Thornton, P. K., Mason-D'Croz, D., Palmer, J., Benton, T. G., Bodirsky, B. L., et al. (2020). Innovation can accelerate the transition towards a sustainable food system. *Nature Food, 1*(5), 266–272.

Honarvar, Z., Hadian, Z., & Mashayekh, M. (2016). Nanocomposites in food packaging applications and their risk assessment for health. *Electronic Physician, 8*(6), 2531–2538.

Horuz, T.İ., & Belibağlı, K. B. (2018). Nanoencapsulation by electrospinning to improve stability and water solubility of carotenoids extracted from tomato peels. *Food Chemistry, 268*, 86–93.

Hosseini, S. F., Rezaei, M., Zandi, M., & Farahmandghavi, F. (2015). Fabrication of bio-nanocomposite films based on fish gelatin reinforced with chitosan nanoparticles. *Food Hydrocolloids, 44*, 172–182.

Hsueh, C. Y., Tsai, M. L., & Liu, T. (2017). Enhancing saltiness perception using chitin nanofibers when curing tilapia fillets. *LWT - Food Science and Technology, 86*, 93–98.

Inguglia, E. S., Zhang, Z., Tiwari, B. K., Kerry, J. P., & Burgess, C. M. (2017). Salt reduction strategies in processed meat products—A review. *Trends in Food Science and Technology, 59*, 70–78.

Itroutwar, P. D., Govindaraju, K., Tamilselvan, S., Kannan, M., Raja, K., & Subramanian, K. S. (2020). Seaweed-based biogenic ZnO nanoparticles for improving agro-morphological characteristics of rice (*Oryza sativa* L.). *Journal of Plant Growth Regulation, 39*(2), 717–728.

Jemaa, M. B., Falleh, H., Serairi, R., Neves, M. A., Snoussi, M., Isoda, H., et al. (2018). Nanoencapsulated Thymus capitatus essential oil as natural preservative. *Innovative Food Science and Emerging Technologies, 45*, 92–97.

Jiang, W. J., Tsai, M. L., & Liu, T. (2017). Chitin nanofiber as a promising candidate for improved salty taste. *LWT - Food Science and Technology, 75*, 65–71.

Keeling, L., Tunón, H., Olmos Antillón, G., Berg, C., Jones, M., Stuardo, L., et al. (2019). Animal welfare and the United Nations sustainable development goals. *Frontiers in Veterinary Science, 6*, 336.

Khaledian, S., Nikkhah, M., Shams-bakhsh, M., & Hoseinzadeh, S. (2017). A sensitive biosensor based on gold nanoparticles to detect Ralstonia solanacearum in soil. *Journal of General Plant Pathology, 83*(4), 231–239.

Kim, K. H., Kabir, E., & Jahan, S. A. (2017). Exposure to pesticides and the associated human health effects. *Science of the Total Environment, 575*, 525–535.

Koltun, P. (2010). Materials and sustainable development. *Progress in Natural Science: Materials International, 20*(1), 16–29.

Kumar, N., Kaur, P., & Bhatia, S. (2017). Advances in bio-nanocomposite materials for food packaging: A review. *Nutrition and Food Science, 47*(4), 591–606.

Kurozawa, L. E., & Hubinger, M. D. (2017). Hydrophilic food compounds encapsulation by ionic gelation. *Current Opinion in Food Science, 15*, 50–55.

Kuswandi, B., Futra, D., & Heng, L. Y. (2017). Nanosensors for the detection of food contaminants. In A. E. Oprea, & A. M. Grumezescu (Eds.), *Nanotechnology applications in food: Flavor, stability, Nutrition and Safety* (pp. 307–333). Elsevier.

Labuschagne, C., Brent, A. C., & Van Erck, R. P. G. (2005). Assessing the sustainability performances of industries. *Journal of Cleaner Production, 13*(4), 373–385.

Lau, H. Y., Wu, H., Wee, E. J. H., Trau, M., Wang, Y., & Botella, J. R. (2017). Specific and sensitive isothermal electrochemical biosensor for plant pathogen DNA detection with colloidal gold nanoparticles as probes. *Scientific Reports, 7*(1), 1–7.

Lee, J. H., Hosseindoust, A., Kim, M. J., Kim, K. Y., Choi, Y. H., Lee, S. H., et al. (2020). Supplemental hot melt extruded nano-selenium increases expression profiles of antioxidant enzymes in the livers and spleens of weanling pigs. *Animal Feed Science and Technology, 262*, 114381.

Lew, T. T. S., Koman, V. B., Gordiichuk, P., Park, M., & Strano, M. S. (2020). The emergence of plant nanobionics and living plants as technology. *Advanced Materials Technologies, 5*(3), 1900657.

Lew, T. T. S., Park, M., Cui, J., & Strano, M. S. (2021). Plant nanobionic sensors for arsenic detection. *Advanced Materials, 33*(1), 2005683.

Li, T. Y., Fu, C. M., & Lien, T. F. (2017). Effects of nanoparticle chromium on chromium absorbability, growth performance, blood parameters and carcass traits of pigs. *Animal Production Science, 57*(6), 1193–1200.

Licciardello, F. (2017). Packaging, blessing in disguise. Review on its diverse contribution to food sustainability. *Trends in Food Science and Technology, 65*, 32–39.

Lin, Y. C., Huang, J. T., Li, M. Z., Cheng, C. Y., & Lien, T. F. (2015). Effects of supplemental nanoparticle trivalent chromium on the nutrient utilization, growth performance and serum traits of broilers. *Journal of Animal Physiology and Animal Nutrition, 99*(1), 59–65.

Liu, Y., Chen, X., Wang, M., & Ma, Q. (2018). A visual electrochemiluminescence resonance energy transfer/surface plasmon coupled electrochemiluminescence nanosensor for Shiga toxin-producing: *Escherichia coli* detection. *Green Chemistry, 20*(24), 5520–5527.

Liu, Y., Yang, G., Zou, D., Hui, Y., Nigam, K., Middelberg, A. P. J., et al. (2020). Formulation of nanoparticles using mixing-induced nanoprecipitation for drug delivery. *Industrial and Engineering Chemistry Research, 59*(9), 4134–4149.

Llanos, J. H. R., & Tadini, C. C. (2018). Preparation and characterization of bio-nanocomposite films based on cassava starch or chitosan, reinforced with montmorillonite or bamboo nanofibers. *International Journal of Biological Macromolecules, 107*(PartA), 371–382.

Lorenzo, J. M., Munekata, P. E. S., Muchenje, V., Saraiva, J. A., Pinto, C. A., Barba, F. J., et al. (2019). Biosensors applied to quantification of ethanol in beverages. In *Engineering tools in the beverage industry* (pp. 447–468).

Lorite, G. S., Rocha, J. M., Miilumäki, N., Saavalainen, P., Selkälä, T., Morales-Cid, G., et al. (2017). Evaluation of physicochemical/microbial properties and life cycle assessment (LCA) of PLA-based nanocomposite active packaging. *LWT - Food Science and Technology, 75*, 305–315.

Lowry, G. V., Avellan, A., & Gilbertson, L. M. (2019). Opportunities and challenges for nanotechnology in the agri-tech revolution. *Nature Nanotechnology, 14*(6), 517–522.

Mahdavi, V., Rafiee-Dastjerdi, H., Asadi, A., Razmjou, J., & Achachlouei, B. F. (2020). Evaluation of *Lippia citriodora* essential oil nanoformulation against the lepidopteran pest *Phthorimaea operculella* Zeller (Gelechiidae). *International Journal of Pest Management*. In press.

Mardalipour, M., Zahedi, H., & Sharghi, Y. (2014). Evaluation of nano biofertilizer efficiency on agronomic traits of spring wheat at different sowing date. *Biological Forum – An International Journal, 6*(2), 349–356. Retrieved from https://www.researchgate.net/publication/287644160.

McKenzie, F. C., & Williams, J. (2015). Sustainable food production: Constraints, challenges and choices by 2050. *Food Security, 7*(2), 221–233.

Medeiros, A. K. D. O. C., Gomes, C. D. C., Amaral, M. L. Q. D. A., de Medeiros, L. D. G., Medeiros, I., Porto, D. L., et al. (2019). Nanoencapsulation improved water solubility and color stability of carotenoids extracted from cantaloupe melon (Cucumis melo L.). *Food Chemistry, 270*, 562–572.

Modarres-Gheisari, S. M. M., Gavagsaz-Ghoachani, R., Malaki, M., Safarpour, P., & Zandi, M. (2019). Ultrasonic nano-emulsification—A review. *Ultrasonics Sonochemistry, 52*, 88–105.

Mohammadi, A., Jafari, S. M., Assadpour, E., & Faridi Esfanjani, A. (2016). Nano-encapsulation of olive leaf phenolic compounds through WPC-pectin complexes and evaluating their release rate. *International Journal of Biological Macromolecules, 82*, 816–822.

Mohammadi, A., Jafari, S. M., Esfanjani, A. F., & Akhavan, S. (2016). Application of nano-encapsulated olive leaf extract in controlling the oxidative stability of soybean oil. *Food Chemistry, 190*, 513–519.

Moncada, M., Astete, C., Sabliov, C., Olson, D., Boeneke, C., & Aryana, K. J. (2015). Nano spray-dried sodium chloride and its effects on the microbiological and sensory characteristics of surface-salted cheese crackers. *Journal of Dairy Science, 98*(9), 5946–5954.

Mondal, A. H., Behera, T., Swain, P., Das, R., Sahoo, S. N., Mishra, S. S., et al. (2020). Nano zinc Vis-à-Vis inorganic zinc as feed additives: Effects on growth, activity of hepatic enzymes and non-specific immunity in rohu, *Labeo rohita* (Hamilton) fingerlings. *Aquaculture Nutrition, 26*(4), 1211–1222.

Muhammad, D. R. A., Sedaghat Doost, A., Gupta, V., bin Sintang, M. D., Van de Walle, D., Van der Meeren, P., et al. (2020). Stability and functionality of xanthan gum–shellac nanoparticles for the encapsulation of cinnamon bark extract. *Food Hydrocolloids, 100*, 105377.

Nabifarkhani, N., Sharifani, M., Garmakhany, A. D., Moghadam, E. G., & Shakeri, A. (2015). Effect of nano-composite and thyme oil (*Thymus vulgaris* L) coating on fruit quality of sweet cherry (Takdaneh Cv) during storage period. *Food Science and Nutrition, 3*(4), 349–354.

Naderi, M., Keyvanshokooh, S., Ghaedi, A., & Salati, A. P. (2019). Interactive effects of dietary Nano selenium and vitamin E on growth, haematology, innate immune responses, antioxidant status and muscle composition of rainbow trout under high rearing density. *Aquaculture Nutrition, 25*(5), 1156–1168.

Nagarajan, M., Benjakul, S., Prodpran, T., & Songtipya, P. (2015). Effects of bio-nanocomposite films from tilapia and squid skin gelatins incorporated with ethanolic extract from coconut husk on storage stability of mackerel meat powder. *Food Packaging and Shelf Life, 6*, 42–52.

Nath, P., Arun, R. K., & Chanda, N. (2015). Smart gold nanosensor for easy sensing of lead and copper ions in solution and using paper strips. *RSC Advances, 5*(84), 69024–69031.

Neamat-Allah, A. N. F., Mahmoud, E. A., & Abd El Hakim, Y. (2019). Efficacy of dietary Nano-selenium on growth, immune response, antioxidant, transcriptomic profile and resistance of Nile tilapia, *Oreochromis niloticus* against Streptococcus iniae infection. *Fish and Shellfish Immunology, 94*, 280–287.

Neethirajan, S., Ragavan, V., Weng, X., & Chand, R. (2018). Biosensors for sustainable food engineering: Challenges and perspectives. *Biosensors, 8*(1), 23.

Nguyen, M. H., Vu, N. B. D., Nguyen, T. H. N., Tran, T. N. M., Le, H. S., Tran, T. T., et al. (2020). Effective biocontrol of nematodes using lipid nanoemulsions co-encapsulating chili oil, cinnamon oil and neem oil. *International Journal of Pest Management*. In press.

Pabast, M., Shariatifar, N., Beikzadeh, S., & Jahed, G. (2018). Effects of chitosan coatings incorporating with free or nano-encapsulated Satureja plant essential oil on quality characteristics of lamb meat. *Food Control, 91*, 185–192.

Pascoli, M., de Albuquerque, F. P., Calzavara, A. K., Tinoco-Nunes, B., Oliveira, W. H. C., Gonçalves, K. C., et al. (2020). The potential of nanobiopesticide based on zein nanoparticles and neem oil for enhanced control of agricultural pests. *Journal of Pest Science, 93*(2), 793–806.

Pateiro, M., Munekata, P. E. S., Cittadini, A., Domínguez, R., & Lorenzo, J. M. (2021). Metallic-based salt substitutes to reduce sodium content in meat products. *Current Opinion in Food Science, 38*, 21–31.

Pei, X., Xiao, Z., Liu, L., Wang, G., Tao, W., Wang, M., et al. (2019). Effects of dietary zinc oxide nanoparticles supplementation on growth performance, zinc status, intestinal morphology, microflora population, and immune response in weaned pigs. *Journal of the Science of Food and Agriculture, 99*(3), 1366–1374.

Pereira, M. C., Oliveira, D. A., Hill, L. E., Zambiazi, R. C., Borges, C. D., Vizzotto, M., et al. (2018). Effect of nano-encapsulation using PLGA on antioxidant and antimicrobial activities of guabiroba fruit phenolic extract. *Food Chemistry, 240*, 396–404.

Peres, M. C., de Souza Costa, G. C., dos Reis, L. E. L., da Silva, L. D., Peixoto, M. F., Alves, C. C. F., et al. (2020). In natura and nanoencapsulated essential oils from *Xylopia aromatica* reduce oviposition of *Bemisia tabaci* in *Phaseolus vulgaris*. *Journal of Pest Science, 93*(2), 807–821.

Perumal, A. B., Sellamuthu, P. S., Nambiar, R. B., & Sadiku, E. R. (2018). Development of polyvinyl alcohol/chitosan bio-nanocomposite films reinforced with cellulose nanocrystals isolated from rice straw. *Applied Surface Science, 449*, 591–602.

Popovic, T., Barbosa-Póvoa, A., Kraslawski, A., & Carvalho, A. (2018). Quantitative indicators for social sustainability assessment of supply chains. *Journal of Cleaner Production, 180*, 748–768.

Qian, Z. S., Shan, X. Y., Chai, L. J., Chen, J. R., & Feng, H. (2015). A fluorescent nanosensor based on graphene quantum dots-aptamer probe and graphene oxide platform for detection of lead (II) ion. *Biosensors and Bioelectronics, 68*, 225–231.

Qin, F., Shen, T., Yang, H., Qian, J., Zou, D., Li, J., et al. (2019). Dietary nano cerium oxide promotes growth, relieves ammonia nitrogen stress, and improves immunity in crab (*Eriocheir sinensis*). *Fish and Shellfish Immunology, 92*, 367–376.

Rahman, M. A., Parvin, A., Khan, M. S. H., Lingaraju, K., Prasad, R., Das, S., et al. (2020). Efficacy of the green synthesized nickel-oxide nanoparticles against pulse beetle, *Callosobruchus maculatus* (F.) in black gram (*Vigna mungo* L.). *International Journal of Pest Management*. In press.

Rajkumar, V., Gunasekaran, C., Dharmaraj, J., Chinnaraj, P., Paul, C. A., & Kanithachristy, I. (2020). Structural characterization of chitosan nanoparticle loaded with *Piper nigrum* essential oil for biological efficacy against the stored grain pest control. *Pesticide Biochemistry and Physiology, 166*, 104566.

Rani, P. U., Madhusudhanamurthy, J., & Sreedhar, B. (2014). Dynamic adsorption of α-pinene and linalool on silica nanoparticles for enhanced antifeedant activity against agricultural pests. *Journal of Pest Science, 87*(1), 191–200.

Sahoo, A., Swain, R. K., Mishra, S. K., Behura, N. C., Beura, S. S., Sahoo, C., et al. (2016). Growth, feed conversion efficiency, and carcass characteristics of broiler chicks fed on inorganic, organic and nano zinc supplemented diets. *Animal Science Reporter, 10*(1), 10–18.

Sahraee, S., Milani, J. M., Ghanbarzadeh, B., & Hamishehkar, H. (2017). Physicochemical and antifungal properties of bio-nanocomposite film based on gelatin-chitin nanoparticles. *International Journal of Biological Macromolecules, 97*, 373–381.

Salcido-Martínez, A., Sánchez, E., Lorena Licón-Trillo, P., Pérez-Álvarez, S., Palacio-Márquez, A., Nubia Amaya-Olivas, I., et al. (2020). Impact of the foliar application of magnesium nanofertilizer on physiological and biochemical parameters and yield in green beans. *Notulae Botanicae Horti Agrobotanici Cluj-Napoca, 48*(4), 2167–2181.

Saleh, A. A., & Ebeid, T. A. (2019). Feeding sodium selenite and nano-selenium stimulates growth and oxidation resistance in broilers. *South African Journal of Animal Sciences, 49*(1), 176–183.

Salerno, G., Rebora, M., Kovalev, A., Gorb, E., & Gorb, S. (2020). Kaolin nano-powder effect on insect attachment ability. *Journal of Pest Science, 93*(1), 315–327.

Salieri, B., Turner, D. A., Nowack, B., & Hischier, R. (2018). Life cycle assessment of manufactured nanomaterials: Where are we? *NanoImpact, 10*, 108–120.

Saminathan, M., Selamat, J., Pirouz, A. A., Abdullah, N., & Zulkifli, I. (2018). Effects of nano-composite adsorbents on the growth performance, serum biochemistry, and organ weights of broilers fed with aflatoxin-contaminated feed. *Toxins, 10*(9), 345.

Sandri, G., Bonferoni, M. C., Rossi, S., Ferrari, F., Aguzzi, C., Viseras, C., et al. (2016). Clay minerals for tissue regeneration, repair, and engineering. In M. S. Ågren (Ed.), *Vol. 2. Wound healing biomaterials* (pp. 385–402). Elsevier.

Sanjay, P. N., Gayithri, K. C., Naveen Kumar, S. K., Krishna, V., & Jagadeesh Prasad, D. (2016). TiO₂-PANI based anti-typhi immobilized nanosensor for *Salmonella* typhi detection. *Materials Today: Proceedings, 3*(6), 1772–1777.

Satapathi, S., Kumar, V., Chini, M. K., Bera, R., Halder, K. K., & Patra, A. (2018). Highly sensitive detection and removal of mercury ion using a multimodal nanosensor. *Nano-Structures and Nano-Objects, 16*, 120–126.

Seifdavati, J., Jahan Ara, M., Seyfzadeh, S., Abdi Benamar, H., Mirzaie Aghjeh Gheshlagh, F., Seyedsharifi, R., et al. (2018). The effects of zinc oxide nano particles on growth performance and blood metabolites and some serum enzymes in Holstein suckling calves. *Iranian Journal of Animal Science Research, 10*(1), 23–33.

Shadjou, R., Hasanzadeh, M., Heidar-poor, M., & Shadjou, N. (2018). Electrochemical monitoring of aflatoxin M1 in milk samples using silver nanoparticles dispersed on α-cyclodextrin-GQDs nanocomposite. *Journal of Molecular Recognition, 31*(6), e2699.

Sharifi, R. S., Khalilzadeh, R., Pirzad, A., & Anwar, S. (2020). Effects of biofertilizers and nano zinc-iron oxide on yield and physicochemical properties of wheat under water deficit conditions. *Communications in Soil Science and Plant Analysis, 51*(19), 2511–2524.

Sheiha, A. M., Abdelnour, S. A., Abd El-Hack, M. E., Khafaga, A. F., Metwally, K. A., Ajarem, J. S., et al. (2020). Effects of dietary biological or chemical-synthesized nano-selenium supplementation on growing rabbits exposed to thermal stress. *Animals, 10*(3), 430.

Shen, Y., Wu, T., Zhang, Y., Ling, N., Zheng, L., Zhang, S. L., et al. (2020). Engineering of a dual-recognition ratiometric fluorescent nanosensor with a remarkably large stokes shift for accurate tracking of pathogenic bacteria at the single-cell level. *Analytical Chemistry, 92*(19), 13396–13404.

Sibbel, A. (2007). The sustainability of functional foods. *Social Science and Medicine, 64*(3), 554–561.

Singh, K. K., Maity, S. B., & Maity, A. (2018). Effect of nano zinc oxide on zinc bioavailability and blood biochemical changes in pre-ruminant lambs. *Indian Journal of Animal Sciences, 88*(7), 805–807. Retrieved from https://www.researchgate.net/publication/326530433.

Singh, T., Shukla, S., Kumar, P., Wahla, V., & Bajpai, V. K. (2017). Application of nanotechnology in food science: Perception and overview. *Frontiers in Microbiology, 8*(AUG), 1501.

Song, H., Yuan, W., Jin, P., Wang, W., Wang, X., Yang, L., et al. (2016). Effects of chitosan/nano-silica on postharvest quality and antioxidant capacity of loquat fruit during cold storage. *Postharvest Biology and Technology, 119*, 41–48.

Sun, Y. B., Xia, T., Wu, H., Zhang, W. J., Zhu, Y. H., Xue, J. X., et al. (2019). Effects of nano zinc oxide as an alternative to pharmacological dose of zinc oxide on growth performance, diarrhea, immune responses, and intestinal microflora profile in weaned piglets. *Animal Feed Science and Technology, 258*, 114312.

Tatlisu, N. B., Yilmaz, M. T., & Arici, M. (2019). Fabrication and characterization of thymol-loaded nanofiber mats as a novel antimould surface material for coating cheese surface. *Food Packaging and Shelf Life, 21*, 100347.

Tijunelyte, I., Betelu, S., Moreau, J., Ignatiadis, I., Berho, C., Lidgi-Guigui, N., et al. (2017). Diazonium salt-based surface-enhanced Raman spectroscopy nanosensor: Detection and quantitation of aromatic hydrocarbons in water samples. *Sensors (Switzerland), 17*(6), 1198.

Umaraw, P., Munekata, P. E. S., Verma, A. K., Barba, F. J., Singh, V. P., Kumar, P., et al. (2020). Edible films/coating with tailored properties for active packaging of meat, fish and derived products. *Trends in Food Science and Technology, 98*, 10–24.

Velarde, A., Fàbrega, E., Blanco-Penedo, I., & Dalmau, A. (2015). Animal welfare towards sustainability in pork meat production. *Meat Science, 109*, 13–17.

Vidal, V. A. S., Lorenzo, J. M., Munekata, P. E. S., & Pollonio, M. A. R. (2020). Challenges to reduce or replace NaCl by chloride salts in meat products made from whole pieces—A review. *Critical Reviews in Food Science and Nutrition*, 1–13.

Vinitha, K., Leena, M. M., Moses, J. A., & Anandharamakrishnan, C. (2021). Size-dependent enhancement in salt perception: Spraying approaches to reduce sodium content in foods. *Powder Technology, 378*, 237–245.

Walia, N., Dasgupta, N., Ranjan, S., Chen, L., & Ramalingam, C. (2017). Fish oil based vitamin D nanoencapsulation by ultrasonication and bioaccessibility analysis in simulated gastro-intestinal tract. *Ultrasonics Sonochemistry, 39*, 623–635.

Walia, N., Dasgupta, N., Ranjan, S., Ramalingam, C., & Gandhi, M. (2019). Methods for nanoemulsion and nanoencapsulation of food bioactives. *Environmental Chemistry Letters, 17*(4), 1471–1483.

Wang, Z., Wei, F., Liu, S. Y., Xu, Q., Huang, J. Y., Dong, X. Y., et al. (2010). Electrocatalytic oxidation of phytohormone salicylic acid at copper nanoparticles-modified gold electrode and its detection in oilseed rape infected with fungal pathogen Sclerotinia sclerotiorum. *Talanta, 80*(3), 1277–1281.

Wang, Y., Zhu, Y., Zhang, S., & Wang, Y. (2018). What could promote farmers to replace chemical fertilizers with organic fertilizers? *Journal of Cleaner Production, 199*, 882–890.

Waseem, N., & Kota, S. (2017). Sustainability definitions — An analysis. *Smart Innovation, Systems and Technologies, 66*, 361–371.

Wu, H., Nißler, R., Morris, V., Herrmann, N., Hu, P., Jeon, S. J., et al. (2020). Monitoring plant health with near-infrared fluorescent H_2O_2 nanosensors. *Nano Letters, 20*(4), 2432–2442.

Xiao, W., Xiao, M., Fu, Q., Yu, S., Shen, H., Bian, H., et al. (2016). A portable smart-phone readout device for the detection of mercury contamination based on an aptamer-assay nanosensor. *Sensors (Switzerland), 16*(11), 1871.

Yu, Z., Alsammarraie, F. K., Nayigiziki, F. X., Wang, W., Vardhanabhuti, B., Mustapha, A., et al. (2017). Effect and mechanism of cellulose nanofibrils on the active functions of biopolymer-based nanocomposite films. *Food Research International, 99*, 166–172.

Yu, H., Park, J. Y., Kwon, C. W., Hong, S. C., Park, K. M., & Chang, P. S. (2018). An overview of nanotechnology in food science: Preparative methods, practical applications, and safety. *Journal of Chemistry, 2018*, 5427978.

Zhang, W. (2018). Global pesticide use: Profile, trend, cost/benefit and more. *Proceedings of the International Academy of Ecology and Environmental Sciences, 8*(1), 1–27. Retrieved from www.iaees.org.

Zheng, W., Luo, B., & Hu, X. (2020). The determinants of farmers' fertilizers and pesticides use behavior in China: An explanation based on label effect. *Journal of Cleaner Production, 272*, 123054.

Zhu, Q., Wang, L., Dong, Q., Chang, S., Wen, K., Jia, S., et al. (2017). FRET-based glucose imaging identifies glucose signalling in response to biotic and abiotic stresses in rice roots. *Journal of Plant Physiology, 215*, 65–72.

CHAPTER 13

Food Legislation: Particularities in Spain for Typical Products of the Mediterranean Diet

MARTINA P. SERRANO[A,B,C] • LOUIS CHONCO[D] • TOMÁS LANDETE-CASTILLEJOS[A,B,C] • ANDRÉS GARCÍA[A,B,C] • JOSÉ MANUEL LORENZO[E,F]
[a]Animal Science Techniques Applied to Wildlife Management Research Group, Hunting Resources Research Institute, Castilla-La Mancha University, Albacete, Spain, [b]Hunting and Livestock Resources Section, Regional Development Institute, Castilla-La Mancha University, Albacete, Spain, [c]Department of Agroforestry Science and Technology and Genetics, Higher Technical School of Agricultural and Forestry Engineers, Castilla-La Mancha University, Albacete, Spain, [d]Research Unit of University Hospital Complex of Albacete, Albacete, Spain, [e]Galician Meat Technology Center, Galicia Technology Park, Ourense, Spain, [f]Food Technology Area, Faculty of Sciences of Ourense, Vigo University, Ourense, Spain

13.1 INTRODUCTION

Current sustainable production systems are insufficient to feed the world. However, recently, Gerten et al. (2020) have concluded that transformation toward more sustainable production and consumption patterns could support 10.2 billion people within the planetary boundaries analyzed. Key prerequisites are spatially redistributed cropland, improved water–nutrient management, food waste reduction, and dietary changes. A challenging question, thus, is whether human development goals such as food security can be met while maintaining multiple planetary boundaries along with their subglobal manifestations. To achieve this food security, it is necessary to have current legislation that regulates the production of each and every one of the foods generated worldwide.

13.2 EUROPEAN UNION FOOD LEGISLATION

The objective of European Union (EU) legislation on agriculture, livestock, and food production is to protect the human health. A comprehensive set of rules regulates the entire food production and processing chain in the EU, including import and export of products. The EU food security policy focuses on four areas of protection:

- Food hygiene: from farm to table, food companies, including food importers, must comply with EU legislation.
- Animal health: controls and sanitary measures for companion animals, farm animals (and their movements), wildlife control, and manage disease.
- Plant health: the detection and eradication of pests at an early stage prevents their spread and guarantees the health of the seeds.
- Contaminants and residues: controls keep contaminants away from food and feed. Acceptable maximum limits apply to both domestic and imported food and feed.

In 2002, the European Parliament and the Council adopted the Regulation (EC) 178/2002 (2002) laying down the general principles and requirements of food Law (General Food Law Regulation, GFLR). The GFLR is the foundation of food and feed Law that ensures a high level of protection of human life and consumers' interests in relation to food, while ensuring the effective functioning of the internal market. It sets out an overarching and coherent framework for the development of food and feed legislation, both at EU and national levels. The GFLR lays down general principles, requirements, and procedures that underpin decision making in matters of food and feed safety, covering all stages of food and feed production, and distribution. It also sets up an independent agency responsible for scientific advice and support: the European Food Safety Authority (EFSA).

Following the approval by the European Parliament on 2019, the Council formally adopted a new Regulation

Sustainable Production Technology in Food. https://doi.org/10.1016/B978-0-12-821233-2.00013-7
Copyright © 2021 Elsevier Inc. All rights reserved.

on the transparency and sustainability of the EU risk assessment in the food chain Regulation (EU) 1381/2019 (2019). This new Regulation amends the GFLR and it aims to increase the transparency of the EU risk assessment in the food chain, on strengthening the reliability, objectivity, and independence of the studies used by EFSA, and revisiting the governance of EFSA in order to ensure its long-term sustainability. The new Transparency Regulation was published in the Official Journal on 6 September 2019. It entered into force 20 days after its publication and will become applicable on 27 March 2021. The Commission and EFSA are working closely to ensure the proper implementation of the new Regulation. The main elements of the Regulation aim to:

– Ensuring more transparency: citizens will have automatic access to all studies and information submitted by industry about the risk assessment process. Stakeholders and the general public will also be consulted on the submitted studies. At the same time, the Regulation will guarantee confidentiality, in duly justified circumstances, by setting out the type of information that may be considered significantly harmful for commercial interests and, therefore, cannot be disclosed.

– Increasing the independence of the studies: the EFSA will be notified of all commissioned studies to guarantee that companies applying for authorizations submit all relevant information and do not hold back unfavorable studies. The authority will also provide general advice to applicants prior to the submission of the dossier. Commission may ask the authority to commission additional studies for purposes verification and may perform fact-finding missions to verify the compliance of laboratories/studies with standards.

– Strengthening the governance and the scientific cooperation: member states, civil society, and European Parliament will be involved in the governance of the Authority by being duly represented in its Management Board. Member States will foster the authority's scientific capacity and engage the best independent experts into its work.

– Developing comprehensive risk communication: a general plan for risk communication will be adopted and will ensure a coherent risk communication strategy throughout the risk analysis process, combined with open dialogue among all interested parts.

13.3 SPANISH FOOD LEGISLATION

As an EU Member State since 1986, Spain observes all EU directives, regulations, and obligations, which are either directly applicable or need to be transposed to national Law. In Spain, the Food Safety and Nutrition Law outlines the basic Spanish food and feed regulations (Law 17/2011, 2011). This Law is based on the EU Regulations and Directives and includes the traditional food safety aspects of detection and removal of physical, chemical, and biological hazards as well as other less conventional issues such as obesity prevention and food advertising rules. It applies equally to domestic and imported products.

The Spanish Food Safety and Nutrition Law establishes basic definitions, goals, and principles for food safety. It also defines procedural rules and coordination mechanisms between the different public administrations responsible for food regulation. It sets out general food safety and health protection rules, regulates inspections and inspection fees, detention, and seizure rules of suspect food, and classifies breaches.

In Spain, the Ministry of Health, Consumption, and Social Welfare (MSCBS) controls imports of agricultural product intended for human consumption, while the Ministry of Agriculture, Fisheries and Food (MAGRAMA) controls imports of animal feed/ingredients and live animals not intended for direct human consumption. The Spanish Consumption, Food Safety and Nutrition Agency (AESAN), under MSCBS, is responsible for Food Safety and coordinates the control of the food chain. The AESAN was established as an independent agency and it is also responsible for risk management.

Spain's food regulations apply to both domestically produced and imported food products. However, food and beverage products from the United States do not require Spain specific permits and are not subject to special rules or regulations for retail sale in Spain (United State Department of Agriculture, 2020). However, all products must comply with the generally applied rules and regulations required for any food and beverage product sold within the EU market.

Currently, EFSA is closely monitoring the situation regarding the outbreak of coronavirus disease (COVID-19), which is affecting a large number of countries around the world. Based on EFSA reports, currently, there is no evidence that food is likely to be a source or likely route of transmission of the virus.

13.4 PARTICULARITIES FOR TYPICAL PRODUCTS OF THE MEDITERRANEAN DIET TYPICALLY PRODUCED IN SPAIN

Ros et al. (2014) conducted the largest nutrition study conducted in Europe, known as PREDIMED. This study showed a reduction of approximately 30% in cerebral or myocardial infarctions with the Mediterranean diet, and also large reductions in the risk of developing diabetes, arrhythmias, and breast cancer, especially when maintains a high consumption of extra virgin olive oil. Additionally, Estruch et al. (2018), in the continuation of the previous study but involving persons at high cardiovascular risk, the incidence of major cardiovascular events was lower among those assigned to a Mediterranean diet supplemented with extra-virgin olive oil or nuts than among those assigned to a reduced-fat diet. Among the typical foods of the Mediterranean diet, some ones that should be highlighted are cereals (bread), olive oil, Iberian cured products as ham, and grapes (wine). Next, this chapter describes the production data of each of these foods as well as the legislation by which they are regulated.

13.4.1 Cereals (Bread)

The estimation of total production and consumption of main cereals in Spain are showed in Table 13.1 (Ministerio de Agricultura, Pesca y Alimentación, 2019a). The highest cereal production in Spain is due to barley (mostly used for animal feed), followed by soft wheat (animal and human nutrition), and corn (mostly used for animal feed). The first legislation of the Spanish food code dates from, 1967 (Law 2484/1967, 1967) and has been consolidated in May 2019 (BOE-A-1967-16485, 2019). This Law contains the chapter XVII related to cereals.

Legislation that regulates the cereal sector in Spain is as follows (state provisions):
- Quality standard for rice packaged for consumption in the domestic market: Order 25297/1980 (1980) corrected by Order 27204/1980 (1980), and modified by Order 9343/1984 (1984).
- Technical-sanitary regulations for the elaboration, circulation, and trade of flaked or expanded cereals: Law 1094/1987 (1987) corrected by Law 25798/1987 (1987) modified by Order 11806/1989 (1989), Law 145/1997 (1997), Law 135/2010 (2010), and Law 176/2013 (2013), and consolidated in March 2013 (BOE-A-1987-21012, 2013).
- Wheat quality legislation: Law 1615/2010 (2010) corrected by Order 12109/2010 (2010) and modified by Law 190/2013 (2013).

The high production of soft wheat in Spain is due to this type of wheat makes almost all types of bread, cookies, breakfast cereals, toasts, and cakes. Bread is one of the typical products consumed in Spain within the Mediterranean diet. Based on data from Ministerio de Agricultura, Pesca y Alimentación (2019b):

(1) the 80.7% of the total volume of bread consumption corresponds to fresh bread, which results from the sum of the fresh whole wheat bread, normal fresh bread, and fresh bread without salt;
(2) despite, the bread consumption decreased 2.0% in, 2018 compared to the previous year consumption, as average, the per capita consumption of bread in Spain is of 31.9 kg/person/year;
(3) on average, Spanish households spend 5.1% of their expenditure on food and beverages for the household buying bread, which implies an average cost of € 76.38/person/year (2.2% less than in, 2017).

With the aim of promoting the consumption of bread, recently, the Law 308/2019 (2019) has been published approving the quality standard for bread. This new Law comes to repeal the previous regulation, which has been in force for more than 35 years in Spain, and incorporates numerous quality innovations and guarantees for the consumer, and several that improve nutritional quality and promote the consumption of more healthy bread containing more whole wheat flour and a limited amount of salt.

13.4.2 Olive Oil

Spain is the first in the world in cultivation area and production of olive oil. Olive oil Spanish production represents approximately 60% of EU production and 45% of world production. The total of olive oils comprises the sum of the three types of olive oils marketed for direct consumption: olive oil, virgin olive oil, and olive oil extra virgin. The one with the greatest presence throughout, 2018 was the olive oil with a volume share of 32.1% (+0.2% compared with, 2017; Ministerio de Agricultura, Pesca y Alimentación, 2019b). The extra virgin olive oil was the next with the highest volume quota within olive oils (+2.5% compared to, 2017). Finally, virgin olive oil is positioned as the variety with the lowest volume within the group of olive oils (10.5%) despite having been the one that has experienced the greatest growth compared to the previous study period (+9.2%). Image 13.1 shows an olive grove in Spain. Fig. 13.1 shows production, exports, and imports data of olive oil in Spain in the last two campaigns (Ministerio de Agricultura, Pesca y Alimentación, 2020a).

The general provisions concerning edible vegetable oils are divided into: (1) community provisions, (2) state provisions, and (3) autonomic provisions. The community provisions are summarized below:
1. Regulation (EEC) 2568/1991 (1991) regarding the characteristics of olive oils and olive-pomace oils and on their analysis methods' modified by 31

TABLE 13.1
Total production and consumption of main cereals during 2019/2020 in Spain (Ministerio de Agricultura, Pesca y Alimentación, 2019a). [a]

Item, thousands of tons	Soft wheat	Durum wheat	Barley	Corn	Rye	Oat	Sorghum	Triticale	Total
Production	5049.7	738.8	7378.5	4111.7	242	811.6	30.09	591.3	18,953.8
Imports	5800	300	950	10,500	60	100	250	100	18,060
Total available	11,818.2	1340.6	9806.2	16,127.4	330.8	994.1	339	720.5	41,476.8
Internal consumption	10,714.8	894	8620.2	14,569.9	297	899.7	301.4	690.8	36,987.3
Animal nutrition	6100	230	7400	12,000	190	790	300	650	27,660
Human nutrition	4200	600	10	100	85	35	–	–	5030
Exports	200	350	100	175	20	50	2	7	904
Total uses	10,914.8	1244	8720.2	14,744.9	317	949.7	303.4	697.8	37,891.3
Ending stocks	903.4	96.6	1086	1382.5	13.8	44.5	35.6	22.7	3585.1

[a] Estimated cereal balance in the 2019/2020 season.

IMAGE 13.1 Olive grove in Spain.(Source: elaceite.net.)

regulations and 8 rectifications until the consolidation in 2016 (Regulation (EEC) 01991R2568, 2016) modified by the Delegated Regulation (EU) 2095/2016 (2016) corrected by DOCE (2017).

2. Regulation (EU) 29/2012 (2012) on the marketing rules of olive oil consolidated in 2016 (Regulation (EEC) 2012R0029, 2016).

3. Regulation (EU) 1308/2013 (2013) creating the common organization of markets for agricultural products. This Regulation was consolidated in, 2017 (Regulation (EEC) 02013R1308, 2012).

On the other hand, the state provisions are summarized below:

1. Law 308/1983 (1983) which approves the technical sanitary regulation of edible vegetable oils consolidated in 2015 (BOE-A-1983-5543, 2015).

2. Order 301/1985 (1985) on delivery of virgin olive oil by the oil mills to their harvesters for self-consumption.

3. Order 265/1989 (1989) which approves the quality standard for heated oils and fats. This Order was consolidated in 2013 (BOE-A-1989-2265, 2013).

4. Law 1431/2003 (2003) which establishes certain marketing measures in the olive oil and olive-pomace oil sector. This Law was consolidated in 2013 (BOE-A-2003-21736, 2013).

5. Order APA/2677/2005 (2005) about accounting and declarations for the control in the sector of olive oil and table olives. The Order was modified by Order ARM/2275/2010 (2010).

6. Law 227/2008 (2008) which establishes the basic regulations regarding panels of tasters of virgin olive oil.

Production 2018/2019, thousand of tons

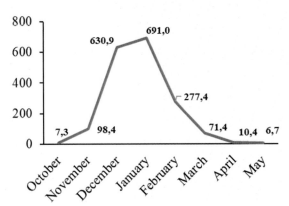

Production 2019/2020ᵃ, thousand of tons

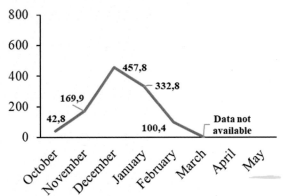

Imports 2018/2019, thousand of tons

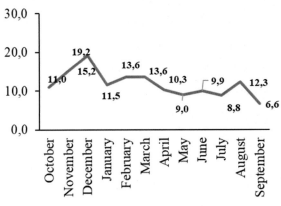

Imports 2019/2020ᵃ, thousand of tons

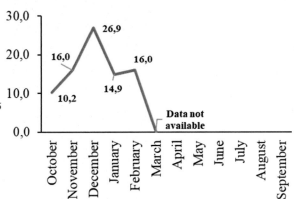

Exports 2018/2019, thousand of tons

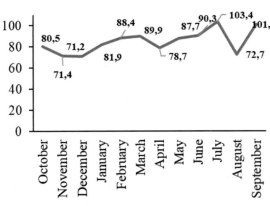

Exports 2019/2020ᵃ, thousand of tons

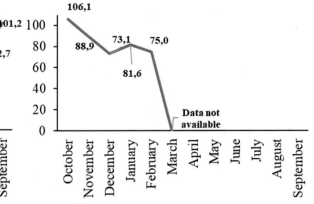

FIG. 13.1 Production, exports, and imports of olive oil in Spain (Ministerio de Agricultura, Pesca y Alimentación, 2020a).

The autonomic provision for olive oil is based in the Law 5/2011 (2011) for Andalucía but, also, olive oil is governed by legislation regarding:

1. Designations of Origin (DO): Spain has 29 DO of olive oils spread throughout the country. The Protected DO recognizes extra virgin olive oils produced in specific areas, with certain varieties (a single variety or several representative of that area), and under specific conditions of production and quality. Only extra virgin olive oils that meet these standards have the possibility of bearing the DO seal on their label.
2. Additives, extraction solvents, flavorings, and technological aids.
3. Provisions related to food information provided to the consumer and other complementary provisions regarding food labeling. General provisions applicable to all food products.
4. Provisions related to food information provided to the consumer and other complementary provisions regarding food labeling. Specific provisions applicable to olive and olive-pomace oils.
5. Nominal quantities for packaged products and control of their effective content. General disposition.
6. Nominal quantities for packaged products and control of their effective content. General and specific provisions.
7. Taking samples.
8. Provisions regarding the designation of origin in extra virgin and virgin olive oils.

All the legislation related to these sections has been summarized by Ministerio de Agricultura, Pesca y Alimentación (2017).

13.4.3 Iberian (IB) Pig Dry-Cured Products

The IB pig (Image 13.2) is a native dark-haired breed of Spain originally reared under free-range conditions, fed on grass and acorns, and slaughtered at heavy weight (160–180 kg of body weight). Most carcasses of IB pigs are destined to the production of high quality dry-cured products mainly hams, shoulders, and loins because of the strong consumer demand for cured products derived from ancestral pigs. In total, in Spain about 3.6 million IB pigs are slaughtered each year, of which this year 733,000 have been fed on acorn and raised in pastures while the rest (almost three million) have been fattened with animal feed, grass, and pasture on farms or outdoors. Unlike the acorn animals, which go to the slaughterhouse from November to March, the rest are slaughtered regularly throughout the year.

Modifications of the traditional rearing system lead to produce IB dry-cured products of different quality and price (Serrano, 2008). In order to avoid commercial frauds and to guarantee the consumers rights, the MAGRAMA has enacted a Law to regulate the market of IB products (Law 4/2014, 2014 consolidated in 2016: BOE-A-2014-318, 2016). Table 13.2 summarizes the Law for Spanish IB pig products according to feeding and handling.

IMAGE 13.2 Iberian pig.(Source: aeceriber.es.)

TABLE 13.2
Law for Spanish Iberian pig products according to feeding and handling (Law 4/2014, 2014 consolidated in 2016) (BOE-A-2014-318, 2016).

Classification	Bellota	Cebo campo[a]	Cebo[b]
Feeding	Concentrate-montanera	Concentrate	Concentrate
Beggining of montanera			
Av. BW[c]/lot, kg	92–115	–	–
Dates	1/10–15/12	–	–
Replacement			
kg BW	>46	–	–
Days	≥60	≥60	–
Slaughter			
Minimum/carcass	108[d]–115	108[d]–115	108[d]–115
Minimum age, months	14	12	10
Dates	15/12–31/03	–	–

[a] Pigs weighing more than 110 kg must have a surface with a minimum total free soil per animal of 100 m^2 during its growing phase.
[b] Pigs weighing more than 110 kg that give rise to products with the designation growing must have a minimum total free floor area per animal of 2 m^2.
[c] Average body weight.
[d] For 100% Iberian pigs.

In addition, IB product labeling must include as a mandatory mention the genetic percentage of IB pig breed:

(1) 100% IB: when it comes to products from animals with 100% of genetic purity of the IB breed, whose parents also have 100% IB racial purity and are registered in the corresponding genealogical book.
(2) IB: products from animals with at least the 50% of its genetic percentage corresponding to the IB pig breed, with parents of the following characteristics:
  2.1. To obtain 75% IB animals, 100% IB breed females will be used enrolled in genealogical book and males from the cross of 100% race mother IB and 100% Duroc breed father, both registered in the corresponding genealogical book of the breed.
  2.2. To obtain 50% IB animals, females of 100% IB breed will be used and 100% Duroc breed males, both registered in the corresponding genealogical book of the race.

13.4.4 Wine

Based on recently data published by Ministerio de Agricultura, Pesca y Alimentación (2020b):
- Of the total production of wine and must in Spain, 33.5 million hectoliters correspond to wine and 3.7 to must.
- Wine production has been mostly red and rosé (a 53%) and the rest of white. Production is located, mainly, in Castilla-La Mancha (54%), Cataluña (9%), Extremadura (8%), Comunidad Valenciana (7%), Castilla y León (5%), La Rioja (5%), and Andalucía (3%).
- The 43% of wine production has been declared as wine with Protected Designation of Origin (PDO), 13% as wine with Protected Geographical Indication (PGI), 19% as varietal wines without Geographical Indication (GI), being the rest of wines the 25% of the production.

Based on data supplied by Ministerio de Agricultura, Pesca y Alimentación (2019b):
(1) as average, the per capita consumption of wine in Spain is 7.89 L/person and year;
(2) Spanish houses spend 1.52% of their budget for food and beverages on wine purchases, representing a per capita expense of 22.82 €/person/year (+3.7% compared to, 2017);
(3) domestic consumption of total wines throughout, 2018 decreased by 2.4% with respect to, 2017;
(4) conversely, the value of the category increased 4% points due to the direct relationship of this variable with the average price, which closes the period at 2.89€/L, which corresponds to an increase of +6.6% over the previous year.

The wine production in Spain is legislated by:

- The state provisions of Vine and Wine Law (Law 24/2003, 2003) consolidated in 2015 (BOE-A-2003-13864, 2015).
- Community provisions for common organization of the wine market: Regulation (EU) 1308/2013 (2013) consolidated in 2019 (BOE 02019R0033, 2019).
- Support programs, trade with third party countries, productive potential, and controls in the sector winemaking: Regulation (EC) 555/2008 (2008) consolidated in, 2018 (BOE 02008R0555, 2018) (community provisions) and Real Decreto 1338/2018 (2018) consolidated in 2020 (BOE-A-2018-14803, 2020) (state provisions).
- Wine product categories, practices, winemaking, and applicable restrictions: Regulation 934/2019 (2019) and Regulation (EU) 935/2019 (2019) (community provisions).
- Designation, denomination, presentation, and protection of wine products: as community provisions, Regulation (EU) 33/2019 (2019) consolidated in 2019 (DOUE 02013R1308, 2019) and corrected in 2019 (DOUE, 2019) and Regulation (EU) 34/2019 (2019). On the other hand, as state provisions, Law 1363/2011 (2011) consolidated in 2015 (BOE-A-2011-17174, 2015).
- Movement and registration of wine products: as community provisions, Regulation (EU) 273/2018 (2018) consolidated by Regulation (EU) 274/2018 (2018). As state provisions, Law 323/1994 (1994) consolidated in 2011 (BOE-A-1994-13633, 2011) and Law 739/2015 (2015).
- Flavored wines, flavored drinks based on wine and cocktails flavored products winegrowing: as community provisions, Regulation (EU) 251/2014 (2014) consolidated in, 2014 (BOE 2014R0251, 2014) complemented regarding to authorize production processes to obtain flavored wine products by Regulation (EU) 670/2017 (2017). As state provisions: (1) BOE (1974) which regulates the elaboration, circulation, and trade of sangria and other beverages derived from wine. This Order was consolidated in 2013 (BOE-A-1974-245, 2013), (2) BOE (1978) for flavored wines and biter-soda. This Order was consolidated in 2013 (BOE-A-1978-5486, 2013), and (3) BOE (1986) for beverage regulation derived from wine. The Order was consolidated in 1997 (BOE-A-1986-33109, 1997).

In conclusion, food legislation in Spain is complex as the European, Spanish, and autonomy guidelines are mandatory. Furthermore, the legislation applies not only for domestically produced products but also for exported ones. It is a dynamic legislation that is frequently modified depending on the particular circumstances of each product. Therefore, it is expected that, in the short and medium term, that new modifications take place given the circumstances that are taking place in 2020 as a consequence of the world coronavirus pandemic.

REFERENCES

BOE. (1974). Orden, de 23 de enero de 1971, por la que se reglamenta la elaboración, circulación y comercio de la sangría y de otras bebidas derivadas del vino. *Boletín Oficial del Estado, 33*, 2376–2381. https://www.boe.es/eli/es/o/1974/01/23/(1)/dof/spa/pdf.

BOE. (1978). Orden, de 22 de febrero de 1978, por la que se designan órganos gestores de las ayudas de fondo nacional de protección al trabajo. *Boletín Oficial del Estado, 47*, 4509–4512. https://www.boe.es/boe/dias/1978/02/24/pdfs/A04509-04512.pdf.

BOE. (1986). Orden, de 11 de diciembre de 1986, sobre reglamentación de bebidas derivadas del vino. *Boletín Oficial del Estado, 304*, 41647. https://www.boe.es/boe/dias/1986/12/20/pdfs/A41647-41647.pdf.

BOE 02008R0555. (2018). Reglamento (CE) 555/2008 de la Comisión de 27 de junio de 2008 por el que se establecen normas de desarrollo del Reglamento (CE) 479/2008 del Consejo, por el que se establece la organización común del mercado vitivinícola, en lo relativo a los programas de apoyo, el comercio con terceros países, el potencial productivo y los controles en el sector vitivinícola. *Boletín Oficial del Estado, 02008R0555*, 1–10. https://www.boe.es/buscar/pdf/2003/BOE-A-2003-13864-consolidado.pdf.

BOE 02019R0033. (2019). Reglamento Delegado (UE) 33/2019 de la Comisión, de 17 de octubre de 2018, por el que se completa el Reglamento (UE) 1308/2013 del Parlamento Europeo y del Consejo en lo que respecta a las solicitudes de protección de denominaciones de origen, indicaciones geográficas y términos tradicionales del sector vitivinícola, al procedimiento de oposición, a las restricciones de utilización, a las modificaciones del pliego de condiciones, a la cancelación de la protección, y al etiquetado y la presentación. *Boletín Oficial del Estado, 02019R0033*, 1–54. https://eur-lex.europa.eu/legal-content/ES/TXT/PDF/?uri=CELEX:02019R0033-20190111&from=EN.

BOE 2014R0251. (2014). Regulation (EU) 251/2014 of the European parliament and of the council of 26 February 2014 on the definition, description, presentation, labelling and the protection of geographical indications of aromatised wine products and repealing Council Regulation (EEC) No 1601/91. *Boletín Oficial del Estado, 2014R0251*, 1–32. https://eur-lex.europa.eu/legal-content/EN/TXT/PDF/?uri=CELEX:02014R0251-20140327&from=ES.

BOE-A-1967-16485. (2019). Decreto 2484/1967, de 21 de Septiembre, por el que se aprueba el texto del Código

Alimentario Español. Texto consolidado última modificación 11 de mayo de 2019. *Boletín Oficial del Estado, 1967-16485*, 1–156. https://www.boe.es/buscar/pdf/1967/BOE-A-1967-16485-consolidado.pdf.

BOE-A-1974-245. (2013). Orden por la que se reglamenta la elaboración, circulación y comercio de la sangría y de otras bebidas derivadas del vino. Texto consolidado última modificación 29 de marzo de 2013. *Boletín Oficial del Estado, 1974-245*, 1–9. https://www.boe.es/buscar/pdf/1974/BOE-A-1974-245-consolidado.pdf.

BOE-A-1978-5486. (2013). Orden de 31 de enero de 1978 por la que se reglamentan los vinos aromatizados y el biter-soda. Texto consolidado última modificación 29 de marzo de 2013. *Boletín Oficial del Estado, 1978-5486*, 1–8. https://www.boe.es/buscar/pdf/1978/BOE-A-1978-5486-consolidado.pdf.

BOE-A-1983-5543. (2015). Real Decreto 308/1983, de 25 de enero, por el que se aprueba la Reglamentación Técnico-Sanitaria de Aceites Vegetales Comestibles. Texto consolidado última modificación de 28 de julio de 2015. *Boletín Oficial del Estado, 1983-5543*, 1–13. https://www.boe.es/buscar/pdf/1983/BOE-A-1983-5543-consolidado.pdf.

BOE-A-1986-33109. (1997). Orden, de 11 de diciembre de 1986, sobre Reglamentación de Bebidas Derivadas del Vino. Texto consolidado última modificación 22 de marzo de 1997. *Boletín Oficial del Estado, 1986-33109*, 1–3. https://www.boe.es/buscar/pdf/1986/BOE-A-1986-33109-consolidado.pdf.

BOE-A-1987-21012. (2013). Real Decreto 1094/1987, de 26 de junio, por el que se aprueba la Reglamentación Técnico-Sanitaria para la elaboración, fabricación, circulación y comercio de cereales en copos o expandidos. Texto consolidado última modificación 29 de marzo de 2013. *Boletín Oficial del Estado, 1987-21012*, 1–7. https://www.boe.es/buscar/pdf/1987/BOE-A-1987-21012-consolidado.pdf.

BOE-A-1989-2265. (2013). Orden, de 26 de enero de 1989, por la que se aprueba la Norma de Calidad para los Aceites y Grasas Calentados. Texto consolidado última modificación 29 de marzo de 2013. *Boletín Oficial del Estado*, 1–6. https://www.boe.es/buscar/pdf/1989/BOE-A-1989-2265-consolidado.pdf.

BOE-A-1994-13633. (2011). Orden de 20 de mayo de 1994, por la que se dictan normas de desarrollo del Real Decreto 323/1994, de 25 de febrero, sobre los documentos que acompañan el transporte de productos vitivinícolas y los registros que se deben llevar en el sector vitivinícola. Texto consolidado última modificación 1 de noviembre de 2011. *Boletín Oficial del Estado, 1994-13633*, 1–6. https://www.boe.es/buscar/pdf/1994/BOE-A-1994-13633-consolidado.pdf.

BOE-A-2003-13864. (2015). Ley 24/2003, de 13 de mayo, de 10 de julio, de la Viña y del Vino. *Boletín Oficial del Estado, 2003-13864*, 1–20. https://www.boe.es/boe/dias/2003/07/11/pdfs/A27165-27179.pdf.

BOE-A-2003-21736. (2013). Real Decreto 1431/2003, de 21 de noviembre, por el que se establecen determinadas medidas de comercialización en el sector de los aceites de oliva y del aceite de orujo de oliva. Texto consolidado última modificación 16 de noviembre de 2013. *Boletín Oficial del Estado*, 1–4. https://www.boe.es/buscar/pdf/2003/BOE-A-2003-21736-consolidado.pdf.

BOE-A-2011-17174. (2015). Real Decreto 1363/2011, de 7 de octubre, por el que se desarrolla la reglamentación comunitaria en materia de etiquetado, presentación e identificación de determinados productos vitivinícolas. Texto consolidado última modificación 29 de enero de 2015. *Boletín Oficial del Estado, 2011-17174*, 1–10. https://www.boe.es/buscar/pdf/2011/BOE-A-2011-17174-consolidado.pdf.

BOE-A-2014-318. (2016). Real Decreto 4/2014, de 10 de enero, por el que se aprueba la norma de calidad para la carne, el jamón, la paleta y la caña de lomo ibérico. Texto consolidado última modificación 11 de junio de 2016. *Boletín Oficial del Estado*, 1–16. https://www.boe.es/buscar/pdf/2014/BOE-A-2014-318-consolidado.pdf.

BOE-A-2018-14803. (2020). Real Decreto 1338/2018, de 29 de octubre, por el que se regula el potencial de producción vitícola. *Boletín Oficial del Estado, 2018-14803*, 1–58. https://www.boe.es/buscar/pdf/2018/BOE-A-2018-14803-consolidado.pdf.

Delegated Regulation (EU) 2095/2016, from 26 September, amending Regulation (EEC) 2568/91, on the characteristics of olive oils and olive-pomace oils and on their methods of analysis. (2016). *Diario Oficial de la Unión Europea, 326*, 1–6. https://eur-lex.europa.eu/legal-content/ES/TXT/PDF/?uri=CELEX:32016R2095&from=ES.

DOCE. (2017). Corrección de errores del Reglamento Delegado (UE) 2016/2095 de la Comisión, de 26 de septiembre de 2016, que modifica el Reglamento (CEE) 2568/91, relativo a las características de los aceites de oliva y de los aceites de orujo de oliva y sobre sus métodos de análisis. *Diario Oficial de la Unión Europea, 211*, 58. https://eur-lex.europa.eu/legal-content/ES/TXT/PDF/?uri=CELEX:32016R2095R(01)&from=ES.

DOUE. (2019). Correction of errors in Commission Delegated Regulation (EU) 33/2019, of 17 October 2018, completing Regulation (EU) 1308/2013 of the European Parliament and of the Council with regard to applications for the protection of appellations of origin, geographical indications and traditional terms of the wine sector, the opposition procedure, the restrictions on use, the modifications to the specification, the cancellation of protection, and the labeling and presentation. *Diario Oficial de la Unión Europea, 269*, 14.

DOUE 02013R1308. (2019). Regulation (UE) 1308/2013 del parlamento europeo y del consejo de 17 de diciembre de 2013 por el que se crea la organización común de mercados de los productos agrarios y por el que se derogan los Reglamentos (CEE) 922/72, (CEE) 234/79, (CE) 1037/2001 y (CE) 1234/2007. *Diario Oficial de la Unión Europea, 02013R1308*, 1–252. https://eur-lex.europa.eu/legal-content/ES/TXT/PDF/?uri=CELEX:02013R1308-20190101&from=EN.

Estruch, R., Ros, E., Salas-Salvadó, J., Covas, M. I., Corella, D., Arós, F., et al. (2018). Primary prevention of cardiovascular disease with a Mediterranean diet supplemented with extra-virgin olive oil or nuts. *The New England Journal of Medicine, 378*(25). https://doi.org/10.1056/NEJMoa1800389. e34(1)–e34(14).

Gerten, D., Heck, V., Jägermeyr, J., Bodirsky, B. L., Fetzer, I., Jalava, M., et al. (2020). Feeding ten billion people is possible within four terrestrial planetary boundaries. *Nature Sustainability, 3*, 200–208. https://doi.org/10.1038/s41893-019-0465-1.

Law 1094/1987, from 26 June, by which the Technical-Sanitary Regulation for the elaboration, manufacture, circulation and trade of flaked or expanded cereals is approved. (1987). *Boletín Oficial del Estado, 215*, 27338–27341. https://www.boe.es/boe/dias/1987/09/08/pdfs/A27338-27341.pdf.

Law 135/2010, from 12 February, repealing provisions relating to the microbiological criteria of foodstuffs. (2010). *Boletín Oficial del Estado, 49*, 18297–18299. https://www.boe.es/boe/dias/2010/02/25/pdfs/BOE-A-2010-3032.pdf.

Law 1363/2011, de 7 de octubre, por el que se desarrolla la reglamentación comunitaria en materia de etiquetado, presentación e identificación de determinados productos vitivinícolas. (2011). *Boletín Oficial del Estado, 263*, 114313–114322. https://www.boe.es/eli/es/rd/2011/10/07/1363.

Law 1431/2003, from 21 November, establishing certain marketing measures in the olive oil and olive-pomace oil sector. (2003). *Boletín Oficial del Estado, 285*, 42415–42416. https://www.boe.es/boe/dias/2003/11/28/pdfs/A42415-42416.pdf.

Law 145/1997, from 31 January, approving the positive list of additives other than colorants and sweeteners for use in the manufacture of food products, as well as their conditions of use. (1997). *Boletín Oficial del Estado, 70*, 9378–9418. https://www.boe.es/boe/dias/1997/03/22/pdfs/A09378-09418.pdf.

Law 1615/2010, from 7 December, by which the wheat quality standard is approved. (2010). *Boletín Oficial del Estado, 301*, 102674–102680. https://www.boe.es/boe/dias/2010/12/11/pdfs/BOE-A-2010-19103.pdf.

Law 17/2011, from 05 July, of food security and nutrition. (2011). *Boletín Oficial del Estado, 160*, 71283–71319. https://www.boe.es/boe/dias/2011/07/06/pdfs/BOE-A-2011-11604.pdf.

Law 176/2013, from 8 de March, whereby certain technical-sanitary regulations and quality standards related to food products are totally or partially repealed. (2013). *Boletín Oficial del Estado, 76*, 24494–24505. https://www.boe.es/boe/dias/2013/03/29/pdfs/BOE-A-2013-3402.pdf.

Law 190/2013, from 15 March, by modifying Royal Decree 1615/2010, of December 7, which approves the wheat quality standard. (2013). *Boletín Oficial del Estado, 82*, 25471–25475. https://www.boe.es/boe/dias/2013/04/05/pdfs/BOE-A-2013-3630.pdf.

Law 227/2008, from 15 February, by which the basic regulations regarding the panels of virgin olive oil tasters are established. (2008). *Boletín Oficial del Estado, 56*, 13323–13325. https://www.boe.es/buscar/pdf/2008/BOE-A-2008-4209-consolidado.pdf.

Law 24/2003, from 10 July, of the Vine and the Wine. (2003). *Boletín Oficial del Estado, 165*, 27165–27179. https://www.boe.es/boe/dias/2003/07/11/pdfs/A27165-27179.pdf.

Law 2484/1967, from 21 September, by which the text of the Spanish Food Code is approved. (1967). *Boletín Oficial del Estado, 251*, 14326–14334. https://www.boe.es/boe/dias/1967/10/20/pdfs/A14326-14334.pdf.

Law 25798/1987, from 26 June, Correction of errata of Royal Decree 1094/1987, which approves the Technical-Sanitary Regulation for the elaboration, manufacture, circulation and trade of cereals in cups or expanded. (1987). *Boletín Oficial del Estado, 276*, 34349. https://www.boe.es/boe/dias/1987/09/08/pdfs/A27338-27341.pdf.

Law 308/1983, from 25 January, by which the technical-sanitary regulation of edible vegetable oils is approved. (1983). *Boletín Oficial del Estado, 44*, 4853–4858. https://www.boe.es/boe/dias/1983/02/21/pdfs/A04853-04858.pdf.

Law 308/2019, from 26 April, by which the quality standard for bread is approved. (2019). *Boletín Oficial del Estado, 113*, 50168–50175. https://www.boe.es/boe/dias/2019/05/11/pdfs/BOE-A-2019-6994.pdf.

Law 323/1994, from 28 February, on the documents accompanying the transport of wine products and the records to be kept in the wine sector. (1994). *Boletín Oficial del Estado, 109*, 14154–14156.

Law 4/2014, from 10 January, by which the quality standard for meat, ham, shoulder and Iberian loin cane is approved. (2014). *Boletín Oficial del Estado, 10*, 1569–1585. https://www.boe.es/boe/dias/2014/01/11/pdfs/BOE-A-2014-318.pdf.

Law 5/2011, from 6 October, from the olive grove of Andalucía. (2011). *Boletín Oficial del Estado, 268*, 116080–116095. https://www.boe.es/boe/dias/2011/11/07/pdfs/BOE-A-2011-17494.pdf.

Law 739/2015, from 31 July, on mandatory declarations in the wine sector. (2015). *Boletín Oficial del Estado, 183*, 66904–66929. https://www.boe.es/eli/es/rd/2015/07/31/739.

Ministerio de Agricultura, Pesca y Alimentación. (2017). Principales disposiciones aplicables a los aceites vegetales comestibles. In *Dirección General de la Industria Alimentaria* S. G. de Control y de Laboratorios Alimentarios. https://www.mapa.gob.es/es/alimentacion/legislacion/recopilaciones-legislativas-monograficas/aceitesvegetalessumariocompleto17082017_tcm30-79051.pdf.

Ministerio de Agricultura, Pesca y Alimentación. (2019a). *Evolución de los balances de cereales en España Campañas 2018/2019 y 2019/2020.* Dirección General de Producciones y Mercados Agrarios S.G. de cultivos herbáceos e industriales y aceite de oliva. https://www.mapa.gob.es/es/ganaderia/estadisticas/evolucionbalancesceralesnov2019_tcm30-135215.pdf.

Ministerio de Agricultura, Pesca y Alimentación. (2019b). *Informe del consumo alimentario en España* (p. 2018). https://www.mapa.gob.es/images/es/20190807_informedeconsumo2018pdf_tcm30-512256.pdf.

Ministerio de Agricultura, Pesca y Alimentación. (2020a). *Datos de producción, movimientos y existencias de aceite de oliva y aceituna de mesa (febrero 2020).* https://www.mapa.gob.es/es/agricultura/temas/producciones-agricolas/datosdeproduccionmovimientosyexistenciasdeaceitedeolivayaceitunademesafebrero2020_tcm30-536663.pdf.

Ministerio de Agricultura, Pesca y Alimentación. (2020b). *La producción de vino y mosto de la campaña 2019/2020 se sitúa en 37,2 millones de hectolitros (enero 2020)*. https://www.mapa.gob.es/es/prensa/200114datosinfovi_tcm30-523945.pdf.

Order 11806/1989, from 24 May, amending the positive list of authorized additives in the production of flaked or expanded cereals. (1989). *Boletín Oficial del Estado, 123*, 15542. https://www.boe.es/boe/dias/1989/05/24/pdfs/A15542-15542.pdf.

Order 12109/2010, from 7 December, correction of errors of the Royal Decree 1615/2010 by which the wheat quality standard is approved. (2010). *Boletín Oficial del Estado, 168*, 78530. https://www.boe.es/boe/dias/2011/07/14/pdfs/BOE-A-2011-12109.pdf.

Order 25297/1980, from 12 November, by which the quality standard is approved for rice packaged for consumption in the internal market. (1980). *Boletín Oficial del Estado, 278*, 25813–25815. https://www.boe.es/boe/dias/1980/11/19/pdfs/A25813-25815.pdf.

Order 265/1989, from 26 January, by which the Quality Standard for Heated Oils and Fats is approved. (1989). *Boletín Oficial del Estado, 26*, 2665–2667. https://www.boe.es/boe/dias/1989/01/31/pdfs/A02665-02667.pdf.

Order 27204/1980, from 18 December. Correction of errors in the Order of 12 November 1980 approving the Quality Standard for rice-packed for destination-consumption in the internal market. (1980). *Boletín Oficial del Estado, 303*, 27893. https://www.boe.es/boe/dias/1980/12/18/pdfs/A27893-27893.pdf.

Order 301/1985, from 12 December, on delivery of virgin olive oil by the oil mills to the harvesting for self-consumption. (1985). *Boletín Oficial del Estado, 5*, 324. https://www.boe.es/boe/dias/1985/01/05/pdfs/A00324-00324.pdf.

Order 9343/1984, from 18 April, by which the quality standard for rice packaged for the domestic market is modified, approved by Order of November 12. (1984). *Boletín Oficial del Estado, 100*, 11391. https://www.boe.es/boe/dias/1984/04/26/pdfs/A11391-11391.pdf.

Order APA/2677/2005, from 8 August, on accounting and declarations for control in the olive oil and table olives sector. (2005). *Boletín Oficial del Estado, 195*, 28566–28585. https://www.boe.es/boe/dias/2005/08/16/pdfs/A28566-28585.pdf.

Order ARM/2275/2010, from 20 August, amending Order APA/2677/2005, of August 8, on accounting and declarations for control in the olive oil and table olives sector. (2010). *Boletín Oficial del Estado, 208*, 74732–74736. https://www.boe.es/boe/dias/2010/08/27/pdfs/BOE-A-2010-13485.pdf.

Real Decreto 1338/2018, de 29 de octubre, por el que se regula el potencial de producción vitícola. (2018). *Boletín Oficial del Estado, 262*, 104825–104885. https://www.boe.es/boe/dias/2018/10/30/pdfs/BOE-A-2018-14803.pdf.

Regulation (EC) 178/2002 of the European Parliament and of the Council of 28 January 2002 laying down the general principles and requirements of food law, establishing the European Food Safety Authority and laying down procedures in matters of food safety. (2002). *Official Journal of the European Community, 31*, 1–24. https://eur-lex.europa.eu/legal-content/EN/TXT/PDF/?uri=CELEX:32002R0178&from=EN.

Regulation (EC) 555/2008. (2008). Regulations commission regulation (EC) 555/2008 of 27 June 2008 laying down detailed rules for implementing Council Regulation (EC) 479/2008 on the common organisation of the market in wine as regards support programmes, trade with third countries, production potential and on controls in the wine sector. *Official Journal of the European Union, 170*, 1–80. https://eur-lex.europa.eu/legal-content/EN/TXT/PDF/?uri=CELEX:32008R0555&from=en.

Regulation (EEC) 01991R2568, from 4 December, relating to the characteristics of olive oils and olive-pomace oils and their methods of analysis. (2016). *Diario Oficial de la Unión Europea, 031.005*, 1–128. https://eur-lex.europa.eu/legal-content/ES/TXT/PDF/?uri=CELEX:01991R2568-20161204&qid=1502366499973&from=ES.

Regulation (EEC) 02013R1308, from 1 August, regulation of execution (EU) 29/2012 of the Commission, of January 13 on the marketing rules of olive oil. (2012). *Diario Oficial de la Unión Europea, 12*, 14–21. https://eur-lex.europa.eu/LexUriServ/LexUriServ.do?uri=OJ:L:2012:012:0014:0021:ES:PDF.

Regulation (EEC) 2012R0029, from 1 January, on the marketing rules for olive oil (codified text). (2016). *Diario Oficial, 005.001*, 1–15. https://eur-lex.europa.eu/legal-content/ES/TXT/PDF/?uri=CELEX:02012R0029-20160101&qid=1457690347909&from=ES.

Regulation (EEC) 2568/1991, from 11 July, relating to the characteristics of olive oils and olive-pomace oils and their methods of analysis. (1991). *Diario Oficial de la Unión Europea, 248*, 1–83. https://eur-lex.europa.eu/LexUriServ/LexUriServ.do?uri=OJ:L:1991:248:0001:0083:ES:PDF.

Regulation (EU) 1308/2013, from 17 December, by which the common organization of markets for agricultural products is created and by which the Regulations are repealed (CEE) 922/72, (CEE) 234/79, (CE) 1037/2001 and (CE) 1234/2007. (2013). *Diario Oficial de la Unión Europea, 347*, 671–854. https://eur-lex.europa.eu/LexUriServ/LexUriServ.do?uri=OJ:L:2013:347:0671:0854:ES:PDF.

Regulation (EU) 1381/2019 of the European Parliament and of the Council of 20 June 2019 on the transparency and sustainability of the EU risk assessment in the food chain and amending regulations (EC) 178/2002, (EC) 1829/2003, (EC) 1831/2003, (EC) 2065/2003, (EC) 1935/2004, (EC) 1331/2008, (EC) 1107/2009, (EU) 2015/2283 and Directive 2001/18/EC. (2019). *Official Journal of the European Community, 231*, 1–28. https://eur-lex.europa.eu/legal-content/EN/TXT/PDF/?uri=CELEX:32019R1381&from=EN.

Regulation (EU) 251/2014 of the European Parliament and of the Council, of 26 February 2014, on the definition, description, presentation, labelling and the protection of geographical indications of aromatised wine products and repealing Council Regulation (EEC) No 1601/91. (2014). *Official Journal of the European Community, 84*, 14–34. https://eur-lex.europa.eu/legal-content/EN/TXT/PDF/?uri=CELEX:32014R0251&from=ES.

Regulation (EU) 273/2018, of 11 December 2017, supplementing Regulation (EU) No 1308/2013 of the European Parliament and of the Council as regards the scheme of authorisations for vine plantings, the vineyard register, accompanying documents and certification, the inward and outward register, compulsory declarations, notifications and publication of notified information, and supplementing Regulation (EU) No 1306/2013 of the European Parliament and of the Council as regards the relevant checks and penalties, amending Commission Regulations (EC) 555/2008, (EC) No 606/2009 and (EC) No 607/2009 and repealing Commission Regulation (EC) No 436/2009 and Commission Delegated Regulation (EU) 2015/560. (2018). *Official Journal of the European Union, 58*, 1–59. https://eur-lex.europa.eu/legal-content/EN/TXT/PDF/?uri=CELEX:32018R0273&from=es.

Regulation (EU) 274/2018, of 11 December 2017, laying down rules for the application of Regulation (EU) 1308/2013 of the European Parliament and of the Council as regards the scheme of authorisations for vine plantings, certification, the inward and outward register, compulsory declarations and notifications, and of Regulation (EU) No 1306/2013 of the European Parliament and of the Council as regards the relevant checks, and repealing Commission Implementing Regulation (EU) 561/2015. (2018). *Official Journal of the European Union, 58*, 60–95. https://eur-lex.europa.eu/legal-content/EN/TXT/PDF/?uri=CELEX:32018R0274&from=ES.

Regulation (EU) 29/2012, from 13 January, on the marketing rules for olive oil (codified text). (2012). *Diario Oficial de la Unión Europea, 12*, 14–21. https://eur-lex.europa.eu/LexUriServ/LexUriServ.do?uri=OJ:L:2012:012:0014:0021:ES:PDF.

Regulation (EU) 33/2019, of 17 October 2018, supplementing Regulation (EU) No 1308/2013 of the European Parliament and of the Council as regards applications for protection of designations of origin, geographical indications and traditional terms in the wine sector, the objection procedure, restrictions of use, amendments to product specifications, cancellation of protection, and labelling and presentation. (2019). *Official Journal of the European Community, 9*, 1–45. https://eur-lex.europa.eu/legal-content/EN/TXT/PDF/?uri=CELEX:32019R0033&from=ES.

Regulation (EU) 34/2019, of 17 October 2018, laying down rules for the application of Regulation (EU) No 1308/2013 of the European Parliament and of the Council as regards applications for protection of designations of origin, geographical indications and traditional terms in the wine sector, the objection procedure, amendments to product specifications, the register of protected names, cancellation of protection and use of symbols, and of Regulation (EU) No 1306/2013 of the European Parliament and of the Council as regards an appropriate system of checks. (2019). *Official Journal of the European Community, 9*, 46–76. https://eur-lex.europa.eu/legal-content/EN/TXT/PDF/?uri=CELEX:32019R0034&from=ES.

Regulation (EU) 670/2017, of 31 January 2017, supplementing Regulation (EU) No 251/2014 of the European Parliament and of the Council as regards the authorised production processes for obtaining aromatised wine products. (2017). *Official Journal of the European Community, 97*, 5–8. https://eur-lex.europa.eu/legal-content/EN/TXT/PDF/?uri=CELEX:32017R0670&from=ES.

Regulation (EU) 935/2019 of 16 April 2019, laying down rules for the application of Regulation (EU) 1308/2013 of the European Parliament and of the Council as regards analysis methods for determining the physical, chemical and organoleptic characteristics of grapevine products and notifications of Member States decisions concerning increases in natural alcoholic strength. (2019). *Official Journal of the European Union, 149*, 53–57. https://eur-lex.europa.eu/legal-content/EN/TXT/PDF/?uri=CELEX:32019R0935&from=ES.

Regulation 934/2019. (2019). Regulations Commission Delegated Regulation (EU) 2019/934 of 12 March 2019 supplementing Regulation (EU) 1308/2013 of the European Parliament and of the Council as regards wine-growing areas where the alcoholic strength may be increased, authorised oenological practices and restrictions applicable to the production and conservation of grapevine products, the minimum percentage of alcohol for by-products and their disposal, and publication of OIV files. *Official Journal of the European Union, 149*, 1–52. https://eur-lex.europa.eu/legal-content/EN/TXT/PDF/?uri=CELEX:32019R0934&from=ES.

Ros, E., Martínez-González, M. A., Estruch, R., Salas-Salvadó, J., Fitó, M., Martínez, J. A., et al. (2014). Mediterranean diet and cardiovascular health: Teachings of the PREDIMED study. *Advances in Nutrition, 5*, 330S–336S. https://doi.org/10.3945/an.113.005389.

Serrano, M. P. (2008). *A study of factors that influence growth performance and carcass and meat quality of iberian pigs reared under intensive management systems* (Tesis Doctoral). E.T.S.I. Agrónomos, UPM. http://oa.upm.es/1676/1/MARTINA_PEREZ_SERRANO.pdf.

United State Department of Agriculture. (2020). *Food and agricultural import regulations and standards country report.* Spain https://apps.fas.usda.gov/newgainapi/api/Report/DownloadReportByFileName?fileName=Food%20and%20Agricultural%20Import%20Regulations%20and%20Standards%20Country%20Report_Madrid_Spain_12-31-2019.

Index

Note: Page numbers followed by *f* indicate figures and *t* indicate tables.

Printed in the United States
by Baker & Taylor Publisher Services